Chironomidae Larvae

Biology and Ecology of the Chironomini

Chironomidae Larvae of the Netherlands and Adjacent Lowlands

Biology and Ecology of the Chironomini

H.K.M. Moller Pillot

KNNV Publishing

CONTENTS

1 INTRODUCTION

This second part of our work on Chironomidae larvae covers the tribes Chironomini and Pseudochironomini. Chironomini play a key role in aquatic systems, especially in stagnant and slowly flowing water, because they often occur in large numbers. The biology and ecology of the species have been given priority because it is not possible to use chironomids for scientific or practical purposes if the data for an important part of the family are not available. We were therefore obliged to delay publication of the keys until the next volume in the series.

As in the previous volume, we limit the scope of our investigation to the Netherlands and adjacent lowlands. This avoided having to devote a lot of attention to fast-flowing streams and the profundal zone of lakes, which made our task much easier. In the descriptions (Chapter 3) and the tables (Chapter 4) all genera and species are treated in alphabetic order. In the tables we omitted species about which we had insufficient data.

Although this volume has only one author, the word 'we' is used. Much information was made available by colleagues and several students did some of the investigations, which delivered valuable information. To refer to them always in the text would cause far too many interruptions. We were not able to go through all the existing literature and all the available data. Nevertheless, we believe that the text and tables are supported by literature research to an extent that has not previously been achieved.

Origin of the data used for this book
As mentioned in Part I of this book series, the information about the biology and ecology of the chironomid species has been taken from the literature, from data supplied by water authorities and from own investigations. Of course, it was impossible to read all the publications, but we collected many articles, theses and books from Western, Central and Eastern Europe and many unpublished Dutch reports. Generally, we have cited only the most important or the best publicly available literature.

For the Netherlands we were able to draw on data from more than 8000 localities (more than 14,500 samples) investigated by water authorities (Nijboer & Verdonschot, 2001; Limnodata.nl). In particular, the results of investigations in the province of Noord-Holland are very well documented (Steenbergen, 1993) and consist of 2774 samples at 1140 sampling sites (stagnant and very slowly flowing water). We also used data from many colleagues and institutes as well as own samples, which together cover 3000 localities. Of course, in many cases information about the influence of many factors (e.g. microhabitat, oxygen content at night) was not available or not useful, and in a number of cases the identification was not reliable. The data from water authorities could rarely be used to establish the occurrence of species in acid or temporary water because such water bodies have rarely been investigated.

In the Netherlands most samples taken for identification are larvae and to a lesser extent also exuviae. If the identification keys raised any doubts this is mentioned in the first paragraph of the text on the relevant species. When identification was doubtful either the information was not used or the reader is warned in the text.

Usefulness of the data outside the Netherlands
The descriptions of the biology and ecology of the species are based partly on the literature from almost the whole of Europe (and sometimes even North America). Data about the life cycle or ecology in

mountain regions or quite different climatic zones are usually not mentioned. We have tried to present all available information relevant to the West European lowlands. Nevertheless, the numerical values are based on the Dutch situation, especially in the matrix tables (Chapter 4). The values in tables 2, 3 and 4 represent the likelihood of the larvae of a species being present in water of the indicated quality that also meets the other requirements of the species. However, some of these values will be different in Denmark or England, for example, because the likelihood of a species being present in other water bodies may be different. In the Netherlands most stagnant water bodies are eutrophic ditches, canals or shallow lakes. A consequence may be that larvae requiring a high oxygen content are rarely found in stagnant water bodies in the Netherlands. Acid water is found mainly in heathland and woodland (and only to a lesser extent in bogs) and oxygen deficit under ice in winter is generally insignificant.
Deep lakes are relatively rare in the Netherlands. The presence of a species in the profundal zone of lakes is mentioned where necessary, but the main text and tables apply to shallow water bodies.

The particular role of food and oxygen
Warwick (1975) suggested that the primary mechanism controlling the sequence of succession of chironomids in Lake Ontario was the availability of food. This thought runs as a continuous thread through our book. We think the importance of food is the main reason why the chironomid species composition tells us something about ecosystem functioning. We have therefore tried to give as much information as possible about the feeding of the different species. But even if we do not know the exact food demands of a species, we have to think about food availability to understand the differences between water bodies.

The Chironomini can probably tell us more about ecosystem functioning than most other chironomids. The decomposition pathways in systems dominated by algae and those dominated by higher vegetation are different. and the intensity and duration of the decomposition processes determines the availability of food for Chironomini larvae. Different species are adapted to differences in the availability of valuable food (see section 2.7) and therefore they depend indirectly on trophic conditions, temperature, pH, etc. For many species the ability to live in running water depends largely on the supply or removal of organic matter (Moller Pillot, 2003) and this feature of a stream or a locality determines the species community. Although this will be important for many species, it is mentioned in this book on only a few occasions where it has been proved.

In polluted water, in the profundal zone of lakes and during long-lasting ice cover in winter, oxygen can be the most important limiting factor. The Chironomini species obviously differ in their degree of susceptibility to oxygen shortage and toxic decomposition products. In this book, therefore, we had to pay more attention to the reasons why certain larvae cannot endure severe saprobity. In sections 2.12 and 2.13 we have tried to give more background information, but for most species the exact reason why they can or cannot occur in highly saprobic water cannot be given. Besides physiological adaptations, the choice of microhabitat and the behaviour of the larvae can be important.

2 GENERAL ASPECTS OF THE SYSTEMATICS, BIOLOGY AND ECOLOGY OF THE CHIRONOMINI

The general biology and ecology of chironomids has been treated in detail by Armitage et al. (1995). Some information of special interest for interpreting samples can be found in Part I of this series (Vallenduuk & Moller Pillot, 2007). The information given in this chapter applies to the Chironomini, although sometimes mention is made of other chironomids.

2.1 SYSTEMATICS

The subfamily Chironominae has been divided into three tribes: Chironomini, Pseudochironomini and Tanytarsini (Saether, 1977a). In this book we treat not only the Chironomini but also the Pseudochironomini because the larvae are not very different and in Europe the Pseudochironomini are represented by only one genus, *Pseudochironomus*. Within the Chironomini the phylogenetic relationships are not clearly expressed in the larval morphology. The *Harnischia* complex, a group of genera around *Harnischia* and *Cryptochironomus*, is an exception and these larvae are usually free living (without larval tubes) and more or less predatory (see Saether, 1971; Saether, 1977; Pinder & Reiss, 1983; Moller Pillot, 1984: 100–101).

2.2 NOMENCLATURE

Many species names are different from those stated in the literature from the 1970s and 1980s (Pinder, 1978; Pinder & Reiss, 1983; Moller Pillot, 1984). Most of these changes have been made as a consequence of the publication by Spies & Saether (2004). In most cases we follow their recommendations. For correct names see also Ashe & Cranston (1990) and Saether & Spies (2004). Particular care has to be given to the endings of species names. Some generic names appeared to be neutral in gender and in this case the species name also has to be neutral in gender if it is adjectival, for example *Cladopelma bicarinatum* and *Tribelos intextum* (Spies & Saether, 2004: 12).

2.3 IDENTIFICATION

Unlike Part I of this book series (Tanypodinae), the identification of larvae could not be treated. For this reason and because there have been many mistaken identifications within the Chironomini, in many cases we give the references which can be used to identify males, exuviae and larvae and state where descriptions or figures can be found. Descriptions of females are also often mentioned. As far as possible we have also tried to warn the reader of cases where the frequently consulted literature has led to mistakes. We do not state under each species that adult males can be identified using Langton & Pinder (2007) and exuviae using Langton (1991). For the exuviae we refer usually only to Langton (1991), unless Langton & Visser (2003) give a different result. The latter publication also contains descriptions of the exuviae of each species.

For many genera identification of larvae to species level is still a problem. The key by Moller Pillot (1984, in Dutch) has to be revised and we hope this will be done in English in the next volume in this series. A provisional English translation of the 1984 publication, with many additions, is used in the Netherlands, but it has no official status. The CD-ROM of Klink & Moller Pillot (2003) is out of print. The keys by Pankratova (1983) in Russian and Nocentini (1985) in Italian are rarely cited because these works are hardly used in northwestern Europe. The key by Biró

(1988) can be used only to a limited extent because much of it is out of date and contains many mistakes. When a special key for one genus exists this is mentioned under the genus heading. A special key has been published for palaeoecological purposes (Brooks et al., 2007). The best descriptions of genera can be found in Pinder & Reiss (1983, 1986; larvae, pupae) and Cranston et al. (1989; adult males).

2.4 DISTRIBUTION

The distribution of all species known from Europe can be found in Saether & Spies (2004), but they are only mentioned under the valid names. For the Netherlands Limnodata.nl gives the distribution of all data supplied by water authorities; other data are not mentioned. Almost all the data are based on identification of the larvae and in some cases, especially in the genus *Chironomus*, this has led to many mistakes. In certain cases the very incomplete maps by Moller Pillot & Buskens (1990) can give an impression.

Reiss (1968a) pointed out that the distribution of chironomids in Europe is mainly determined by temperature and other ecological factors and hardly by historical factors. This will apply more strongly to lowland species. However, it is possible that some species confined to large rivers and estuaries have survived mainly in Eastern Europe and the Netherlands and currently have a disjunct distribution, for example *Lipiniella araenicola* and *Paratendipes intermedius*. Some of these species may disperse to other countries in the future. The reverse is also possible: many species are only known from the first half of the twentieth century or are possibly retreating to more northerly countries in response to climate change. An example of a new species in the Netherlands is *Polypedilum nubifer*. In future it will be important to give species distributions for different periods.

2.5 LIFE CYCLE

The general aspects of the life cycle and development of Chironomidae have been described in Vallenduuk & Moller Pillot (2007: 9–14). Some aspects of special significance for the tribe Chironomini are summarised here.

Influence of temperature and feeding

The Chironomini are thermophilous and most species emerge later in spring than other chironomids. The number of generations in the Netherlands varies from 1 to 7 per year, depending on the species. The characteristics of the life cycle of all the species, as far as they are known, are given in Chapter 4, table 1. As temperature and feeding have much influence on the duration of development (Mundie, 1957; Laville, 1971; Sankamperumal & Pandian, 1991), very early emergences have been reported in urban water bodies and the number of generations in these water bodies are most probably higher (e.g. in *Polypedilum nubeculosum*). Even within a species the instar composition of overwintering larvae may vary substantially from year to year according to climatic conditions (Armitage et al., 1995: 246).
To give an impression of the influence of temperature and day length many data given in Chapter 3 are from outside the region covered by this book. Moreover, in some cases only data from Northern, Eastern or Southern Europe were available. Trophic conditions also influence the life cycle: in several species Orendt (1993: 179) found more generations in more eutrophic lakes.

Differences between spring and summer generation

Many authors report a large spring generation and a small summer generation, or the opposite (e.g. Mundie, 1957; Schleuter, 1985; Matěna, 1989; Otto, 1991). A small summer generation may be only a partial second (or third) generation because development was too slow to permit emergence before autumn. However, pre-

G

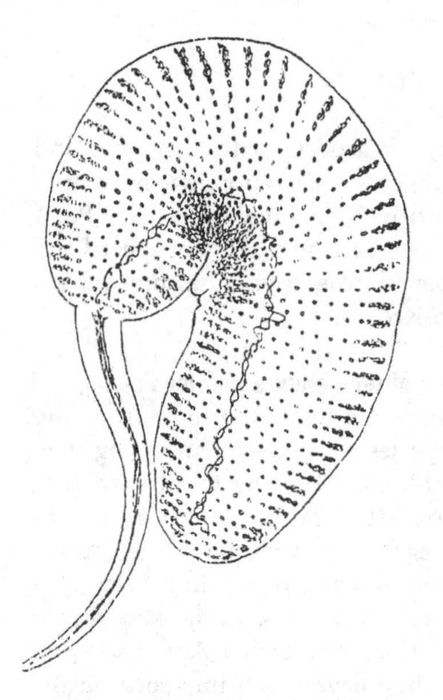

1 *Egg mass of Chironomus (from Munsterhjelm, 1920)*

dation by fishes or invertebrate predators is a factor in many cases (Matěna, 1989). Kajak et al. (1972) stated that an increase in the fish stock led to a decrease in benthos biomass and an increase in the biomass of fauna living on aquatic plants. In a number of cases we reported that species were present in spring in a water body where the oxygen content approached zero during summer nights. Although the absence of these species in summer was not proven, the oxygen content may be a cause of differences between spring and summer generations. A survey of all possible factors for *Chironomus plumosus* is given by Sokolova (1983: 223–244). Another consideration is that the spring generation is often better synchronised than later generations.

In brooks and streams losses in winter are often more important because fast currents transport the animals and their food sources downstream, resulting in a small spring generation (Pinder, 1983; Moller Pillot, 2003: 42). In many cases the cause of the reduced numbers in spring or summer is not known (e.g. in *Polypedilum sordens*). The possible impact of differences in food availability between seasons has been investigated for only a few species.

Eggs

The egg masses and eggs of the Chironomini are described in Nolte (1993: 45–51), which also contains the most important data from Munsterhjelm (1920) and much other literature. The egg masses of many Chironomini resemble a straight or slightly bent cylinder (see fig. 1), but globular, club-shaped, bale-shaped and string-shaped egg masses also occur. The number of eggs varies between species, but is also influenced by various factors, including temperature: higher numbers can be expected at lower temperatures (Dettinger-Klemm, 2003). Larger species, such as *Chironomus plumosus*, can produce egg masses with more than 2000 eggs, whereas within the genus *Polypedilum* sometimes no more than 100 eggs have been recorded. The mean length to width ratio is often about 3 or lower, which is a little lower than in the Tanypodinae. The colour of the eggs is very variable.

Larvulae

The first instar larvae or larvulae of the Chironomini usually have a trifid median tooth (Kalugina, 1959; Soponis & Russell, 1982), which is often different from the shape of this tooth in later instars (see fig. 2). The latter authors also give other characteristics of the larvulae.

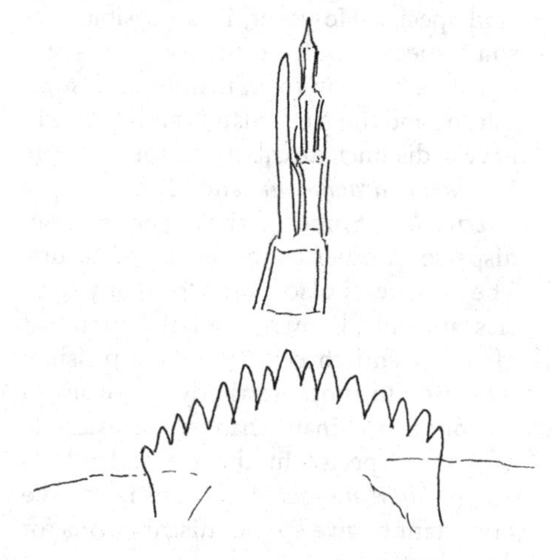

2 *Antenna and mentum of first instar larva of Endochironomus impar (after Kalugina, 1959)*

The larvulae are free swimming and feed on small particles. Gripekoven (1913: 204) found diatoms in the gut as well. Usually after two or more days they begin to settle and build tubes. Reist & Fischer (1987) stated that the larvulae of some *Chironomus* species already settled and built tubes within a few hours after hatching. Diapause in first instar has been reported only in *Paratendipes albimanus* (Ward & Cummins, 1978).

Older larvae

As a rule, the second instar larvae already exhibit all the most species-specific characters (other than sizes). Exceptions are the second instar larvae of *Endochironomus dispar* and *Paratendipes* species. The different instars can be identified by the size of the sclerotised parts: head length and head width of third instar larvae are 60% of the lengths found in fourth instar larvae, and this 60% ratio also applies to the second and third instar. The wear of the mentum is often strongest in the fourth instar and varies between individuals, depending on feeding behaviour. A very strong wear often occurs in the genus *Stictochironomus*.

Most Chironomini larvae in northwestern Europe have a winter diapause induced by short day length in late summer. Hibernation takes place in second, third and/or fourth instar and is more or less characteristic of each species (Goddeeris, 1983, 1986).

Pupae

The pupae of most Chironomini live in a tube. In many cases this is the larval tube, in other cases (*Harnischia* complex) the larvae build a pupal case. The thoracic horns function as a tracheal gill. Within their tube the pupae perform rhythmic undulations of the abdomen to continually draw in fresh water from outside the tube. As they have to swim to the water surface to emerge, many fall prey to fish and water fowl.

Adults

The lifespan of the adult depends on temperature and humidity. The large species dry out more slowly and live longer. In *Chironomus plumosus* Hilsenhoff (1966) found a maximum of eleven days at 16 °C.

2.6 MICROHABITAT

Many Chironomini species build a larval tube from fine particulate material using products from their salivary glands. Some species mine in plants (e.g. in the genera *Endochironomus* and *Glyptotendipes*); a few species live in colonies of sponges or Bryozoa or as parasites on snails. Most species of the *Harnischia* complex are free living.

The majority of the Chironomini are bottom dwellers. Another group of species lives on plants temporarily and at other times in or near the bottom, sometimes in later instars (e.g. *Microtendipes*) or during winter (e.g. *Endochironomus albipennis*).

The species of one genus may have very different microhabitat requirements (e.g. *Polypedilum*, *Glyptotendipes*), while the species in other genera occupy similar microhabitats (e.g. *Chironomus*). As a rule, small differences in the substrate can be very important. Many authors have stated an obvious preference for more or less fine particulate bottom material (e.g. McLachlan & McLachlan, 1975; Tolkamp, 1980). Higler (1977) reported great differences between submerged and emersed *Stratiotes* plants and Tokeshi & Pinder (1985) found contrasting patterns of microdistribution on different parts of plants.

Bottom dwellers

The vertical distribution of the larvae within the sediment is important for interpreting sampling results. If only the superficial layers of the sediment are investigated, not all the larvae will be caught. The percentage of larvae in the upper layers depends on the substrate type, the oxygen content, the species, the time of the year and the larval instars. For instance, the larvae of

Polypedilum were found by Chernovskij (1938) most numerously in the upper 2 cm of the muddy bottom (detritus + sand) of lake Chainoe and never deeper than 8 cm. However, in the sand-gravel bottom of the river Morava, Holzer (1980) found *Polypedilum* larvae in fairly high numbers at a depth of 20 cm and some specimens at a depth of more than 50 cm.

The differences between larvae of different species at the same site are considerable. In muddy bottoms with no oxygen Chernovskij (1938) found only *Chironomus plumosus* in large numbers deeper than 2 cm. In summer van de Bund (1994) found fourth instar larvae of *Stictochironomus sticticus* most numerously in the 2–5 cm layer – but the two first larval instars inhabited mainly the top 1 cm – and larvae of *Polypedilum* in the top 1 cm of the sandy sediment. By burrowing to greater depths parts of the populations of the later instars escape from predation by bream (*Abramis brama*). Olafsson (1992) proposed that different distributional patterns among instars is likely to reflect differences in diet, feeding strategies and haemoglobin content. Van de Bund (1994) suggested that *Stictochironomus* and *Polypedilum* larvae rely on very similar (!) food resources, leading to spatial segregation.

According to Olafsson (1992) third and especially fourth instar larvae burrowing deeper into the sediment is a general phenomenon in chironomids. This author observed no regular seasonal differences in the vertical distribution between larvae of any instar. However, many larvae move deeper into the bottom in late summer and in winter, especially in northern regions (Danks, 1971).

Inhabitants of plants, wood and stones

In contrast to other chironomids, a significant proportion of the plant-inhabiting Chironomini larvae mine in living or decaying stems and leaves. Some of them are facultative miners and can also be found on wood and stones (e.g. *Glyptotendipes pallens*). Mining larvae can eat plant tissue, but in most cases they use the mine to filter food from the water, which they draw through the tube by undulating their abdomens.

The inhabitants of plants and other firm substrates are less affected by harmful decomposition products and hypoxia than species living in the bottom. The matrix tables in Chapter 4 show that many of them seem to be more resistant to pollution. The reverse can be the case for pH because the pH in the bottom is often less extreme.

2.7 FEEDING

The specific differences between genera and species of Tanytarsini lie mainly in the characters of the antennae and setae on the head and abdomen, but in Chironomini they are as a rule found in the structure of the mouth parts. Webb (1980) and Webb & Scholl (1987) found differences in the structure of the ventromental plates between almost all the species of the genus *Chironomus*. The microarchitecture of the ventromental plates is involved in the processes of silk spinning and is especially important for filter feeding (Webb et al., 1981). The presence of a dorsal tooth may be an adaptation that increases the scraping area of a spoon shaped mandible (Olafsson, 1992a). Kurazhskovskaya (1969) found obvious differences between genera and species in the structure of the salivary glands, especially in the tribe Chironomini. For example, the salivary glands of *Glyptotendipes signatus* are conspicuously different from those of other *Glyptotendipes* species (ibid. p. 189, fig. 3). In most cases, the function of the differences are still unknown, but these examples indicate that feeding has played an important role in the evolution of genera and species.

The tribe Chironomini can be divided into the *Harnischia* complex, which displays predominantly carnivorous feeding behaviour, and the other genera, in which the larvae are phytodetritophages. The species composition of the latter group in particular can be useful for assessing water quality because the larvae of different species require different types of decomposing material and decomposition pathways. These aspects of feeding are examined more closely below.

Carnivorous species

The mouthparts of larvae of the *Harnischia* complex display obvious characters of carnivorous species, especially the sickle-shaped mandibles. Larvae associated with scraping, deposit feeding or filter feeding tend to have more triangular and stoutly formed mandibles (Chernovskij, 1949; Olafsson, 1992a). However, not all genera of the *Harnischia* complex appear to be carnivores: species of *Cladopelma* and *Microchironomus* feed on algae and detritus. Morozova (2005) observed that smaller *Cryptochironomus* species and younger larvae of this genus also eat more detritus. Some *Parachironomus* species are parasites on leeches.

Carnivorous species are free living. The prey is usually swallowed whole; complete head capsules of devoured small chironomids can often be seen in their intestines. The larvae of *Xenochironomus xenolabis* and *Demeijerea rufipes* feed in an exceptional manner, living inside the bodies of sponges and feeding on their tissues.

Phytodetritophagous species – food selectivity

Very few Chironomini feed on living tissues of aquatic plants (e.g. some *Endochironomus*, fide Gripekoven, 1913, and van der Velde & Hiddink, 1987). Gaevskaya (1969: 96) stated that the larvae of *Endochironomus tendens* and *Glyptotendipes cauliginellus* are obligate phytophages feeding mainly on the living tissue of higher aquatic plants, but this was based on incomplete observations.

Most larvae eat decaying plant material, fine particulate organic material, filamentous algae, diatoms, etc. Many species are able to exploit several different types of food and according to Monakov (2003: 328) they easily switch from one food type to another, selecting the type most readily available. Brundin (1956: 190) stated that chironomids in the profundal zone of lakes utilise plankton of very different species and in different stages of decomposition. However, this does not apply to all species, because obvious selectivity has sometimes been observed. For example, Rasmussen (1985) reported that *Chironomus riparius* and *Glyptotendipes paripes* utilised totally different food. Moore (1979) observed that the larvae of *Polypedilum nubeculosum* selected algae of a well-defined small size, such as *Scenedesmus*, and therefore ate very little filamentous algae. However, some algae like *Scenedesmus* were hardly or not digested.

Specialisation within a genus reduces the concurrence of species. Some species of *Chironomus* depend primarily on decomposing phytoplankton (e.g. *C. anthracinus*), whereas other species (e.g. *C. commutatus*) live mainly in water bodies with little phytoplankton (see the descriptions of the ecology of these species in Chapter 3). These differences explain much of the relation between species composition and ecosystem functioning.

Tokeshi & Townsend (1987) pointed out that interspecific competition plays a largely insignificant role in dynamic systems, like the epiphytic chironomid community in streams, because most species are irregularly present. Such epiphytic systems are mainly inhabited by Orthocladiinae, which display little feeding specialisation. Concurrence is probably most important in more stable systems, for instance on the bottom of lakes, where mainly Chironomini are found. It is possible that species in oligotrophic systems usually benefit most from feeding specialisation, as Goedkoop & Johnson (1992) suggested for profundal chironomids. In any case, oligotrophic and mesotrophic systems can contain very high numbers of species, which stimulates avoidance of concurrence.

The characteristics found in Naididae are probably also present in chironomids. For example, Bowker et al. (1983) stated that *Nais elinguis* selected unicellular algae and discriminated against multicellular algae, and also that *Chaetogaster setosus* specialised in very elongate diatoms. These seemed to be connected with the morphometry of pharynx and intestines.
For more information about food selectivity, see Armitage et al. (1995: 160 et seq.).

Decomposition – nutritive value
Decomposing higher plants can be an important source of food. Barnes (1983) reported a strong rise in the number of Chironomini larvae in new ball-clay ponds when vegetation increased. Izvekova (2000) reported a strong increase in the number of chironomids during and after the decay of inundated terrestrial plants. The larvae fed on decomposed plant tissues as well as on algae.

Many authors have stated that organic material is most valuable for chironomids when there is much bacterial activity (Fischer, 1969: 39; Izvekova & Lvova-Katchanova, 1972; McLachlan et al., 1979; van de Bund, 1994). Both uninfected material and old detritus have no value for chironomids (Ward & Cummins, 1979; Kajak, 1987). Baker & Bradnam (1976) found that *Chironomus* digest at least half the bacteria they ingest in situ. Bacteria may sometimes be the most important food for *Polypedilum nubeculosum* (Moore, 1979a). Goedkoop & Johnson (1992) suggested that enhancing bacterial production by burrow ventilation (microbial gardening) can be important for deposit feeders.

According to Bjelke et al. (2005) shredders can be divided in two groups – summer and winter growing species – which feed on the more easily degradable material and the more resistant matter respectively. Some tree leaves and most grasses, herbs and water plants decompose quickly and Izvekova (1980: 76–77) stated that decaying *Cladophora* lost much of its nutritive value after just a few days. Without doubt the availability of valuable decomposition products during the seasons is an important aspect of ecosystem functioning.

Adaptation of life cycles is (after food selectivity) the second type of specialisation which is the reason that species composition can be used to gain insight into ecosystem functioning. Ward & Cummins (1979) concluded that the life cycle of *Paratendipes albimanus* is adapted to the availability of detritus with a rich microbial flora. It is evident that species like *Chironomus anthracinus* that depend on phytoplankton cannot emerge very early in spring, as can be seen in *Chironomus dorsalis*. Temperature and pH are especially important factors. See also Moller Pillot & Buskens (1990: 8) and Moller Pillot (2003: 29 et seq.).

The digestion and ingestion of algae depends on the algal group, the condition of the algae and the chironomid species concerned. These aspects were investigated for the orthoclad *Cricotopus sylvestris* by Sorokin (1968), who found that green algae were eaten vigorously, but hardly digested; bacteria were digested a little better and diatoms and blue-green algae very well. Moore (1979a) also noted that many chironomids have problems digesting algae and Kajak (1987) found that *Chironomus plumosus* digested diatoms much better than other algae. We need to know more about such questions before we can explain the differences in ecology of many Chironomini species.

2.8 SWARMING AND OVIPOSITION

Chironomini females deposit their eggs only after mating; parthenogenesis is very exceptional (see *Zavreliella marmorata*). The males of all species swarm, but some species, such as *Chironomus riparius*, can mate in cages less than 50 cm high and can be reared for many generations (Strenzke, 1959: 31).

The majority of species deposit their eggs on firm substrates like macrophytes, wood or lake shores. Relatively few Chironomini, including *Chironomus plumosus*, lay their eggs directly on the water surface. Sokolova (1983) and Shilova (1965b) observed oviposition by females flying low over the water surface. Izvekova (1980: 98–99) stated that swarming and oviposition was severely reduced in strong wind and wind direction influenced the places where egg masses were deposited.
The females of some (or most?) species are able to actively select habitats for laying their eggs, a mechanism that can play a role in the distribution of species across habitats (Strenzke, 1959: 33; 1960: 121; Matěna, 1990: 55).

2.9 WATER TYPE

Current

Current velocity is important for different reasons. If the current is too strong (for shorter or longer periods) the larvae risk being carried away; if the current is too slow the oxygen content may become too low or too much silt may accumulate.

The Chironomini inhabit stagnant and slow-flowing water more than other subfamilies and the tribe Tanytarsini. Some Chironomini genera, for example *Paracladopelma* and *Polypedilum*, contain a greater proportion of species that inhabit running water. Very few Chironomini species can live in fast currents. Although publications on fast-flowing streams mention many species, in most cases these are larvae in more lentic stretches of the stream or at the under side of stones, or even the drift of larvae and exuviae from stagnant waters in contact with the stream. Species of stagnant water were found drifting in the stream more often than living at the sampling site.

Many species characteristic of flowing water also live in the littoral zone of lakes. Water movement in the lakes causes a flow of oxygen around the animals living near the shore, which can also be carried away by wave action. The velocities here are not fully comparable with current velocities in streams. For the interpretation of the current velocities given in this book, see Chapter 4.

Depth

For species living in lakes we also mention the depth at which the larvae live. This is considered to be especially important in relation to food and oxygen availability. Luferov (1972) stressed the role of light: in very clear lakes the larvae penetrate much deeper than in turbid lakes. Many larvae travel down in late afternoon and travel back when dawn breaks.

Temperature

The Chironomini are more thermophilous than other chironomids and this delays the emergence of the first generation in spring. Some genera, for example *Glyptotendipes*, *Cryptochironomus* and *Parachironomus*, are absent from alpine lakes above 1000 m (Reiss, 1968a).

Shade

Brooks, ditches and pools in woodland are often entirely shaded. This can influence the occurrence of species in different ways. However, for most species there is little information about these factors. Four main processes are at work:

1. In contrast to many terrestrial species, most aquatic chironomids avoid flying in woodland, and not only in summer (Delettre et al., 1992). The chance of egg deposition is therefore lower. There is no information about differences between species.
2. Shaded water bodies are poor in water plants and marsh plants, which provide a substrate and food for chironomid larvae.
3. Many Chlorophyceae require a fairly high light intensity and are scarce or absent in heavily shaded woodland (Hynes, 1970: 61), so that these algae are less available as food for chironomids. Many diatoms appear to be fairly

indifferent to this light reduction. In the Slinge stream near Winterswijk van Iersel (1977) reported a sharp fall in chlorophyll-a-content over five kilometres.

4. Production and consumption of oxygen depend on the quantities of plants and algae. The oxygen content during daytime is lower in shaded water bodies and the differences between day and night are smaller (van Iersel, 1977).

Permanence

In principle the text and tables about permanence apply to isolated water bodies falling dry in summer (see Vallenduuk & Moller Pillot, 2007: 29). Much of the literature on Chironomini is from studies of fish ponds which dry out at other times of the year, of dry beds in estuaries during low tide, or in rivers during periods of low water. Such situations are only treated in some cases in the text; the tables in Chapter 4 do not apply to such water bodies. In Central Europe and northern regions the complete freezing of many temporary water bodies may have more influence on the occurrence of species in these waters than in the Netherlands.

In general, mainly bivoltine and multivoltine species (flying until autumn) are able to live in temporary water because most Chironomini (in contrast to many Orthocladiinae) cannot survive, or can barely survive, in completely desiccated bottoms. The occurrence of most species in temporary water therefore depends more than in permanent water on the presence of egg-laying females in the surroundings. Noticeable exceptions are *Polypedilum nubifer* and *Polypedilum uncinatum*, which can survive in mud with a very low water content (Tourenq, 1975; Dettinger-Klemm, 2002, 2003).

2.10 pH

The influence of an environmental factor on the occurrence of a species is always complicated, even in the case of pH. Our pH data are based on measurements taken in the water column, often sampled when taking a sample of the fauna. It has not been possible to take into account other possibly important factors (see below) because the required data were rarely available.

Season

The pH in a water body varies during the season. In winter the pH value can be 1.5 to 2 lower than in summer, especially when the alkalinity (buffering capacity) is low. The tolerance of a species is also not constant. In some cases only the emergence of the adult is inhibited (Bell, 1970) and larvae in diapause or larvae in other instars can be more tolerant of acidity. During the winter many larvae usually stay in the bottom, where the pH in organic sediments is not as extreme as in the water.

If the preference for a higher or lower pH is based on the decomposition by bacteria or fungi, it is possible that only the duration of development is changed.

Humic level

Sensitivity to low pH is higher in clear water than in humic water, as has been proved for many invertebrates by Schartau et al. (2008). These authors also indicated that high levels of humic substances may ameliorate the toxicity of metals, primarily aluminium.

NH_3 and H_2S

The dissociation of NH_4^+ decreases at low pH values, but the toxicity of H_2S is much increased and mortality is possible, especially in silt layers without oxygen (Caspers, 1972). This means that the occurrence of a species in water with a high or low pH may depend on the NH_3 or H_2S content.

Other stress factors

It is self-evident that different stress factors reinforce each other, especially because in toxic environments larvae need more oxygen. This question has no particular implications for the use of the matrix keys, because all the values apply only to water bodies that are suitable for the species.

Predation
A rise in numbers of chironomids during acidification can be the result of a cessation of fish predation (Henrikson et al., 1982).

2.11 TROPHIC CONDITIONS

Trophic lake type systems
Using the occurrence of chironomids to estimate the intensity of production processes in lakes has a long history. It began with August Thienemann's investigations in the Eifel area of Germany. In one of his first articles (Thienemann, 1913) he made a distinction between lakes characterised by the presence of *Chironomus* and others where *Tanytarsus* was the dominant genus. The oxygen content in the profundal zone appeared to be the decisive factor: in eutrophic lakes hardly or no oxygen was present in the profundal zone for at least part of the year. The history of the trophic lake type systems has been summarised by Brundin (1949: 616 et seq.).

Saether (1979) stressed that different members of indicator communities have very different indicative properties and all biotopes are a mosaic of different habitats. He gives two tables of indicator species in oligohumic lakes, one for the profundal zone and one for the sublittoral and littoral zone, showing the range of occurrence in oligotrophic to very eutrophic lakes for every species. In the profundal zone of lakes the phytoplankton is practically the only source of food. This means that the process of decomposition on the bottom correlates well with the production and phosphate content in the water column. The production and decomposition of the vegetation are additional factors in the littoral zone and these are more dispersed in space and time. Sedimentation of silt is not uniform either. Langdon et al. (2006) stated that the presence of some species in shallow eutrophic lakes in England depended more on plant structure than on total phosphorus or chlorophyll-a.
Brodersen et al. (1998), who developed a method for using macroinvertebrates as trophic indicators in Danish lakes, got around the structure problem in the littoral zone by only using the fauna on stones. In lakes with a more extensive vegetation, however, the relation between the production in the lake and the presence of chironomids will never be direct and simple.

Trophic relations in Dutch water bodies
As the chironomid associations of Saether (1979) cannot be used to estimate trophic conditions in mesohumic and polyhumic lakes and the trophic systems are not useful in lakes with extensive vegetation, the problems in the pools, canals and ditches in the Netherlands are immense. But this does not mean the primary production has no influence on the presence of species. Production and decomposition of phytoplankton provides food for chironomid species and the oxygen regime depends on the nature of the autotrophic organisms and their lifespan. In water without much sedimentation of newly decaying phytoplankton, the species will have to be adapted to other food sources (Goedkoop & Johnson, 1992). Many species of chironomids are adapted to eat and digest specific food or their life cycle is adapted to match the progress of the decomposition process (see under Feeding). Besides quality, the quantity of food can also be important. *Chironomus riparius* displayed very low survival and growth rates in poor sediments (Vos, 2001), whereas other species, like *Pagastiella orophila*, can live in oligotrophic conditions (Saether, 1979).

In this book we try to give as much information as possible on these relations. We mention when a correlation between trophic parameters and the occurrence of species is given in the literature, but this will never be a linear relation in all types of water. We cannot give such relations in table form in Chapter 4 as we do for saprobity.

2.12 SAPROBITY

Our starting point is that assessments of saprobity are used to reflect the situation of the ecosystem as a whole: the bottom and the water layer. We confine ourselves mainly to undeep water and the chironomids living there.

As we stated in Part I (Vallenduuk & Moller Pillot, 2007: 20) we follow Sládeček (1973) in his interpretation of saprobity. However, it remains a rather vague concept if used for stagnant and running water bodies, acid and non-acid, etc. The main reason for giving separate data for saprobity and oxygen in our matrix tables is that we are concerned with different factors. These differences are significant, especially for Chironomini, which are mainly bottom dwellers and detritus feeders. There are three important factors:

1. Decaying organic matter is an important food source for many Chironomini.
2. The presence of ammonia, sulphide and nitrite can be the reason for the absence of species (see Neumann et al., 2001).
3. The larvae need a certain supply of oxygen that depends on the oxygen content of the water column and substrate, water movements and on their own physiology and behaviour (haemoglobin content, ventilation, etc.).

The first two factors mainly reflect the situation on the bottom; for the oxygen content we use the situation in the water column on the understanding that we use the regime during the 24 hour day. It will be clear that larvae living on plants have a higher tolerance to these factors (defined in this way) than obligate bottom dwellers.

We call a system **oligosaprobic** when there is so little decomposition that hardly any oxygen is used and the bottom is also rich in oxygen, even in places where dead leaves or wood lie on the bottom. In a **polysaprobic** environment the accumulation of decaying material is visible everywhere, the bottom conditions are almost anaerobic and decomposition products are present in reduced form (NH_3, H_2S); see also Chapter 4, introduction to table 2.

A limited number of Chironomini are inhabitants of plants, wood and stones in the water column or even near the surface. Such species (e.g. *Parachironomus arcuatus)* can live in water with a relatively poor oxygen regime and are also less influenced by toxic decomposition products, not because of physiological adaptation, but by choice of microhabitat. This can be seen in table 2 in Chapter 4.

Our figures for saprobity for the separate species (Chapter 4, table 2) are often different from those in the tables by Moog (1995). This is mainly because we take oxygen separately and because of differences between Austria and the Netherlands. Deposition of decaying material, the toxicity of chemicals and the flow through the bottom are quite different in mountain and lowland streams.

2.13 OXYGEN

Haemoglobin – ventral tubules

All or almost all species have some quantity of haemoglobin in the body fluid. This factor may partly account for the dominance of Chironomini in most tropical and subtropical standing waters with relatively high temperature and low oxygen concentrations. Species living mainly on plants near the surface (e.g. *Polypedilum cultellatum*, *Parachironomus arcuatus* and *Endochironomus albipennis*) usually have noticeably less haemoglobin than other species of these genera. Some species which burrow deep into the sediment have more haemoglobin per mg dry weight, for example *Polypedilum bicrenatum* (Heinis, 1993: 53). For the different types of haemoglobin see Heinis (1993: 60).

The haemoglobin concentrations in the larvae are variable and the larvae can probably change their haemoglobin level in response to environmental changes (Rossaro et al., 2007: 338). Moreover, the larvae have a lower haemoglobin production when their food does not contain enough iron (Neumann, 1961). Rossaro et al. (2007) found no correlation between haemoglobin content and the relative size of the ring organ of the pupa, although both the development of the pupal thoracic horn (measured as the area of the ring organ) and the haemoglobin content of the larva are thought to be good predictors of the oxygen tolerance of a species.

Fourth instar larvae develop much larger **ventral tubules** in oxygen poor conditions than in oxygen saturated water, but such differences are hardly observed in younger instars (Haas, 1956). Toxic chemicals can also influence the length of ventral tubules (see below). The **anal tubules** become conspicuously enlarged in electrolyte poor conditions and are not influenced by anoxia (Haas & Strenzke, 1957; Strenzke, 1960).

Anoxia

Walshe (1948) showed that larvae of the genus *Chironomus* survive longer under anaerobic conditions than many other chironomids. She found that under experimental conditions 50% of *Chironomus longistylus* larvae were still alive after 68 hours, much higher than in most other species with haemoglobin, for example *Paratendipes albimanus* and *Polypedilum nubeculosum*. Heinis (1993) reported that in Lake Maarsseveen I in the Netherlands a large population of *C. anthracinus* survived 2.5 months of anoxia, but this species did not survive a 5 month deoxygenation period in Lake Maarsseveen II. Hamburger et al. (1995) found that traces of oxygen in the environment prolong the survival time of the larvae considerably. Moller Pillot (1971) found much larger numbers of larvae in lowland brooks where oxygen was not completely absent.

It is highly likely that all species of the genus are able to survive anoxia. We found *C. riparius*, *C. annularius*, *C. plumosus* and *C. obtusidens* in ditches at Bergambacht, the Netherlands, where almost every night in summer the oxygen content dropped for some hours below 0.5 mg/l. However, during the whole night traces of oxygen were also present near the bottom of the ditches, most probably because of convection. Only *C. riparius* was present in the most polluted ditch and in winter these larvae survived more than a month of ice cover before dying.

C. riparius is not the most resistant species of the genus. In general, small larvae are more sensitive to anoxia than large larvae (Heinis, 1993: 73; Rossaro et al., 2007). The much larger larvae of *C. plumosus* are therefore more resistant to anoxia than the smaller *C. riparius* and the presence of the latter in more polluted ditches may be due to better tolerance of pollution. Larger larvae also survive anoxia much longer than smaller larvae of the same species (Hamburger et al., 1995; Int Panis et al., 1996). In conditions with less oxygen Heinis (1993) found considerably larger and older larvae than in more favourable oxygen conditions. Entz (1965: 134, 136) noted that in autumn the largest larvae of *Chironomus plumosus* (> 20 mm) were darker red than larvae with a length of 17–18 mm.

As mentioned in Chapter 1, anoxia under ice cover in winter is largely irrelevant in the Netherlands because the winters are usually very mild. In regions with colder winters, however, this may be why some species can be absent from shallow eutrophic water bodies.

2.14 IRON

Rasmussen & Lindegaard (1988) stated that Fe^{2+} ions have a considerable influence on the occurrence of chironomid species. However, high concentrations of Fe^{2+} ions can be expected only if there is seepage of groundwater through the bottom, especially where the sediment is rich in iron sulphide. The presence of Fe^{2+} ions may prevent chironomids from retreating into the bottom sediment and surviving there. Because water authorities mainly measure the total amount of iron and not the concentration of Fe^{2+} ions very little information about the influence of iron on macroinvertebrates is available.

2.15 TRACE METALS AND OTHER CHEMICALS

Much research has been done on the influence of toxic chemicals on chironomids. However, in this book make little mention of toxic chemicals because hardly anything is known about differences in response between species. Much of the work on the toxicity of trace metals and other chemicals has been on the genus *Chironomus*. Williams et al. (1986) stated that the first instar larvae in particular are very sensitive to cadmium. Heinis et al. (1990) found that food uptake by *Glyptotendipes pallens* ceased completely when fourth instar larvae were exposed to high cadmium concentrations. For further information about heavy metal contamination see e.g. Pinder (1986: 14) or Heinis (1993: 115 et seq.).

According to Neumann et al. (2001) the presence of sulphide and nitrite can be the cause of the absence of *Chironomus* larvae at many places, despite their ability to detoxify these chemicals. These authors also stated that the length of the ventral tubules depends on the nitrite concentration in the environment. The length of the anal tubules was not influenced.

2.16 SALINITY

For basic information about the influence of salinity see Vallenduuk & Moller Pillot (2007). In brackish water the majority of the chironomids usually belong to the Chironomini, especially to the genus *Chironomus*. Tourenq (1975) stated that *Chironomus annularius* is able to live in water with a higher chloride content when the oxygen content is higher. This is probably true for all species that are not very well adapted to brackish environment. Most authors do not take this into account and this effect of oxygen content should be borne in mind when using data on the tolerances of different species. It may also partly explain why Baltic and Scandinavian publications often report higher chloride contents than Dutch publications (see Vallenduuk & Moller Pillot, 2007: 20).

To get around the problem of irregular seasonal influences (and also the influence of oxygen) Krebs (1982, also published in Moller Pillot & Krebs, 1981) made a classification of chironomid communities according to increasing salinity of water bodies. In the literature sometimes salinity is used and in other cases chlorinity; the reader should note this difference when consulting the species descriptions in Chapter 3. In the tables in Chapter 4 we use only chlorinity (1‰ salinity = 540 mg Cl/l).

2.17 PIONEER SITUATIONS

It is a well-known phenomenon that during the first few years after a fish pond or reservoir has been filled a succession of species can be observed. *Chironomus* species are prominent colonisers of new habitat. Morduchai-Boltovskoi (1961) mentioned *C. plumosus* in particular. Matěna (1990) reported that in fish ponds other *Chironomus* species like *C. melanescens* and *C. annularius* can be still earlier. Species of other genera can also appear within a short time, for example species of the genera *Procladius*, *Cryptochironomus* and *Psectrocladius* (Brown & Oldham, 1984; Buskens, 1989; own data). Differences in species composition between the initial and later succession stages of a water body may be due to:

- better dispersal capacities of species;
- decomposition of terrestrial organic material (e.g. Buskens, 1989);
- absence of many parasites and predators;
- less concurrence.

The larvae of pioneer species can be found in habitats with very different saprobity levels than in later succession stages. This phenomenon has been taken into account in table 2 in Chapter 4. Pioneer situations have been excluded positively only for *Chironomus luridus* and *Chironomus riparius*. The phenomenon also occurs in most temporary waters.

2.18 DISPERSAL

Adults

The dispersal of adults has been treated at some length in Part I (Vallenduuk & Moller Pillot, 2007). In general, larger species can fly further than smaller species and many Chironomini species probably cover many kilometres. However, dispersal capacity does not only depend on the size of the species (see Part I: p. 9).

Larvae

The larvulae, and to a lesser extent second instar larvae, swim around in the water column before settling. Older larvae often leave their tubes when disturbed and in some species this occurs even after the slightest disturbance, for example in *Endochironomus* and *Glyptotendipes* species (Kalugina, 1959). Swimming by larvae is an important dispersal mechanism in the genus *Parachironomus* (see under the genus). Kalugina (1959) supposed that seasonal migration in search of a better habitat might be a dispersal mechanism in many species. Many larvae leave their tubes in oxygen poor conditions (Thienemann, 1954:133). Vos (2001) observed that *Chironomus* larvae in third instar creep or swim around when there is food shortage. In running water such larvae are often transported downstream (Moller Pillot, 2003: 30, 42 et seq.).

3 SYSTEMATICS, BIOLOGY AND ECOLOGY OF GENERA AND SPECIES

This Chapter contains written descriptions. For a numerical evaluation see the tables in Chapter 4.

Beckidia zabolotzkyi (Goetghebuer, 1938)

Cryptochironomus zabolotzkii Chernovskij, 1949: 54, fig. 13
Beckiella zabolotzkyi Saether, 1977: 120

SYSTEMATICS AND IDENTIFICATION

Saether (1979) proposed the name *Beckidia* as a replacement for *Beckiella* Saether nec Grandjean. Two species are known from Europe, one of which has been collected in the Netherlands as a subfossil in sediments of the river Rhine (Klink, 1989). Adult males and larvae of both species can be identified using Saether (1977). The pupa of the genus has been described and illustrated by Pinder & Reiss (1986) and Langton (1991). The larva of *Beckidia zabolotzkyi* has been described and illustrated by Chernovskij (1949), Pankratova (1983), Nocentini (1985) and Klink & Moller Pillot (2003). *B. tethys* has been illustrated by Pinder & Reiss (1983).

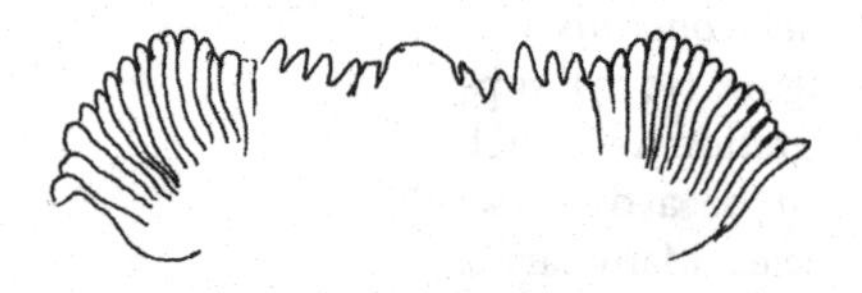

3 *Beckidia: mentum (after Chernovskij, 1949)*

DISTRIBUTION IN EUROPE

B. zabolotzkyi has been collected in Eastern and Southeastern Europe, including Poland and Italy, and as subfossil in the Rhine in the Netherlands (Klink, 1989). *B. tethys* is known from Italy and the Balkan peninsula (Saether & Spies, 2004).

MICROHABITAT

The larvae live in clean sandy bottoms in large rivers (Chernovskij, 1949).

Chernovskiia Saether, 1977

Chernovskiia Saether, 1977: 107–115; Pankratova, 1983: 150–152
Cryptochironomus monstrosus Chernovskij, 1949: 56, fig. 14
Orthocladiinae gen.? l. *macrocera* Chernovskij, 1949: 94, fig. 71–72

SYSTEMATICS

According to Saether (1977) and Saether & Spies (2004) two species of this genus occur in Europe: *C. orbicus* (syn. *Cryptochironomus monstrosus* Chernovskij) and *C. macrocera* (syn. Orthocladiinae gen.? l. *macrocera* Chernovskij). Pankratova (1983) renamed the first species *C. ra* Ulomskij 1941, but this is most probably a nomen nudum. She also writes that *C. macrocera* belongs to the Orthocladiinae, but probably overlooked the note by Saether (1977: 109) which states that what appears to be the mentum in the description and fig. 71 in Chernovskij (1949) in reality represents the collapsed ventromental plates.

IDENTIFICATION

The adult male of *C. macrocera* is unknown, but the name is possibly a synonym of *C. amphitrite*. Under this name, adult males and pupae can be identified using Saether (1977), who also keyed both species as a larva (see also Klink & Moller Pillot, 2003). The larva of *C. orbicus* has also been described and illustrated by Pinder & Reiss (1983). The larva of *C. macrocera* has been described and illustrated by Nocentini (1985).

DISTRIBUTION IN EUROPE

C. macrocera has been collected in Russia and Italy and as subfossil in the Rhine in the Netherlands (Klink, 1989). *C. orbicus* is known from Russia and the Balkan peninsula (Saether & Spies, 2004).

FEEDING

According to Chernovskij (1949) the larvae are carnivores, have a flexible body and creep through the sandy bottom.

MICROHABITAT

The larvae live in clean river sand, where the current is swift (Chernovskij, 1949; Pankratova, 1983). They belong to the psammorheobiontic fauna.

Chironomus Meigen, 1803

SYSTEMATICS

At present a division of the genus in four subgenera is generally accepted (e.g. Wülker, 1999; Saether & Spies, 2004; Langton & Pinder, 2007):
Camptochironomus
Chaetolabis
Lobochironomus
Chironomus

A number of complexes have been discerned cytologically within the subgenus *Chironomus* (see Keyl, 1962; Michailova, 1989; Kiknadze et al., 1992; Wülker, 1999), including:
C. thummi complex
C. pseudothummi complex
C. parathummi complex
C. lacunarius complex

Each complex can be divided into groups of species, but this division into groups is not the same in different publications. These groups have nothing to do with the larval groups defined by Lenz (1954–62) and copied by later authors (e.g. Moller Pillot, 1984). These larval groups have no systematic or ecological value.

Some species earlier attributed to the genus *Einfeldia* have now been placed by many authors under *Chironomus (Lobochironomus)*. In this book we have not adopted this reclassification for *E. carbonaria* and *E. dissidens*, pending further investigations. The larval type of these two species is quite different from *Chironomus* larvae and we think these species should be placed in a separate genus (see *Einfeldia*).

Hybridisation between different species of *Chironomus* has been stated at least within the *plumosus* group (Ryser et al., 1983; Rychen Bangerter & Fischer, 1989). In the laboratory *C. riparius* and *C. piger* can also hybridise, but no hybrids have been found in nature (Strenzke, 1960).

Kalugina (1972) stated that adults of this genus of hibernated larvae are usually darker in colour and larger than the adults of summer generations. Such seasonal forms are sometimes described as individual species.

IDENTIFICATION

According to Kiknadze et al. (1991) more than 100 species can be discerned worldwide on the basis of karyotype characteristics. It is not possible to identify all these species reliably from the adult male, pupa or larva. For karyological characteristics and a key to some species see also Michailóva (1989).

Strenzke (1959) described and keyed 15 species as adult male and female based on morphological characters. Pinder's key to adult males (1978) includes 20 species. Langton & Pinder (2007) extended this to 32 species. In the same manner, Langton (1991) keyed the pupal exuviae of 41 species (some of them without a valid name) and expanded this key in Langton & Visser (2003) by adding more than 20 European species. Because the identification of this genus is very difficult and extensive descriptions are only present in the 2003 publication, it is hardly possible to identify this genus without consulting the latter publication. In many cases identification using the provisional matrix table by Wilson & Ruse (2005: 154) is unreliable. At present the most useful key for identifying the larvae is Vallenduuk et al. (1995). However, this key is still very incomplete and should be improved as soon as possible. Where identification using Webb & Scholl (1985) gives different results, this is mentioned below.

LIFE CYCLE

All *Chironomus* species probably have a diapause in winter in fourth (rarely also in third) instar larvae. Second instar larvae are very rarely collected in winter, probably as a result of late oviposition. Pupation takes place in the larval burrow (Hilsenhoff, 1966; Jónasson & Kristiansen, 1967; own observations). When development is complete, the pupa swims to the surface and emergence takes less than one minute. In summer most specimens emerge in the evening (Fischer, 1969; Hilsenhoff, 1966).

In a temperate climate, as in the Netherlands, most species have two or three generations a year. However, four to seven generations can occur, for example in *C. riparius*. Univoltinism is an exception (e.g. *C. pilicornis*), except in cold lakes and boreal and alpine regions. Some females deposit two egg masses (e.g. *C. nuditarsis*, see Fischer, 1969); the second egg mass is deposited two or three days after the first and contains fewer eggs. Larval food and temperature influence the size of the female and the number of eggs (Fischer: 1969: 37).

EGGS

Nolte (1993) described the characteristics of the eggs and egg masses of many *Chironomus* species.

OVIPOSITION

The species of *Chironomus* show obvious differences regarding the place of oviposition. Egg masses can be thrown off on the open water surface, but most species choose a more or less firm substrate, often in a species-specific habitat. These differences are treated for every species separately where information is available. Strenzke (1960) stresses the importance of the active selection of habitats by female *Chironomus* for laying their eggs (*aktive Verteilungsregulation*). He suggests that the larvae are able to live in a greater variety of water types than selected by the females.

MICROHABITAT

Larvae of the genus *Chironomus* are typical bottom dwellers, living in tubes of different forms. They are scarce or absent on stones without mud (Mol et al., 1982; Brodersen et al., 1998). Large numbers will settle on artificial substrates only if much organic silt becomes lodged between the pieces of firm substrate (own unpublished observations). *Chironomus* larvae are also found on plants that are more or less covered with mud (Higler, 1977; Drake, 1982). In silty bottoms the larvae can be found to a depth of 12 and rarely 20 cm (Sokolova, 1983: 194). Šterba & Holzer (1977) reported the presence of many *Chironomus* larvae in sand/gravel substrate in the bottom of rivers at greater depths, down to more than 70 cm below the surface.

PARASITISM

Chironomus larvae are often parasitised, possibly more than other genera. Thienemann (1954) presented a survey which indicates that especially Protozoa, viruses, Trematoda and Mermithidae can be important. Fischer (1969) observed a severe decline in population numbers because nearly 80% of the older larvae were infected with Mermithidae. Sokolova et al. (1992) reported infestation rates of *C. piger* by the mermithid *Hydromermis contorta* ranging from 29% to 96% in one year and from 2% to 59% in another year.

D

Chironomus aberratus Keyl, 1961

IDENTIFICATION

The adult males cannot be reliably distinguished from those of *C. sororius* (Wülker, 1973). It is also very difficult to distinguish between the larvae of *C. aberratus* and *C. sororius*. Geiger et al. (1978) illustrated a difference in the colouration of the frontal apotome, but this was not copied by Webb & Scholl (1985) or Vallenduuk et al. (1995). The recorded observations of this species given below were karyologically verified.

DISTRIBUTION IN EUROPE AND THE NETHERLANDS

C. aberratus is known only from Central Europe, Bulgaria and Russia (Saether & Spies, 2004). Some larvae from the Netherlands identified as *C. aberratus* (see Nijboer & Verdonschot, 2001: 53) have yet to be verified.

WATER TYPE

Wülker (1973) found *C. aberratus* together with *C. sororius* in pools in the Black Forest. Wülker & Klötzli (1973: 476, 481) supposed that the larvae are not strongly cold stenothermous because some larvae were found in a meadow pool which showed large fluctuations in temperature. Consistent with this supposition, Michailova (1989) collected the larvae in pools near the coast of the Black Sea in Bulgaria. Ryser et al. (1980) collected larvae from pools in the Alps near Bern. Matěna & Frouz (2000) also collected this species from small water bodies (sandpits, forest pools) in the Czech Republic, often together with *C. sororius*.

Permanence

Michailova (1989) reported that the larvae lived in temporary pools.

Chironomus acidophilus Keyl, 1960

C. acidophilus has been collected in Germany, Central Europe and Sweden (Saether & Spies, 2004). In the Netherlands larvae from several localities have been identified as *C. acidophilus*, but none of them have been cytologically verified. Strenzke (1960: 120) suggested that *C. acidophilus* is a stenotope inhabitant of mineral acid water.

Chironomus acutiventris Wülker, Ryser & Scholl, 1983

Chironomus gr. *fluviatilis* p.p. Lenz, 1954–62: 153, 161, fig. 73

SYSTEMATICS AND IDENTIFICATION

The adult male is absent from the key by Langton & Pinder (2007). Wülker et al. (1983) described the karyotype and the morphology of the adult male and the larva. The *fluviatilis* group classified by Lenz (1954–62) and copied by Moller Pillot (1984) applies mainly to this species, especially because Lenz added 'river inhabitants' to his key. The acute points of the ventral tubules are rather characteristic. The larva can be distinguished from *C. obtusidens* by the larger mental size (distance between first lateral teeth), although there is a small overlap in this character between both species. The number of striae on the ventromental plates, as noted by Vallenduuk et al. (1995) and Istomina et al. (2000), is not reliable, but as a rule *C. acutiventris* can be also identified by the darker ventromental plates and the more acutely pointed ventral tubules. Istomina et al. also reported some differences with other species of the *obtusidens* group (especially *C. sokolovae*). The exuviae are absent from the key by Langton (1991), but can be identified using Langton & Visser (2003).

Wülker et al. (1983) distinguished two subspecies on the base of their karyotypes. Their larval morphology is hardly different.

DISTRIBUTION IN EUROPE AND THE NETHERLANDS

The occurrence of *C. acutiventris* has been stated definitively in only a few countries in Central and Western Europe (Saether & Spies, 2004). However, the species is most probably not rare in the greater part of Europe. In the Netherlands the species has been collected in the floodplains of the large rivers and at scattered sites in the Pleistocene sandy regions (Nijboer & Verdonschot, unpublished data.; H. Cuppen, unpublished data; own data). In at least one case (Beerze stream, Westelbeers) the identity of the larvae has been cytologically verified.

LIFE CYCLE

Wülker et al. (1983) found females depositing eggs from early May until September. The species was here possibly multivoltine; development in the laboratory was quicker than in *C. plumosus*. Smit et al. (1996) reported two generations, but their data are not very clear and nothing is mentioned about the generation emerging in May. One generation emerged in early July, followed by increasing numbers of small larvae in mid July. In September/October they again found a small peak in the numbers of small larvae.

FEEDING

Smit et al. (1996: 505) noted that the larvae show an opportunistic feeding behaviour (without further explanation). We observed larvae gathering organic detritus particles as food. In the estuaries they probably feed more on dying algae and in the lowland brooks more on fine detritus.

MICROHABITAT AND SOIL TYPE

In Dutch estuaries the species seems to be especially present in dynamic situations, possibly as a result of oviposition activities by females in recently submerged sediments (Smit et al., 1996). They are found here on silty and sandy sediments (Smit et al., 1994). Peeters (1988) collected larvae of gr. *fluviatilis* locally on fine sand in the river Meuse. In dead river arms the larvae also live on clay (own data).
In lowland brooks we collected the larvae at sheltered places in sandy sediments with some organic silt. During rearing experiments they lived in heaps of gathered organic particles. Klink (1991) collected the larvae scarcely on stones in the river Meuse.

WATER TYPE

The larvae are found in a number of European rivers (Lenz, 1954–62). In the freshwater lakes of the Rhine-Scheldt estuary the species is locally numerous in the 'upper eulittoral' zone (Smit et al., 1994; 1996) The larvae are scarce in the Dutch part of the river Meuse (Klink, 1991; Peeters, 1988). They are not rare in lowland brooks in the Netherlands (Limnodata.nl; H. Cuppen, pers. comm.; own data). The species is rarely mentioned in the literature on faster flowing streams (e.g. Orendt, 2002a). In stagnant water the species is most probably confined to lakes (especially estuaries) and dead river arms, where it is less rare than the matrix table for current indicates. Wülker et al. (1983) collected the species in Swiss lakes close to the shoreline. Records of finds in ditches have yet to be verified.

D

pH

The species has not been collected in acid water.

TROPHIC CONDITIONS AND SAPROBITY

In lowland brooks the larvae are characteristic of eutrophic water with little organic pollution (H. Cuppen, unpublished data; own data). In the Rhine-Meuse estuaries the larvae live in highly productive areas with very high chlorophyll-A contents (Smit et al., 1996). In both situations the oxygen content near the bottom is high.

SALINITY

Lenz (1954–62: 161) reported presence of larvae of group *fluviatilis* in brackish lakes in Siberia, but it is uncertain whether this actually applies to *C. acutiventris*.

Chironomus agilis Shobanov & Dyomin, 1988

SYSTEMATICS AND IDENTIFICATION

C. agilis belongs to the *plumosus* group. Identification of the adult is not possible. The larvae can be identified using Shobanov (1989, 1989a) and Vallenduuk et al. (1995). See also the figures in Kiknadze et al. (1991).

DISTRIBUTION IN EUROPE

According to Saether & Spies (2004) the species is only known to be present in Russia. A recorded find in the Biesbosch basins in the Netherlands by Kuijpers (pers. comm.) needs to be verified.

Chironomus annularius Meigen, 1818

NOMENCLATURE AND IDENTIFICATION

C. annularius presents some problems when identifying adult males. Pinder (1978) supposed that his species described by Degeer (sensu Edwards) was not the same as the species of Meigen (sensu Strenzke). Lindeberg & Wiederholm (1979) give the differences between them. Spies & Saether (2004) argued that the conspecifity of all material reported as '*C. annularius*' appears highly unlikely. Langton & Pinder (2007) made no decision on this point and called the species *C. annularius* auct. Moreover, it appeared to be difficult to differentiate *C. annularius* from *C. entis*, a species only recently collected in Western Europe.

When identifying larvae it seems to be no problem to distinguish *C. annularius* from *C. entis* because the latter is much larger (head width >770 μm) and belongs to gr. *plumosus* sensu Shobanov, 1989. The differences between *C. annularius*, *C. melanotus* and *C. cingulatus* as given in Vallenduuk et al. (1995) are not reliable (see under *C. cingulatus*). These authors group the three species together as *C. annularius* agg. For the time being we treat *C. annularius* as a species, although it is actually an aggregate, but we think it is not appropriate to add a new name or category.

DISTRIBUTION IN EUROPE AND THE NETHERLANDS

C. annularius lives in nearly the whole of Europe (Saether & Spies, 2004). In the Netherlands the species can be found throughout the country (Nijboer & Verdonschot, unpublished data).

LIFE CYCLE

Jónsson (1987) recorded only one flight period of *C. annularius* at Lake Esrom in Denmark, in June. Kalugina (1972) found four generations a year in Russian fish ponds. In the Czech Republic Matěna (1990) observed three clear emergence periods from April to September. Palmén & Aho (1966) noted that a third generation is likely in Southern Finland, depending on the temperature conditions.

Krebs (1978) found larvae in third and fourth instar in winter. In the Camargue the adults emerge the whole year round, with a maximum in May and June (Tourenq, 1975). Smith & Young (1973) also found much less emergence in summer than in spring. Dettinger-Klemm (2003: 135) reported experimental results showing that total development is delayed by long days, but quickened by higher temperatures between 5 and 30 °C. In experiments with short days and low temperatures the development of *C. annularius* was faster than that of *C. dorsalis* (sensu Strenzke) and *C. luridus*. At 15 °C, however, the development of *C. dorsalis* was faster (Dettinger-Klemm, 2003: 136). Matěna (1990) also noted relatively slow growth of *C. annularius* larvae. The role of food in all these cases remains unclear.

FEEDING

Matěna (1990) observed an increase in the numbers of larvae after an increase in the amount of decaying filamentous algae. Ferrarese (1992) found the larvae damaging the roots of rice seedlings; the larvae are most likely to be solely or mainly root growth disturbers. In most cases the larvae probably feed on fine particulate organic material, but periphyton may make up an important part of the diet (see Microhabitat). The larvae do not appear to depend on the presence of phytoplankton.

OVIPOSITION

Strenzke (1959) saw some egg masses of *C. annularius* deposited on a firm substrate, but supposed that the egg masses were often deposited on the water surface. Matěna (1990) suggested that the egg masses are deposited along the water's edge. Strenzke (1960: 121) observed egg deposition along water edges without macrophytes. The number of eggs per egg mass is 940–1250 (Strenzke, 1959); Konstantinov (1956, cited by Nolte, 1993) found only 700 eggs per mass.

MICROHABITAT

The larvae live in places with much detritus. According to Palmén & Aho (1966) a bottom sediment with coarse detritus is preferred (see also Palmén, 1962). Tourenq (1975) noted larvae on plant stems and leaves covered with periphyton. However, Matěna (1990) stated that a clean muddy bottom is preferred to coarse organic material and Moldován (1987) found the species only scarcely in litter-rich habitats.

DENSITIES

Palmén & Aho (1966) obtained 2750 emergent adults/m^2 from one locality, but in most other cases fewer than 50/m^2. Moldován (1987) found maximum densities up to 6000 larvae/m^2 in a reservoir of a sewage treatment plant in Hungary.

WATER TYPE

Current

In the literature *C. annularius* is usually mentioned as a stagnant water species and only occasionally collected in running water (Lehmann, 1971; Krieger-Wolff & Wülker, 1971; Fittkau & Reiss, 1978; Orendt, 2002a; Limnodata.nl). However, in moderately polluted slow-flowing lowland brooks the larvae are not rare, especially if the velocity of the current is less than 15 cm/s (GWL, unpublished data; own data). Caspers (1991) collected the exuviae of this species in the upper parts of the river Rhine, but it seems to be unlikely that the larvae live in the river itself (see also Caspers, 1980; Becker, 1994).

Dimensions

C. annularius has often been collected in small pools or ponds (Remmert, 1955; Geiger et al., 1978; Matěna, 1990). In the Netherlands, Krebs (1981, 1984) found the larvae (province of Zeeland) commonly in small and medium-sized water bodies up to 50 m wide and 3 m deep. The species is absent in the very large estuary lakes in the western part of the country (van der Velden et al., 1995; Smit et al., 1996), where the soil is possibly also too poor in detritus. Palmén & Aho (1966) stated that the species is absent at depths of more than 4 m and abundant only in shore localities at 1–2 m depth.

Shade
The larvae are rather common in woodland pools (own data).

Permanence
Dettinger-Klemm (2003: 135) called *C. annularius* a typical species of permanent ponds. Schleuter (1985) and Matěna (1990) did not find the species in temporary pools. However, the larvae are often collected in temporary water such as pools, the upper courses of lowland brooks (own data), rice fields (Ferrarese, 1992) and marshes (Tourenq, 1975). Their presence in temporary water will be the result of oviposition in pioneer situations (see above) and depends on the presence of water during the flying period of the species.

pH

The larvae are especially common in more or less alkaline water with relatively fast decomposition. In acid water they are usually absent (e.g. Moller Pillot, 2003). However, there are some records of larvae and exuviae in acid water with pH values of about 4.4 (van Kleef, unpublished).

TROPHIC CONDITIONS AND SAPROBITY

C. annularius is a rather small species within the genus and therefore probably less resistant to anoxia (see under the genus). Nevertheless, the larvae (at least belonging to *C. annularius* agg.) are found in polysaprobic conditions more often than most other species of *Chironomus*, but they tolerate less severe pollution than *C. riparius* (unpublished own results). The larvae are relatively often collected in rather severely polluted lowland streams (GWL, unpublished data). Based on the identification of adult males, Krebs found the species regularly at NH_4^+ levels in the order of 1.5 to 2 mg/l, and on two occasions at levels as high as 4.7 and 15.3 mg/l (Krebs & Moller Pillot, in prep.).

SALINITY

The larvae are most common in fresh water, but endure also slightly brackish water up to 4000 or 5000 mg chloride/l (Remmert, 1955; Tourenq, 1975; Krebs & Moller Pillot, in prep.). When the oxygen content is high (12 mg/l or more) the larvae can live in water with a chloride content of 8000 mg/l (Tourenq, 1975).

DISPERSAL

Palmén & Aho (1966) suggested an offshore migration of larvae in autumn or winter in Finnish lakes. Matěna (1990) reported rapid colonisation by oviposition of newly filled fish ponds. Krebs (1978) noted rapid colonisation of a brackish ditch after refreshment in autumn (after a few weeks).

Chironomus anthracinus Zetterstedt, 1860

IDENTIFICATION

The adult males can be identified using Strenzke (1959) and Langton & Pinder (2007). However, Strenzke gives a beard ratio of 6.8–8.2, whereas Langton & Pinder give 4.5–5.7. Misidentifications are probably not rare (see also Wülker, 1999 and below under Water type). The female has been described and illustrated by Strenzke (1959) and Rodova (1978). The exuviae have been keyed by Langton (1991), the larvae by Vallenduuk et al. (1995); it is probable that another species also keys out as *C. anthracinus* in the latter key. The larvae are absent from the key by Webb & Scholl (1985). Shobanov (1996) gives a description, with figures, of the adult male, pupa and larva in fourth instar.

DISTRIBUTION IN EUROPE AND THE NETHERLANDS

The species has been recorded in many countries scattered over Europe (Saether & Spies, 2004). Only a few reliable records are known from the Netherlands (Lake Maarsseveen, Biesbosch reservoirs) because there are few deep lakes and the profundal zone of sandpits have been rarely investigated. All records mentioned by Limnodata.nl are doubtful (see Identification).

LIFE CYCLE

C. anthracinus has a one-year or two-year life cycle, but in the latter case part of the population may emerge after one year (Jónasson, 1972; Johnson & Pejler, 1987; Heinis, 1993). The larvae winter in third and fourth instar (Johnson & Pejler, 1987; Heinis 1993: 25). Emergence takes place mainly in spring (mainly in May), but a smaller emergence peak was observed by Kouwets & Davids (1987) and Heinis (1993) from July to September. There is no information about the life cycle in the Biesbosch reservoirs, where a high oxygen content is artificially maintained in the profundal zone.

FEEDING

Jónasson & Kristiansen (1967) describe the tube as a conical tube in the mud and a cylindrical 'chimney' above the mud surface. The presence of the chimney depends on the oxygen content of the water column and the quality of food. Stagnation of food uptake in summer can be the result of oxygen shortage (Jónasson & Kristiansen, 1967). The food is collected from the mud surrounding the tube. According to Johnson (1986) *C. anthracinus* is a deposit feeder, scraping particles (detritus, diatoms and flagellates) from the sediment surface. Walshe (1951) noted that the larvae were sometimes seen feeding on the mud well below the mud surface. However, when the surface layers are absent, the larvae have hardly any food in the gut and hardly survive (Kajak, 1987).

The food base for this species is the decaying phytoplankton that reaches the bottom of the lake (Jónasson & Kristiansen, 1967; Sokolova, 1971). In Lake Erken the growth rate was highest during the autumn diatom bloom (Johnson & Pejler, 1987). Jónasson (1972) also recorded a higher growth rate in Lake Esrom after the spring maximum of phytoplankton. Izvekova (1980) proved that the nutritional value of algae decaying during some days was much higher than that of fresh material. Decaying particles from vascular plants were also much consumed and had a high nutritional value. The larvae also filter the water.

MICROHABITAT

Johnson (1984) found a strong preference for muddy sediments with a small particle size (< 20 μm), but the species is less dependent on sediment composition than *C. plumosus*.

OVIPOSITION

Pelagic egg masses are thrown off above the profundal zone and after swelling the egg mass sinks to the bottom (Sokolova, 1971). Sokolova found 400–1000 eggs per egg mass.

DENSITIES

Johnson & Pejler (1987) recorded maximum densities of fourth instar larvae of about 1000 larvae/m^2 in Lake Erken. Jónasson (1972) found an average number of about 9000 larvae/m^2 (of all instars) in Lake Esrom. When the density of larvae was higher than 2000/m^2, these larvae cleared the mud surface of newly laid eggs, preventing the development of a new generation.

WATER TYPE

Dimensions

Fittkau & Reiss (1978) reported the occurrence of *C. anthracinus* in small and large, stagnant and flowing water bodies. Records from running water are scarce and usually based on identification of adult males (e.g. some males caught by Lehmann, 1971 and Becker, 1994). All cytologically verified material of this species is from more or less large lakes and reservoirs (Krieger-Wolff & Wülker. 1971; Matěna & Frouz, 2000). The emergence of *C. anthracinus* from small shallow rock pools (Lindeberg, 1958) and the recorded presence in a pond in Germany (Schleuter, 1985) are improbable. Shobanov (1996) called all reports of *C. anthracinus* in undeep water bodies doubtful. Whiteside & Lindegaard (1982) found the larvae mainly in the deeper parts (> 9 m) of the Danish Lake Grane Langsø and never less than 6 m deep. In Lake Erken the larvae live at a depth of 12–20 m (Johnson, 1984) and in Lake Maarsseveen at a depth of 8–20 m (Heinis & Swain, 1986).

pH

Most records are from alkaline water, but there are a few exceptions (e.g. Koskenniemi & Paasivirta, 1987: pH 5.5–6). Shobanov (1996) called all reports of *C. anthracinus* in acid water doubtful.

TROPHIC CONDITIONS AND SAPROBITY

According to Saether (1979) and Wiederholm (1980) *C. anthracinus* larvae live in the profundal and sublittoral zone of lakes, mainly in oligotrophic and mesotrophic water, in less eutrophic conditions than the larger larvae of *C. plumosus* (see under this species). However, Matěna & Frouz (2000) and Kuijpers (pers. comm.) found fewer or even no larvae in water with more scanty food supplies. *C. anthracinus* is very tolerant to hypoxia: according to Heinis & Crommentuijn (1992) the larvae are still able to feed at oxygen concentrations as low as 2.5 mg/l. Jónasson (1972) noted that growth stops once the oxygen content has fallen below 1 mg/l. The larvae live in mesotrophic lakes at great depth and also in environments relatively rich in food. They can survive a period of 2.5 months of anoxia (Jónasson & Kristiansen, 1967; Heinis, 1993). However, they did not survive a 5 month deoxygenation period in Lake Maarsseveen II (Heinis, 1993). According to Mundie (1965), the larvae of this species may survive long periods of anoxia partly because they (and not only early instars) often move to the water mass of the lake, especially in late summer. The matrix table for saprobity and oxygen has not been filled in for this species because the saprobity and oxygen regime in the profundal zone cannot be compared with other situations.

SALINITY

There is one unverified record of *C. anthracinus* from slightly brackish water in the Netherlands.

Chironomus aprilinus Meigen, 1830

Chironomus halophilus Strenzke, 1959: 25 et seq., fig. 14; 1960: 121 et seq.; Palmén & Aho, 1966: 217–244; Tourenq, 1975: 180 et seq.

NOMENCLATURE AND IDENTIFICATION

In the past there have been problems with the species name *aprilinus* (see Spies & Saether, 2004: 34). At the moment the name is generally accepted. The interpretation of the name by Tourenq (1975: 178) is unclear.

Identification of adult males has never been a problem. The species has been keyed by Strenzke (1959) and Pinder (1978). The female has been described and illustrated by Strenzke (1959: 25). Parma & Krebs (1977) gives a comparison of measurements of Dutch, German and Finnish populations of males and females and of their wing lengths at different times of the year. The exuviae have been keyed by Langton (1991) and the larvae have been keyed by Webb & Scholl (1985) and Vallenduuk et al. (1995). Descriptions and figures of the larval morphology and karyology can be found in Kiknadze et al. (1991).

DISTRIBUTION IN EUROPE AND THE NETHERLANDS

The species has been collected in many countries in Western, Northern and Southern Europe (Saether & Spies, 2004). In the Netherlands the larvae have been found mainly in the Delta region, but also in the brackish regions in the northern part of the country (Limnodata.nl).

LIFE CYCLE

Palmén & Aho (1966) reported two generations in southern Finland, but in cold environments only one and in warm years possibly a partial third generation. In the delta region of the Netherlands Krebs (1978) observed two generations a year (sometimes three?), emerging from the end of March until early November. Larvae overwinter in third and fourth instar. Strenzke (1959: 39) stated that rearing and copulation is possible in the laboratory.

OVIPOSITION

The eggs are deposited on the water surface and sink within a short time to the bottom (Strenzke, 1959). Sometimes this author found egg masses deposited on a firm substrate. He counted 840 to 1250 eggs per egg mass.

MICROHABITAT

The larvae are bottom dwellers. Palmén & Aho (1966) found the species at localities with a large amount of organic detritus and an overlying layer of filiform algae covering the hard mineral bottom. The larvae were scarce on fine sand sediments poor in organic detritus and on very fine mud. However, Steenbergen (1993) found the larvae scarcely between submersed vegetation, mostly on clay and more scarcely on sandy bottoms. Krebs (1978) stated that the densities on the bottom of a ditch were higher in places where fine sand presents a firm substrate, and the larvae were absent from places with soft black anaerobic silt. The larvae also live on rocks and stones (Palmén & Aho, 1966; own observations). The larvae often live in tubes (Tourenq, 1975).

DENSITIES

Palmén & Aho (1966) observed never more than 1200 emergents/m^2 in one season. Krebs (1978) recorded a maximum of about 11,000 larvae/m^2 in a brackish ditch in December, a minimum of 300/m^2 at the end of April and again a maximum of 6000/m^2 at the end of July.

WATER TYPE

Dimensions

C. aprilinus has been found in very large brackish water lakes (Palmén & Aho, 1966). Steenbergen (1993) and Krebs (1981, 1984) found the larvae mainly in medium-sized water bodies, but also in pools and narrow ditches.

D

Permanence
According to Tourenq (1975) the species can be present in temporary water, mainly in winter.

TROPHIC CONDITIONS AND SAPROBITY

We have few data on the occurrence of the larvae in polluted water. In some cases we found larvae in polluted ditches and they are without doubt no less resistant to anoxia than other species of the genus. Nevertheless, Palmén & Aho (1966) and Krebs (1978) suggested that the larvae avoid bottoms consisting of anaerobic silt. Tourenq (1975) mentioned that the larvae are absent from polluted water.

SALINITY

Strenzke (1960) reported that the females only deposited eggs in brackish water with a chloride content of 4000–13,700 mg/l. In nature he found the species in Germany at 1200–16,100 mg/l. Krebs & Moller Pillot (in prep.) and Tourenq (1975) reported incidental occurrence in fresh water, but give 1500–6000 mg/l as the normal range. Tourenq even found larvae in places where the chloride content rose to 15,000–22,000 mg/l. Steenbergen (1993) gave a mean of 5800 mg/l and found the larvae rarely below 1000 mg/l. The species disappeared within a year after the enclosure and freshening of Lake Volkerak-Zoommeer, a part of the Rhine-Scheldt estuary in the Netherlands (van der Velden et al., 1995).

Chironomus balatonicus Dévai, Wülker & Scholl, 1983

Chironomus muratensis Smit et al., 1992, 1994; Smit, 1995; van Nes & Smit, 1993 (misidentification)

SYSTEMATICS AND IDENTIFICATION

The species can be identified almost only by karyological analysis, although there are some morphological differences between the larvae of *C. plumosus* and *C. balatonicus* (Shobanov, 1989, 1989a; Vallenduuk et al., 1995). Hybridisation between these two species is possible, but only in one direction (Michailova, 1989: 12). Both species appear to be closely related.

DISTRIBUTION IN EUROPE AND THE NETHERLANDS

C. balatonicus has been identified in only a few European countries (Saether & Spies, 2004), but this is probably only because of the difficulty with identification. Larvae from the Rhine-Meuse Delta (Haringvliet) in the Netherlands, where the species was common, have been cytologically verified (van der Velden et al., 1996). According to Limnodata.nl, *C. balatonicus* is present in several places, especially in the western part of the country.

LIFE CYCLE

Data collected by Smit et al. (1996) indicate two generations a year, emerging in May and late summer. According to Specziár & Vöros (2001), the species can become multivoltine when the primary production of algae is permanently high.

FEEDING

The larvae feed by filtering and non-selective grazing of planktonic and benthic algae (Smit et al., 1992). On the Ventjagers flats benthic algae contributed to more than 30% of the total number of algae found in the gut. As a rule Specziár & Vöros (2001) found a higher production of *C. balatonicus* in Lake Balaton when the phytoplankton concentration increased. After the algal bloom in late summer the population remains high until spring. The next generation develops when the required phytoplankton level is reached again.

MICROHABITAT

Michailova (1989) collected the larvae (cytologically investigated) in muddy substrata. In contrast to *C. plumosus*, Smit (1995: 162) found the larvae in bottoms with a high sand content, but not in soft mud. However, Smit et al. (1992) reported a positive correlation with the silt fraction.

DENSITY

In Lake Volkerak-Zoommeer van der Velden et al. (1995) found mean densities of 600 larvae/m^2 in the most optimal year.

WATER TYPE

The larvae are found mostly in large lakes and estuaries, but also in carp ponds (Michailova, 1989; van der Velden et al., 1996; Matěna & Frouz, 2000; Specziár & Vörös, 2001). In the Netherlands larvae have been found (not cytologically verified) in larger ditches and canals, sometimes slowly flowing (Limnodata.nl).

pH

Polukonova (2000) stated in Russia that *C. balatonicus* prefers more mineralised water with higher Na, K and Ca contents. The studied populations in Hungary (Specziár & Vörös, 2001) and the Netherlands (e.g. van der Velden et al., 1996) lived in strongly alkaline water.

TROPHIC CONDITIONS AND SAPROBITY

All cytologically verified records are from strongly eutrophicated water. However, the large lakes and estuaries where the larvae have been found will never have very low oxygen contents. The larvae of *C. balatonicus* can tolerate mild hypoxia (Smit et al., 1992). The time spent in ventilation is more or less constant over a range of 1–9 mg O_2/l.

SALINITY

Van Nes & Smit (1993: 25) mentioned the occurrence of larvae of this species in waters with up to 3100 mg Cl/l. However, the larvae colonised the more saline zone of Lake Volkerak much more slowly than the deeper zone, suggesting that the larvae prefer less saline water, although the authors thought it possible that the presence of *Nereis diversicolor* prevented colonisation by chironomids.

Chironomus beljaninae Wülker, 1991

C. beljaninae has been collected in Northern Europe (Scandinavia, Russia). Wülker (1991, 1999) supposed that its ecology is identical to that of *C. fraternus*.

Chironomus bernensis Klötzli, 1973

SYSTEMATICS

Wülker & Klötzli (1973) attributed *C. bernensis* to the *lacunarius* complex within the subgenus *Chironomus*. The species is related to *C. commutatus*.

D

IDENTIFICATION

Fully reliable identification is at present only possible by karyological investigation. However, the adult male can be identified using Langton & Pinder (2007), the exuviae using Langton (1991) and the larvae using Vallenduuk et al. (1995). Nearly all Dutch data are based on identification of larvae.

DISTRIBUTION IN EUROPE AND THE NETHERLANDS

C. bernensis has been recorded in only a few countries, mainly in the central and western parts of Europe (Saether & Spies, 2004). In the Netherlands the species has been collected mainly in the Pleistocene areas in the eastern and southern provinces (Limnodata.nl), but it also occurs in the floodplains of the large rivers (see Water type).

LIFE CYCLE

C. bernensis develops in laboratory cultures much faster than *C. plumosus* (Reist & Fischer, 1987). However, nothing is known about the development time and number of generations in nature.

FEEDING

In the laboratory Reist & Fischer (1987) reared the larvae on organic detritus, but the nutritional value of the detritus from the bottom of the lake appeared to be too low. Rieradevall & Prat (1989) supposed that the larvae in Lake Banyoles fed on detritus.

MICROHABITAT

Wülker & Klötzli (1973) collected the larvae on thick layers of polluted organic silt, mainly along a flowing channel in a lake. Michailova (1989) found them in mud. In the Netherlands the larvae have also often been found on silt (e.g. Smit et al., 1994: 206), but they appear also to live on sandy bottoms, for example in floodplain pools (Klink, pers. comm.) and in lowland brooks (own data). In Lake Banyoles the larvae lived on a sediment mainly composed of limestones and clay (Rieradevall & Prat, 1989).

WATER TYPE

Karyologically verified material has been collected from lakes (Wülker & Klötzli, 1973; Rieradevall & Prat, 1989), from fish ponds (Michailova, 1989) and from running water (Michailova, 1989; Matěna & Frouz, 2000). In Lake Banyoles the larvae lived in the profundal zone. Orendt (2002a) mentioned two faster running streams in Bavaria where the species has been collected. In the Netherlands most records of larvae (identified using Vallenduuk et al., 1995) are from lowland brooks and small rivers or from floodplain pools; in a few cases the species has been collected in estuarine lakes (Smit et al., 1994), in the lower parts of large rivers or in ditches and pools (Limnodata.nl; Klink, pers. comm.).

pH

The species has not been collected in very acid water, but the larvae have sometimes been collected in lowland brooks with pH values between 6 and 7.

TROPHIC CONDITIONS AND SAPROBITY

The lakes, fish ponds, pools and streams from which the species has been recorded are without doubt mostly eutrophic to hypertrophic. An exception is the oligotrophic Lake Banyoles (Rieradevall & Prat, 1989). In a number of cases a layer of organic silt has been reported; in Lake Wohlen near Bern this silt had an anthropogenic origin (Wülker & Klötzli, 1973). In contrast, in the Netherlands larvae have been found mainly in brooks with good water quality. We have not found the species in severely polluted water with low oxygen content and the larvae have not been reported from the profundal zone of eutrophic lakes. Limnodata.nl gives some records from moderately polluted waters with lowered oxygen content and BOD5 values up to 8mg/l, but the species has been collected most often in brooks with better quality. The characteristic habitats (lakes and flowing water) also suggest that the larvae often live in more or less saprobic bottoms, but only when the water column has a relative good oxygen regime.

SALINITY

There is only one record from slightly brackish water with 200–1000 mg Cl/l (Limnodata.nl).

Chironomus calipterus Kieffer, 1908

C. calipterus occurs only in the Mediterranean area (Saether & Spies, 2004) and will not be treated here. The species has been studied by Tourenq (1975).

Chironomus cingulatus Meigen, 1830

IDENTIFICATION

Adult males can be identified using Langton & Pinder (2007), but *C. melanotus* is absent from this key and the differences between this and some other species (*C. bernensis, C. nuditarsis, C. acutiventris*) are small (see also Klötzli, 1974). A problem is also that Strenze (1959) and Wülker et al. (1983) give a beard ratio of 4.4, whereas the key by Langton & Pinder leads to *C. cingulatus* only after choosing < 4.0. Identification using older literature is not reliable at all because more species have been described recently. Several species (including *C. commutatus*) that differ from *C. cingulatus* mainly in colour are taken together or considered as varieties in Kruseman (1933), Goetghebuer (1937–54) and Pinder (1978). Identifying exuviae using Langton (1991) is similarly problematic, but this has been partly resolved in Langton & Visser (2003).

The larvae resemble those of *C. annularius* and *C. melanotus*. The L1/L2 ratio given as a discriminating character by Webb & Scholl (1985) and Vallenduuk et al. (1995) shows a big overlap and is usually greater than 3.65 in all three species, as can be seen in Webb & Scholl, table 6 and fig. 9. It is no wonder that in the Netherlands *C. cingulatus* is rarely found on the basis of larval identification (e.g. Nijboer & Verdonschot, 2001; Limnodata.nl) and appears to be very common based on the identification of adults (Kruseman, 1933; Krebs, 1981, 1984; Kouwets & Davids, 1984) or exuviae (van Kleef, unpublished, identified by P. Langton).

In treating this species we assume that data in the literature based mainly on the identification of adults or exuviae are correct, although some of them may apply to another species.

DISTRIBUTION IN EUROPE AND THE NETHERLANDS

C. cingulatus has been found in many European countries and from almost all areas of the European mainland (Saether & Spies, 2004). The species is also most probably common in all parts of the Netherlands, but there are few reliable data (see under Identification).

LIFE CYCLE

Brundin (1949), Lehmann (1971) and Otto (1991) collected the adults emerging from lakes or running water from early May to September. Schleuter (1985) observed emergence in a pool near Bonn in early April. Schleuter (1985), Otto (1991) and Orendt (1993) found 2 to 3 generations a year in Germany.

FEEDING

Titmus & Badcock (1981) found mainly detritus in the guts of larvae, but also many diatoms and other algae, and conclude that the larvae are non-selective feeders.

OVIPOSITION

Strenzke (1959) stated that the egg masses are most probably deposited on firm substrates. He found 910 to 1200 eggs per egg mass.

MICROHABITAT

Shilova (1976) collected the larvae on soft mud and muddy sand. Lehmann (1971) found the larvae on mud in lentic habitats, but also on branches and plants covered with mud. Drake (1982) collected the species emerging from detritus on the leaves of the bulrush *Schoenoplectus lacustris*.

WATER TYPE

Shilova (1976) collected *C. cingulatus* mainly in small and shallow pools, ditches and slow flowing brooks, more rarely in the littoral zone of lakes. Other authors report the species also being present in pools (Schleuter, 1985) or in more or less large lakes (Humphries, 1936; Brundin, 1949; Otto, 1991; Orendt, 1993). In Lake Windermere the occurrence of the larvae was almost entirely confined to the sublittoral zone (Humphries, 1936). However, Moog (1995) suggested the species occurs sometimes also in the profundal zone. Records from brooks and streams (e.g. Drake, 1982; Orendt, 2002a) most probably refer to finds in lentic stretches (cf. Lehmann, 1971; Shilova, 1976).

Permanence

Schleuter (1985) observed that the species was present in a pool which dried out in some years. The presence of the larvae in temporary water is probably not rare (cf. Shilova, 1976).

pH

Most lakes, ditches and streams where the larvae were reported in the literature and for which the pH was stated had a pH higher than 7 (e.g. Krebs, 1981, 1984; Otto, 1991; Orendt, 1993). Schleuter (1985) collected the species at pH 5.0. In Dutch moorland pools the exuviae were collected twice at a pH around 4 and often at pH 5 or 6 (van Kleef, unpublished, identified by P. Langton).

TROPHIC CONDITIONS AND SAPROBITY

Humphries (1936) and Brundin (1949) collected the species in oligotrophic to mesotrophic lakes, Otto (1991) and Orendt (1983) in eutrophic lakes. The last author suggested that this discrepancy was because it does not apply to the same species. Cytological examination is required in order to resolve this question. Moog (1995) indicated that *C. cingulatus*, more than other species in this genus, lives in oligosaprobic water. This is consistent with the frequent occurrence in Dutch moorland pools (see under pH).

SALINITY

Kruseman (1933: 126) called *C. cingulatus* euryhaline. Ruse (2002) collected the species mainly in brackish water. Krebs (1981, 1984) found it only in fresh and slightly brackish water.

Chironomus commutatus Keyl, 1960

IDENTIFICATION

The adult male is absent from Pinder (1978), but has been keyed by Langton & Pinder (2007). Klötzli (1974) expressed doubts about distinguishing between *C. commutatus* and *C. bernensis* and also gives some differences between *C. commutatus* and other species. The female has been described by Laville (1971). Exuviae can be identified using Langton (1991) and larvae using Webb & Scholl (1985) or Vallenduuk et al. (1995). According to Laville (1971) the anterior ventral tubules are straight; this is not consistent with the Dutch material. However, Dutch material from one locality has been verified cytologically by Dr. I. Kiknadze.

DISTRIBUTION IN EUROPE AND THE NETHERLANDS

C. commutatus is widespread in Europe, but is possibly absent from some parts of the Mediterranean area (Saether & Spies, 2004). In the Netherlands *C. commutatus* is one of the most common and widespread species of the genus, collected in about 500 localities, a little more frequently in the Holocene regions (Limnodata.nl).

LIFE CYCLE

There is no information about the life cycle.

FEEDING

On several occasions the larvae have been found in water with hardly any phytoplankton and feeding mainly on detritus and decaying filamentous algae (own data).

MICROHABITAT

The larvae live mainly on organic or clayish silt and only scarcely on sand with little organic material (own data).

WATER TYPE

Current

There are many records from slowly to moderately flowing lowland brooks in the Netherlands, although the species seems to be most common in stagnant water (Limnodata.nl). In the literature the larvae are most frequently reported from lakes and pools, but there are at least two cytologically verified records from flowing water (Krieger-Wolff & Wülker, 1971; Matěna & Frouz, 2000).

Dimensions

The species is common in small water-bodies as pools and ditches as well as in large lakes and canals (e.g. Krieger-Wolff & Wülker, 1971; Limnodata.nl). In the Biesbosch-reservoirs the larvae live in goodish numbers at a depth of 16 to 26 m (A. Kuijpers, unpublished).

Permanence

There are only very few records from temporary water, possibly because these waters are now less investigated and this genus was not previously identified in the Netherlands.

pH

In the Netherlands there are only a few records from acid water with a pH lower than 6. All of them apply to identified larvae. The species is rarely collected in acid moorland pools and was absent from the moorland pools investigated by van Kleef (identification of exuviae).

TROPHIC CONDITIONS AND SAPROBITY

The records from the Netherlands as well as in the literature are rarely from mesotrophic lakes or pools and mainly from eutrophic ponds, ditches and canals. In many cases a layer of organic silt is present, but the larvae seem to be rare in severely polluted water (H. Cuppen, pers. comm.). Limnodata.nl and GWL (unpublished data) give a few cases of a BOD5 of 10 or more and/or very low oxygen content (0.1–1 mg O_2/l in the daytime). In the reservoirs in the Biesbosch, where the larvae lived at a depth of 16–26 m, the oxygen content was always high (A. Kuijpers, unpublished data).

SALINITY

The only records from brackish water apply to water bodies with less than 800 mg Cl/l (Limnodata.nl). Krebs (1981, 1984), who investigated brackish water and identified only adults, could not identify this species.

Chironomus crassimanus Strenzke, 1959

SYSTEMATICS

Saether & Spies (2004) placed the species in the subgenus *Chironomus* and according to Michailova (1989) the species belongs to the '*pseudothummi*' complex. In Langton & Pinder (2007) the species has been placed mistakenly under the subgenus *Lobochironomus* on p. 159 (sub 5), whereas it should be on p. 163 (sub 22). The same mistake has to be corrected in volume 2, p. 159. Although the pupa has a break in the hook row of segment 2 (which is a character of *Lobochironomus*) the species most probably belongs to *Chironomus* s.s. (Langton, pers. comm.).

IDENTIFICATION

The adult male can be identified using Strenzke (1959) and Langton & Pinder (2007) (see above regarding the mistake in the latter key). Strenzke also described and illustrated the female. The exuviae cannot be identified using Langton (1991) and Langton & Visser (2003) because the break in the hook row of segment 2 had not been noticed in the material of Strenzke (Langton, pers. comm.). The larvae can be identified provisionally using Vallenduuk et al. (1995); however, the larvae of some related species are still unknown.

DISTRIBUTION IN EUROPE

C. crassimanus has been collected in only very few European countries (Saether & Spies, 2004), including Germany (near Hamburg, see Strenzke, 1959) and the British Isles (Langton & Pinder, 2007).

ECOLOGY

Strenzke (1960: 120) suggested that *C. crassimanus* is a stenotope inhabitant of mineral acid water. The British records are also from very acid water (Langton, pers. comm.). The specimens collected by Michailova (1989) in salt ponds in Bulgaria possibly belong to another species.

Chironomus dorsalis sensu Strenzke, 1959

Chironomus dorsalis auct. Pinder, 1978: 114, fig. 55D, 145A; Shobanov et al., 1996: 54; Langton & Pinder, 2007: 163, fig. 84D, 203B
nec *Chironomus (Lobochironomus) dorsalis* Meigen, 1818; Langton & Pinder, 2007: 159, fig. 82H, 199B
? *Chironomus dorsalis* Kiknadze et al., 1991: 18, fig. 20 (see below)
Tendipes viridicollis Kruseman, 1933: 160, 169
nec *Tendipes dorsalis* Kruseman, 1933: 160, 167

NOMENCLATURE AND IDENTIFICATION

Spies & Saether (2004) argued that the name *Chironomus dorsalis* Meigen, 1818 has to be used for the species named by most authors as *Chironomus longipes*. However, specimens falling within *C. dorsalis* sensu Strenzke, 1959 cannot be linked with any valid name. For this reason we have to continue the use of this name, and as long as there is no other name for this species we continue to use the name *C. longipes* for the true *C. dorsalis* Meigen.

Another problem is that the larva of *C. dorsalis* in Kiknadze et al. (1991) differs from the species keyed by Vallenduuk et al. (1995). The larval description in Vallenduuk et al. was based on rearing of larvae and identification of the adults using Strenzke (1959) and Pinder (1978). These larvae have a pigmented frontal apotome and a small dark patch on the gula, in contrast to the larvae of Kiknadze et al. (1991: 18). Dettinger-Klemm (2003: 109) also reared the species and his description corresponds with the Dutch specimens. This author stated that specimens that developed at high water temperatures had a weak colouration of the head. We therefore think that Kiknadze et al. described the same species.

DISTRIBUTION IN EUROPE AND THE NETHERLANDS

C. dorsalis has been recorded nearly everywhere in Europe (Saether & Spies, 2004). In the Netherlands there are only about 30 records because temporary waters have been badly investigated. The records given by Limnodata.nl are scattered over the whole country.

DENSITY

Tourenq (1975) noted local densities of larvae up to more than 50,000/m^2 . In this case many larvae were parasitised by *Hydromermis contorta*.

MICROHABITAT

Tourenq (1975) found the larvae in bottoms rich in organic material. We collected the species in different water types, usually on thin layers of organic silt or dead leaves.

LIFE CYCLE

Kalugina (1972) reported four generations a year in Russia, from April until August. In the Camargue the adults emerge from January until November, with maximum in March and April (Tourenq, 1975). Brundin (1949) collected adults from the end of April until the middle of October in Sweden. Dettinger-Klemm (2003) observed emergence in Germany from May until late September. We caught an adult male in the Netherlands in an urban environment as early as March.

Dettinger-Klemm (2000) investigated the mean duration of development of this species in culture vessels. At 30 °C they developed in about 17 days and at 15 °C within

D

35 days. Males emerged earlier than females. In natural conditions development takes longer. Wotton et al. (1992) found no emergence of adults of this species in sand filter beds temporary filled with water, although these containers had a bed run from 16 to 77 days. In winter the larvae are in oligopause in fourth instar (Dettinger-Klemm, 2003).

OVIPOSITION

Strenzke (1959) supposed that the eggs are deposited on firm substrates, for instance along the water edge. This author mentioned 800–910 eggs per egg mass and Nolte (1993) 900 to 1102; Dettinger-Klemm found only 150 to 631 eggs per mass. Some of the numbers given by Nolte, may refer to another species.

FEEDING

Walshe (1951) stated that the larvae do not make a net for filtration; they only gather detritus particles. Rusina (1956) stated that blue-green algae were digested and assimilated, but only when these were already dying.

WATER TYPE

Current

C. dorsalis has been rarely mentioned in literature as an inhabitant of flowing water. However, in the Netherlands there are a number of records in slow-flowing lowland brooks, mainly in upper courses (Limnodata.nl; own data).

Dimensions

In most cases the larvae are found in small to very small water bodies such as ponds, pools, flat roofs and rain puddles, both in nature and near houses (Krieger-Wolff & Wülker, 1971; Schleuter, 1985, 1986; Dettinger-Klemm, 2003; own data). However, the species also lives (scarcely) in lakes, larger pools and ditches, and even in canals and small lowland streams (Brundin, 1949; Limnodata.nl; van Kleef, unpublished).

Permanence

C. dorsalis is a typical inhabitant of temporary water (Schleuter, 1985; Dettinger-Klemm, 2000; Matěna & Frouz, 2000). However, the species also lives in permanent water bodies such as lakes (Brundin, 1949) or canals (Limnodata.nl). Dettinger-Klemm (2000) argued that in dry summers and severe winters the species cannot survive in small temporary pools, but that surviving in permanent water can be difficult for this species because of concurrence and predation. The species does not tolerate total desiccation or temperatures higher than 33 °C (Dettinger-Klemm, 2000a, 2003). McLachlan (1974) stated that *C. dorsalis* was one of the first colonisers in a newly filled basin, followed later by other species of macroinvertebrates more typical for this type of water.

pH

The larvae live in water with a range of pH values (Thienemann, 1954). Limnodata.nl gives a mean pH of 7.5. The larvae have often been collected in water with a pH around 5 (e.g. Schleuter, 1985; Moller Pillot, 2003). Van Kleef (unpublished) found the exuviae in three moorland pools with a pH of 4.2–4.5.

TROPHIC CONDITIONS AND SAPROBITY

The larvae live scarcely in the Swedish oligotrophic lakes (Brundin, 1949) and the records in the Netherlands are mainly from eutrophic, hardly polluted water (Limnodata.nl). In a moorland pool the ammonium content was very high (van Kleef, unpublished). Very little information exists about the ability of the species to live in severely polluted water. In any case, the larvae have no problem surviving in water with a low oxygen content (Tourenq, 1975; Limnodata.nl) or higher temperatures up to 30 °C. (Dettinger-Klemm, 2000a). Wülker & Klötzli (1973) noted that the larvae occur in meadow pools eutrophicated by cattle dung.

SALINITY

The larvae are euryhaline: according to Tourenq (1975: 180) and the literature cited by him they live in fresh and brackish water with a chloride content up to 10,000 mg/l. However, Krebs (1981,1984, 1985) did not collect the species at all in brackish water in the province of Zeeland, although he investigated many small pools.

Chironomus entis Shobanov, 1989

SYSTEMATICS AND IDENTIFICATION

C. entis is very similar to *C. plumosus* and is almost impossible to identify without cytological investigation. The adult male resembles *C. annularius* and *C. prasinus* (Langton & Pinder, 2007 vol. 2: 160). The exuviae have been described and keyed by Langton & Visser (2003). Shobanov (1989, 1989a) described and keyed the larva according to a longer first antennal segment, but the sizes of antennal segments are most probably not the same in different regions. For figures see also Kiknadze et al. (1991: fig. 9).

DISTRIBUTION IN EUROPE

The species has been recorded in lakes and ponds in Scandinavia and Central and Eastern Europe (Shilova & Shobanov, 1996; Wülker, 1999; Saether & Spies, 2004). Matěna & Frouz (2000) collected the species once in a carp pond in the Czech Republic. The species may occur in Western Europe.

Chironomus fraternus Wülker, 1991

IDENTIFICATION

The adult male and female, larva and pupa have been described by Wülker (1991). The Dutch material has been (cytologically) identified by I. I. Kiknadze. The exuviae can be identified using Langton & Visser (2003), who also give a description. The larva is still undescribed.

D

DISTRIBUTION IN EUROPE AND THE NETHERLANDS

C. fraternus has been collected in Finland, Sweden, Norway and Russia (Wülker, 1991, 1999) and in the Netherlands (Bargerveen, leg. H. Vallenduuk).

MICROHABITAT

Wülker (1991) noted that in Scandinavia the larvae live in *Drepanocladus* mosses and sometimes on organic bottoms. In the Netherlands the larvae were collected in a small peat cutting without or with very little *Sphagnum*; the bottom consisted of a thick layer of mud.

WATER TYPE

The species has been collected in Scandinavia in lakes and in the bog pond Siikaneva in Finland. In the Netherlands the larvae lived in a peat cutting (see Microhabitat).

Chironomus heteropilicornis Wülker, 1996

This species is treated under *C. pilicornis*.

Chironomus holomelas Keyl, 1961

IDENTIFICATION

The adult male cannot be identified with certainty (see Langton & Pinder vol. 2: 161) and the larvae can hardly be identified without cytological investigation (see Webb & Scholl, 1985; Vallenduuk et al., 1995). The exuviae have been described and keyed by Langton & Visser (2003).

DISTRIBUTION IN EUROPE AND THE NETHERLANDS

According to Saether & Spies (2004) the species has been collected only in Central Europe. Recently the species has also been found in the British Isles (Langton & Pinder, 2007 vol. 2: 161). In the Netherlands the species has been reported three times. These records have to be confirmed by karyological investigation of (new) material.

WATER TYPE

Ryser et al. (1980) noted the occurrence of the larvae in pools in the Jura (Switzerland). Matěna & Frouz (2000) collected them once in running water and a number of times in natural pools in peatland in the Czech Republic. Matěna (1990) also found them in newly filled ponds.

Chironomus inermifrons Goetghebuer, 1921

IDENTIFICATION

The adult male can be identified using Pinder (1978) and Langton & Pinder (2007). The exuviae and larvae have not been described.

DISTRIBUTION IN EUROPE AND THE NETHERLANDS

The species has been found only in some countries in Western and Northern Europe (Saether & Spies, 2004).

LIFE CYCLE

In the Three Dubs Tarn in the English Lake District the adults emerged only from the middle of July until early September, suggesting only one generation a year (Macan, 1949).

ECOLOGY

Macan (1949) collected the adults emerging from the centre of a mesotrophic tarn in England with a pH of 6.7. This tarn was about 3 m deep.

Chironomus jonmartini Lindeberg, 1979

Chironomus neglectus Wülker, 1973: 369

DISTRIBUTION IN EUROPE

C. jonmartini is only known from Scandinavia (Saether & Spies, 2004).

WATER TYPE

The species is known from rock pools (Palmén & Aho, 1966; Wülker, 1999), bog ponds and a reservoir (Koskenniemi & Paasivirta, 1987).

Chironomus lacunarius Wülker, 1973

IDENTIFICATION

The adult male can be identified using Langton & Pinder (2007), the exuviae using Langton (1991) and the larva using Vallenduuk et al. (1995).

DISTRIBUTION IN EUROPE

C. lacunarius has been collected in a few countries scattered over Europe (Saether & Spies, 2004). The species has not been found in the Netherlands.

WATER TYPE

The larvae have been collected in alpine meadow pools (Wülker & Klötzli, 1973; Geiger et al., 1978; Ryser et al., 1980).

Chironomus longipes Staeger, 1839

Chironomus (Lobochironomus) dorsalis Meigen, 1818; Saether & Spies, 2004; Langton & Pinder, 2007: 159, fig. 82H, 199B
Tendipes longipes Kruseman, 1933: 172, fig. 34
Einfeldia longipes Pinder, 1978: 120, fig. 151C; Shilova, 1980: 176–180, figs. 9A3, B3, 10

SYSTEMATICS AND IDENTIFICATION

Webb & Scholl (1987) stated that '*Einfeldia longipes*' had to be placed in the genus *Chironomus*, subgenus *Lobochironomus*. Exuviae and larvae have the normal characters of this genus. Identification of larvae has so far been problematic, mainly because there is a large overlap in the characters given by Vallenduuk et al. (1995). Most probably all fourth instar larvae with head width 350–430 µm belong to *Lobochironomus*, but the species of this subgenus can be identified only by studying the ventromental plates at high magnification (see Webb & Scholl, 1987). In the Netherlands and adjacent lowlands *C. longipes* is probably the only species within this subgenus.

DISTRIBUTION IN EUROPE AND THE NETHERLANDS

C. longipes has been recorded nearly everywhere in Europe (Saether & Spies, 2004). An unknown part of the Dutch records to date were based on identification of the larvae and are not reliable (see above). However, Kruseman (1933) collected the characteristic adult males in the Holocene western as well as in the Pleistocene eastern and southern parts of the country. Langton identified the exuviae from a number of moorland pools (coll. van Kleef).

LIFE CYCLE

Exuviae and adults have been collected from May to August (Kruseman, 1933; van Kleef, unpublished; own data).

WATER TYPE

Pinder & Reiss (1983) suggested that the larvae of *Einfeldia* spec. gr. C (in the majority of cases *C. longipes*?) occur mainly in dystrophic small water bodies. However, some records from Kruseman (1933: e.g. Groote Lindt), from different types of moorland pools (van Kleef, unpublished) and our own records from the floodplain of the river Pripyat in Belarus show that *C. longipes* lives in different types of pools and small lakes. The species seems to be rare in large lakes: Brundin (1949) collected the adult males rarely from Swedish lakes and most other publications on lake chironomids do not mention the species at all.

Limnodata.nl gives many records from eutrophic ditches in the Holocene part of the Netherlands and also from some lowland brooks in the Pleistocene areas. Unless what has been stated under Systematic and identification above, many of these records will indeed apply to *C. longipes*. No author reports the species from flowing water in brooks or streams. The adult males caught by Klink (1985a) near the river IJssel will have emerged from stagnant water bodies near the river. Schleuter (1985) collected the species in some nearly desiccated pools, but there is no record from actual temporary water.

pH

The exuviae have been collected in ponds, moorland pools and small lakes with a pH from 3.9 to 7.5 (Schleuter, 1985; van Kleef, unpublished; own data).

SALINITY

There are no records from brackish water. Krebs (1981, 1984, 1985, 1990) did not find the species at all in the province of Zeeland, based on identification of adults.

Chironomus longistylus Goetghebuer, 1921

NOMENCLATURE AND IDENTIFICATION

According to Langton (unpublished) *C. longistylus* in Belgium and Great Britain (including the type of Goetghebuer) is not the same species as described by Wülker (1999 etc.). If this is true, the text below applies to two different species. The adult male can be identified using Pinder (1978) and Langton & Pinder (2007). The exuviae have been described and keyed by Langton (1991) and Langton & Visser (2003).

DISTRIBUTION IN EUROPE AND THE NETHERLANDS

C. longistylus has been collected in several countries scattered over large parts of Europe (Saether & Spies, 2004). In the Netherlands there are three records identified from exuviae by P. Langton.

MICROHABITAT

Griffiths (1973) collected the larvae much more often on fine detritus than on coarse detritus and rarely or not on plants. Paasivirta & Koskenniemi (1980) found *C. cf. longistylus* to be rather common on moss-covered littoral sites in two Finnish lakes which dried up and froze in late winter.

WATER TYPE

According to Langton (1991) the larvae live in ponds. Griffiths (1973) collected the species from an acid moorland pond (pH 4.4). Wülker (1999) mentioned several ponds and lakes in Scandinavia. In the Netherlands the exuviae have been collected in three acid moorland pools (van Kleef, unpublished).

Chironomus lugubris Zetterstedt, 1850

SYSTEMATICS AND IDENTIFICATION

Pinder (1978) stated that this may be no more than a dark form of *C. riparius*. Many authors (e.g. Wülker, 1999) do not mention the name at all. Langton & Visser give a description of the exuviae. The larvae have no lateral tubules (Lindeberg, 1959).

DISTRIBUTION IN EUROPE AND THE NETHERLANDS

C. lugubris has been collected in a few European countries, mainly in Northern and Central Europe and in the British Isles (Saether & Spies, 2004). In the Netherlands one specimen has been collected as exuviae in a moorland pool (van Kleef, unpublished, det. P.H. Langton).

FEEDING

McLachlan et al. (1979) stated that the larvae feed on small peat particles with many bacteria which had settled from suspension, partly after having been broken down by the cladoceran *Chydorus sphaericus*. The latter can then feed on the pellets of the chironomid larva.

OVIPOSITION

Munsterhjelm (1920) mentioned egg deposition in August in Southern Finland and about 550 eggs per egg mass.

MICROHABITAT

According to McLachlan & McLachlan (1975) the larvae are strictly associated with the fine fraction of the peat mud on the bottom of the lake.

WATER TYPE

C. lugubris has been collected in bog lakes (McLachlan et al., 1979), a moorland pool (van Kleef, unpublished) and rock pools (Lindeberg, 1959).

pH

McLachlan & McLachlan (1975) collected the larvae in a bog lake with pH values between 3 and 4. In the Netherlands van Kleef (unpublished) found one specimen as exuviae in a moorland pool at pH 4.2.

D

Chironomus luridus Strenzke, 1959

SYSTEMATICS AND IDENTIFICATION

Karyologically *C. luridus* belongs to the *pseudothummi* complex (see e.g. Michailova, 1989; Kiknadze et al., 1991). Within this complex at least four species cannot be identified from larva; they are taken together by Vallenduuk et al. (1995) as *C. luridus* agg. In Limnodata.nl these species seem to be mentioned together as *C. luridus*. Detailed figures of the larval head and karyotype is given by Kiknadze et al. (1991: fig. 21, photo 20). Identification of adults and exuviae, especially in combination with larvae, is possible, but not fully reliable, because the species group is still insufficiently described.

DISTRIBUTION IN EUROPE AND THE NETHERLANDS

C. luridus occurs nearly everywhere in Europe (Saether & Spies, 2004). The species is also widespread in the Netherlands.

LIFE CYCLE

We caught adults in nature from April to October. There are at least two or three generations a year. In winter the larvae are in third and fourth instar (own data).

MICROHABITAT

Moldován (1987) supposed that the occurrence was associated with habitats with allochthonous litter (mainly of *Populus* trees). Matěna (1990) found the larvae of *C. luridus* more commonly in places with plant remnants than on plain muddy bottoms. Krebs & Moller Pillot (in prep.) found the species in the province of Zeeland mainly in water with a well-developed vegetation. Klink (unpublished) stated the same in pools along the large rivers. We also collected the species in small water bodies in shaded and sheltered places without vegetation, often between dead leaves, but sometimes on mud.

OVIPOSITION

Strenzke (1959) noted that the egg masses are deposited on firm substrates. He found 910–1030 eggs per egg mass. Strenzke (1960: 121) observed that the preferred locations for egg deposition were shaded by trees (see also under Microhabitat).

WATER TYPE

Current

The species is mentioned in the literature as occurring mainly in astatic water, ponds and pools (e.g. Krieger-Wolff & Wülker, 1971; Ryser et al., 1980; Matěna & Frouz, 2000), but has also been found in running water. In the Netherlands most records of *C. luridus* agg. from lowland brooks seem to be for this species. The species is rarely reported from faster running streams (Orendt, 2002a; cf. Moog, 1995).

Dimensions

Although most records are from small pools or relatively narrow brooks and ditches, the larvae can be found in lower numbers in wider canals with well-developed vegetation (Krebs, 1984). However, *C. luridus* was a pioneer species in carp ponds (Matěna & Frouz, 2000).

Permanence

The species is a common inhabitant of temporary pools (Dettinger-Klemm & Bohle, 1996, as *C.* cf. *luridus*). Matěna & Frouz (2000) found the species in astatic water like rain pools. We reared *C. luridus* from a flat roof and a rain tub.

pH

Orendt (1999) recorded the occurrence of the species in streams with pH values ranging from 3.8 to 6.6 (he did not test water with a pH higher than 6.7). However, as a rule this species has not been found in acid water and possibly the material found by Orendt was not correctly identified.

TROPHIC CONDITIONS AND SAPROBITY

In pioneer situations the larvae can be found in water without much organic material. However, they usually live in an environment of decomposing dead plant material, often in water without anthropogenic pollution. In some cases the larvae have been collected from severely polluted ditches or brooks, sometimes with hardly any oxygen (Grontmij | Aqua Sense, unpublished data; own data).

SALINITY

Krebs reared *C. luridus* regularly from slightly brackish water with a maximum of 1600 mg Cl/l (Krebs & Moller Pillot, in prep.). Tourenq (1975: 182) found *C. luridus semicinctus* in the Camargue in fresh and slightly brackish water up to 6000 mg Cl/l. According to Limnodata.nl the species is relatively scarce in brackish water.

Chironomus macani Freeman, 1948

SYSTEMATICS

C. macani is placed in a separate subgenus *Chaetolabis* Townes, 1945 (see Ashe & Cranston, 1990).

IDENTIFICATION

The adult male can be identified using Langton & Pinder (2007), the exuviae using Langton (1991). Males and females from northern Scandinavia are described by Wiederholm (1979).

The larva has been keyed only by Pankratova (1983). The larva resembles *C. riparius* (or *C. bernensis* ?), but has mentum type II or III and a somewhat darker frontal apotome.

DISTRIBUTION IN EUROPE

C. macani has been found in only a few countries, mainly in the western and northern parts of Europe (Saether & Spies, 2004).

MICROHABITAT

Paasivirta & Koskenniemi (1980) found the larvae of *C. macani* in the moss-covered littoral zone of two polyhumic lakes in Finland. These sites dried up and froze in late winter. The adult males collected by Macan (1949) emerged from the centre of a tarn.

WATER TYPE

The species has been collected in small and large lakes (see Wülker, 1999).

Chironomus melanescens Keyl, 1961

D

IDENTIFICATION

The adult male cannot be identified using Langton & Pinder (2007). The exuviae are absent from the keys by Langton (1991) and Langton and Visser (2003). A description of the larva with figures showing details can be found in Kiknadze et al. (1991). The data below are based on cytologically verified material (Wülker, Matěna) or on morphological identification of larvae (Dutch data).

DISTRIBUTION IN EUROPE

C. melanescens has been collected in a few European countries, including some in Central and Western Europe (Saether & Spies, 2004). Relatively few data (about twenty localities) are available from the Netherlands. The species is probably not rare, but the suitable water bodies have hardly been investigated.

LIFE CYCLE

Matěna (1990) mentions relatively fast growth of the larvae compared with those of *C. annularius* and found predominantly mature larvae after only 11 days after filling a carp pond. The adults emerged from April to October.

OVIPOSITION

From observations of Matěna (1990) it seems certain that the egg masses are deposited along the water's edge.

MICROHABITAT

Matěna (1990) found the larvae significantly more frequently in places with plant remnants than on plain muddy bottoms. This corresponds with our own records.

WATER TYPE

Matěna (1990) and Matěna & Frouz (2000) stated that the species is characteristic of stagnant waters in early stages of succession. The species colonised all the carp ponds they investigated immediately after filling, notwithstanding the season (from April to October). Wülker (1999) mentioned many pools, but also one record in the river Jenissei. We have some records from water bodies in the floodplain area of the Netherlands and in Belarus, which are temporarily inundated in winter or spring (Klink, pers. comm.; own data). Moller Pillot (2003) collected the larvae scarcely in the Roodloop, a temporary upper course of a lowland brook. Other Dutch records are from permanent lowland brooks, a ditch and some moorland pools.

pH

In the Roodloop the larvae lived at pH 5–6 (Moller Pillot, 2003). The larvae have also been collected in some acid moorland pools (Limnodata.nl; H. Cuppen, unpublished). The other records are from non-acid water.

TROPHIC CONDITIONS AND SAPROBITY

The larvae live in mesotrophic moorland pools and upper courses, but also in very eutrophic water. One Dutch record is from a polluted ditch with hardly any oxygen at night during summer.

Chironomus melanotus Keyl, 1961

IDENTIFICATION

The adult male is absent from the key by Langton & Pinder (2007). According to Wülker (1973) the species resembles *C. anthracinus*, but is smaller, with a lower AR. Identification of the exuviae using Langton (1991) is possible, but very difficult. The larva resembles that of *C. annularius* and *C. cingulatus*. The differences given by Webb & Scholl (19850 and Vallenduuk et al. (1995) make no reliable identification possible. According to the figures in Kiknadze et al. (1991: fig. 15) the shape of the first antennal segment is different, being more conical in *C. melanotus*. The records mentioned below (unless otherwise stated) are based on cytological identification.

DISTRIBUTION IN EUROPE AND THE NETHERLANDS

The species has been recorded in several European countries scattered over the European mainland (Saether & Spies, 2004). In the Netherlands the species has been reported from about 20 localities, for the greater part in the eastern and southern provinces. However, all these records are for larvae identified from morphological characters, with the exception of one record based on identification of exuviae.

WATER TYPE

Current

The larvae have been collected in a subalpine river in Norway (Wülker, 1999). Matěna

& Frouz (2000) noted the species as typical for diverting ditches. Other records in the literature are from stagnant water (e.g. Ryser et al., 1980; Schleuter, 1985). In the Netherlands the larvae have been also reported from some lowland brooks (identified only from morphological characters).

Shade
Many authors found the species in woodland, probably at least partly in shady water (Krieger-Wolff & Wülker, 1971; Schleuter, 1985; Matĕna & Frouz, 2000).

Dimensions
All records of *C. melanotus* are from small shallow pools, ponds, ditches, dead river arms, etc. In the Netherlands the species (not cytologically verified) has sometimes been collected in canals and lowland brooks (Limnodata.nl).

Permanence
As far as mentioned in the literature, the pools and ditches where the larvae have been found were permanent. Most probably the larvae also occur in temporary water.

TROPHIC CONDITIONS AND pH

The larvae live in humic, non-humic and sometimes also in eutrophic water (e.g. Wülker, 1973). Schleuter (1985) collected them in acid water with pH around 5 (based on the identification of adults). One Dutch record based on identified exuviae is from a moorland pool with pH 4.2 (identified by P. Langton). The larvae have also been identified from other acid moorland pools (Limnodata.nl; van Kleef, unpublished).

Chironomus mendax Storå, 1936

SYSTEMATICS

C. mendax belongs to the subgenus *Lobochironomus* (Webb & Scholl, 1987).

DISTRIBUTION IN EUROPE

C. mendax has been collected in lakes and pools in Scandinavia, Russia, the Alps and the Black Forest (Wülker, 1999; Saether & Spies, 2004). The species will not be treated here.

D

Chironomus montuosus Ryser, Wülker & Scholl, 1985

SYSTEMATICS

C. montuosus belongs to the subgenus *Lobochironomus* (Webb & Scholl, 1987; Shilova & Shobanov, 1996).

DISTRIBUTION IN EUROPE

C. montuosus has been collected in Central Europe and Russia (Saether & Spies, 2004). The species will not be treated here, although in Russia the species occurs in lowland bogs (Shilova & Shobanov, 1996).

Chironomus muratensis Ryser, Scholl & Wülker, 1983

nec *Chironomus muratensis* Smit et al., 1992, 1994 (misidentification, = *C. balatonicus*)

SYSTEMATICS

C. muratensis belongs to gr. *plumosus* (Ryser et al., 1983). The authors found several cases of hybridisation with *C. plumosus* and *C. nudiventris*. However, according to Michailova (1989: 13) hybridisation between *C. muratensis* and *C. plumosus* is not possible because these species seem to have diverged further than *C. plumosus* and *C. balatonicus*.

IDENTIFICATION

Adults of *C. muratensis* cannot be distinguished from those of *C. plumosus* and *C. nudiventris* (Ryser et al., 1983). The exuviae are still unknown. In their key Webb & Scholl (1985) distinguish between larvae of *C. muratensis* and *C. plumosus* by the length of the ventral tubules. This character is not reliable and has led to misidentifications. As proven by Kiknadze et al. (1991), short ventral tubules can also occur in *C. plumosus* and *C. balatonicus*. Although there seem to be slight differences in the mean length of the tubules, this character is not used in later keys (e.g. Shobanov, 1989a; Vallenduuk et al., 1995).

Dutch material from one locality has been verified cytologically by Dr. I. Kiknadze.

DISTRIBUTION IN EUROPE AND THE NETHERLANDS

The species has been recorded in several countries scattered over Europe (Saether & Spies, 2004). In the Netherlands the species seems to be most common in the Holocene western part of the country (Limnodata.nl). However, we also collected larvae in the eastern part of the country.

LIFE CYCLE AND OVIPOSITION

Ryser et al. (1983) observed egg deposition in Switzerland from May until early October, most numerously in June and July. They suppose that the species is at least partly bivoltine. The authors observed egg deposition on wood at the shoreline. In winter we found most (or all) larvae in fourth instar.

FEEDING

We found much fine detritus, many diatoms and some filamentous algae in the guts of the larvae. Ryser et al. (1983) reared the larvae successfully with Tetraphyll.

MICROHABITAT

In the Murten See, Ryser et al. (1983) found the larvae mainly on sandy bottoms without much organic silt. Michailova (1989) also noted that the larvae occur in sandy substrata. We collected *C. muratensis* mainly from sandy bottoms too, and in lower numbers on organic silt.

WATER TYPE

C. muratensis is an inhabitant of larger water bodies such as lakes and the lower parts of rivers and freshwater estuaries (Ryser et al., 1983). The larvae have been found especially in relatively undeep water. Matěna & Frouz (2000) collected the species only in larger reservoirs in the Czech Republic. However, we found the larvae in ponds no more than 50 m wide and sometimes even in ponds less than 20 m wide. Polukonova (2000) also reports the occurrence of *C. muratensis* in a pond and Limnodata.nl gives records in ditches.

pH

All Dutch larvae identified as *C. muratensis* using Vallenduuk et al.(1995) lived in water with a pH > 7 (Limnodata.nl). The species probably does not occur in acid water.

However, Polukonova (2000) stated that in Russia *C. muratensis*, as distinct from *C. plumosus* and *C. balatonicus*, also lived in water with lower Na, K and Ca contents.

TROPHIC CONDITIONS AND SAPROBITY

Int Panis et al. (1996) found that the larvae of *C. muratensis* had on average a lower haemoglobin concentration than larvae of *C. plumosus*. However, the haemoglobin concentrations within this genus seem to be very variable, also within one species (see Rossaro et al., 2007). In view of the preference for bottoms with less organic silt, it seems probable that the larvae need a better oxygen regime. However, Limnodata.nl gives one record from water with a very bad oxygen regime and some cases of a rather high BOD5 (about 10 mg O_2/l).

SALINITY

There is only one record from brackish water (Barend creek, prov. of Noord-Brabant, > 1000 mg Cl/l).

Chironomus nuditarsis Keyl, 1962

Chironomus nuditarsus Limnodata.nl (misspelling)

SYSTEMATICS AND IDENTIFICATION

Michailova (1989) placed this species in the '*thummi*' complex. The larvae more or less resemble *C. plumosus*, but do not belong to gr. *plumosus* sensu Shobanov (1989). Identification of the adult male is difficult; the hypopygium resembles that of *C. cingulatus*, *C. acutiventris*, *C. commutatus* and *C. bernensis*. The species is absent from Pinder (1978), but can be identified using Langton & Pinder (2007). For additional characters see Klötzli (1974) and Lindeberg & Wiederholm (1979). Klötzli also describes the female. The exuviae can be identified using Langton (1991), the larva using Vallenduuk et al. (1995). Polukonova (2005) described the differences between the larvae of *C. nuditarsis* and *C. curabilis*. The latter species is only known from Russia.

DISTRIBUTION IN EUROPE AND THE NETHERLANDS

C. nuditarsis is widely distributed in Western, Central and Northern Europe (Saether & Spies, 2004). Based on the morphological identification of larvae, the species has been recorded in the Netherlands from more than 100 localities scattered over nearly the whole country (Nijboer & Verdonschot, 2001; Limnodata.nl; GWL, pers. comm.; own data). The species is most probably more common in the Holocene areas than in the sandy Pleistocene areas.

D

LIFE CYCLE

Fischer (1969: 26) supposed about four generations a year. In Lake Wohlen, Switzerland, the species was most abundant in late summer. Eggs were found from mid April to mid November. According to Reist & Fischer (1987) and Rychen Bangerter & Fischer (1989) *C. nuditarsis* develops faster than *C. plumosus*. The larvae show a dormancy response somewhat later in the year (at day length 12 h or less) and are supposed to have a longer reproduction period.

FEEDING

Reist & Fischer (1987) reared the larvae in the laboratory on organic detritus, but the nutritional value of the detritus from the bottom of the lake appeared to be too low.

OVIPOSITION

C. nuditarsis deposits the eggs on the surface of open water (Fischer, 1969: 31).

MICROHABITAT

Fischer (1969) found numerous larvae in thick layers of anaerobic silt. Michailova (1989) collected the larvae in muddy substrata. In the Netherlands the larvae occur on anaerobic silt as well as on sandy bottoms without much silt.

WATER TYPE

Current

The larvae are collected relatively scarcely in flowing water, but in the Netherlands there are many records from lowland brooks. Specimens collected in fast flowing water (e.g. Orendt, 2002a) probably originated from localities with a weaker current.

Dimensions

C. nuditarsis is mainly reported as occurring in lakes, but also lives locally in pools (Fischer, 1969; Fittkau & Reiss, 1978; Geiger et al., 1978). In the Netherlands many records are from small lakes and canals, but the larvae are also not rare in shallow and narrow ditches.

Permanence

There are no records from temporary water, but in view of the life cycle the species may settle in some sites after temporary desiccation.

pH

The larvae have been collected only rarely in acid water with pH values between 4 and 6 (Limnodata.nl; van Kleef, unpublished; GWL, unpublished).

SAPROBITY, OXYGEN

Fischer (1969) reported abundant larvae in anaerobic silt in Lake Wohlen near Bern, which was strongly polluted but had a high oxygen content due to flow in the water column. Langton (1991) characterised the preferred habitat as organically polluted ponds, but according to Moldován (1987) it is not a characteristic species of sewage water. However, they can be numerous in ditches in helophyte filter systems (GWL, unpublished). Many records in the Netherlands are from water with low oxygen contents, but where larvae were found in anaerobic silt as a rule there seems to be no severe anthropogenic pollution. In many cases the larvae were even collected in water bodies with a rather good water quality. The larvae seem to prefer places with much decomposition of organic material. They do not appear to be dependent on the presence of phytoplankton if the nutritional value of the detritus is sufficient.

SALINITY

The species is only known from fresh and slightly brackish water. The highest chloride contents lie between 1000 and 2000 mg Cl/l (Limnodata.nl).

PARASITISM

In Lake Wohlen near Bern the population of *C. nuditarsis* appeared to be severely parasitised by *Mermis*. The species recovered with greater difficulty than *C. plumosus*.

Chironomus nudiventris Ryser et al., 1983

Tendipes gr. *reductus* Lenz, 1954–62: 153, 161
Tendipes plumosus f.l. *reductus* Pankratova, 1964: 198
Chironomus f.l. *reductus* Moller Pillot, 1984: 218

SYSTEMATICS

C. nudiventris belongs to gr. *plumosus* (Ryser et al., 1983). These authors found two instances of hybridisation with *C. muratensis*.

IDENTIFICATION

The adult males cannot be distinguished from *C. plumosus* and other species of the group *plumosus* (Ryser et al., 1983). Therefore the species is often treated under the name *plumosus* (e.g. Pankratova, 1964). The exuviae are absent from Langton (1991), but can be identified using Langton & Visser (2003). The larva is conspicuous because of the absence of ventral tubules; for differences from *C. agilis* see Shobanov (1989a). A description of the morphology and karyotype of the larva, with figures, is given by Kiknadze et al. (1991).

DISTRIBUTION IN EUROPE AND THE NETHERLANDS

C. nudiventris is widespread in Europe, but is possibly absent from Scandinavia (Saether & Spies, 2004). In the Netherlands the larvae occur in nearly all parts of the country.

LIFE CYCLE

Ryser et al. (1983) stated egg deposition in Switzerland from May until early October. They supposed that the species is at least partly bivoltine or multivoltine. The data presented by Smit et al. (1996) seems to confirm this. The emergence in May was probably somewhat later than in other *Chironomus* species. In winter nearly all larvae are in fourth instar (own data).

FEEDING

The larvae feed by non-selective grazing on all algal groups present on the sediment (Smit et al., 1992). We also found much detritus in the gut.

MICROHABITAT

The species occurs rarely in clean sand, but is most common on silty sand (Smit et al., 1994; 1996). Pankratova (1964) also found the larvae on silty sand bottoms and more scarcely on clay. According to Klink (1994) the species is an indicator for sandy riparian or slightly dynamic zones of rivers with a moderate silt content. In the river Meuse the larvae were found almost entirely on sandy bottoms (Peeters 1988). In sandy bottoms in the Haringvliet Klaren (1987) collected more larvae in the upper 5 cm of the sediment than at a depth of 5–10 cm. The larvae are rarely collected on stones.

DENSITY

Pankratova (1964) recorded maximal densities of about 100 larvae/dm^2 in the river Oka in Russia. In the river Meuse rarely more than 20 larvae/dm^2 were found (own data). Smit et al. (1994) recorded average densities of 2 to 3 larvae/dm^2 in optimal habitats in the Rhine-Meuse delta.

WATER TYPE

Current

The larvae have been found in lakes and rivers (Lenz, 1954–62; Ryser et al., 1983). Pankratova (1964) observed that the larvae occurred in water with different currents. In the Netherlands the larvae have been found in the slow-flowing parts of the large rivers, in oxbow lakes and in the estuarine lakes, and more scarcely in small streams with a slow current. Records from other lakes or canals are scarce (Limnodata.nl). In Germany M. Siebert (pers. comm.) collected larvae in the lower stretches of rivers. The records from lakes in Switzerland (Ryser et al., 1983) concern only small populations.

Dimensions

As mentioned above the larvae are mainly collected in large rivers and more rarely in small rivers. The species has also been found in stagnant water, especially in lakes more than 50 m wide. In the Biesbosch reservoirs the larvae were locally present at depths of 16 to 24 m (A. Kuijpers, pers. comm.). Ryser et al. (1983) could not find them in small pools in Switzerland. In the river Meuse the larvae were found numerously at depths down to 8 m, rarely in the shallow areas (Peeters, 1988). Smit et al. (1996) found higher densities in the deeper (less exposed) parts of a tidal sandy flat in the enclosed Rhine-Meuse delta, probably because of higher predation by waders in the more exposed parts.

pH

The pH in large rivers and estuaries is nearly always higher than 7.5. A lower pH probably presents no direct problem for larvae or pupae.

TROPHIC CONDITIONS AND SAPROBITY

The larvae are tolerant of mild hypoxia, showing no higher ventilation behaviour at oxygen concentrations of 3 mg/l (Smit et al., 1992). They were locally numerous in the deeper parts of the Meuse in 1971, when the water quality was bad (own data). Most of these samples contained no other chironomid species. However, in that year the larvae were absent in the Rhine, where the minimum oxygen content was then 1 mg/l and the average BOD nearly 50 mg/l (bij de Vaate, 2003: 17). In small rivers the larvae seems to be limited to less polluted sites (H. Cuppen, pers. comm.). They are always absent in thick layers of organic silt. The species rarely lives in sites subject to anoxia.

SALINITY

There are no records from brackish water. In 1973 the species was very scarce in the Haringvliet, where the chloride content was then about 300 mg/l (own data).

Chironomus obtusidens Goetghebuer, 1921

SYSTEMATICS AND IDENTIFICATION

C. obtusidens belongs within the *thummi* complex of the *obtusidens* group, together with *C. acutiventris* (see Wülker et al., 1983; Istomina et al., 2000). Identification of the adult male delivers no problems because of the broad anal point, although in *C. acutiventris* the anal point is also a little broadened. The exuviae are absent from the key by Langton (1991), but can be identified using Langton & Visser (2003). The larvae of the two species can usually be identified by the mental size; the other characters given by Webb & Scholl (1985) and Vallenduuk et al. (1995) are less reliable (see under *C. acutiventris*).

DISTRIBUTION IN EUROPE AND THE NETHERLANDS

The species is widespread in Europe (Saether & Spies, 2004). In the Netherlands the species is widespread with an unmistakable preference for the Pleistocene sandy soil regions (Limnodata.nl).

LIFE CYCLE

In winter nearly all larvae are in fourth instar (own data).

FEEDING

According to Moog (1995) the larvae feed on fine particulate organic material. They gather this mainly from the sediment, but to a lesser extent they obtain their food by active filtration (see also Monakov, 2003: 331).

MICROHABITAT

The larvae are found on sand, clay or peat bottoms with variable organic silt content.

WATER TYPE

C. obtusidens seems to be absent from fast-flowing brooks and streams (Lehmann, 1971; Orendt, 2002a). Wülker et al. (1983) reviewed the records in the literature; the species prefers flowing water or the littoral zone of lakes where there is at least some water movement. Matěna (1990) found the species more in running waters and in inflow and outflow ditches than in the ponds he investigated. Most of the more than 200 Dutch records (based on identification of larvae) are also from lowland brooks and slow-flowing canals, but the larvae are not rare in stagnant ditches and ponds (e.g. Limnodata.nl; own data). In larger water bodies the larvae seem to be confined to the littoral zone.

Permanence

Michailova (1989) collected the larvae in temporary water. Moller Pillot (2003) reported a rather frequent presence in a temporary upper course of a lowland brook.

pH

Brodin & Gransberg (1993) stated that *C. obtusidens* became more common in a Scottish lake after acidification (based on identification of head capsules); the larvae were most common (around 1900 individuals) at a pH of about 5.2. However, the identification is doubtful, also because the larvae must have lived in the profundal zone of the lake. There are no Dutch records from very acid water, but the larvae have been frequently collected in slightly acid water with a pH of around 6 (Moller Pillot, 2003; Limnodata.nl).

TROPHIC CONDITIONS AND SAPROBITY

C. obtusidens was nearly absent from the oligotrophic lakes in Sweden (Brundin, 1949). The larvae appear to be most common in very eutrophic to moderately polluted water and have been collected rarely in severely polluted water (Limnodata.nl; own data). In many cases the larvae lived in places where polluted water had drained away or on organic silt in more or less flowing water, indicating a preference for decomposing matter if enough oxygen is available (Limnodata.nl; H. Cuppen, pers. comm.; own data). However, in two cases larvae were found in water without any oxygen during the whole night.

SALINITY

The larvae have not been collected in brackish water. B. Krebs, who investigated many brackish pools and ditches, has never found the species.

Chironomus pallidivittatus Edwards, 1929

Camptochironomus pallidivittatus Shilova, 1976: 103; Pinder, 1978: 111, fig. 142B; Kiknadze et al., 1991: 20, 90

SYSTEMATICS AND NOMENCLATURE

The species belongs to the subgenus *Camptochironomus*. See under *C. tentans*. The problems with nomenclature are treated by Spies & Saether (2004). Most publications give Malloch as the author of the species name.

DISTRIBUTION IN EUROPE AND THE NETHERLANDS

The species is widespread in Europe and also in the Nearctic (Saether & Spies, 2004). Limnodata.nl contains records based on larval identification mainly from the Holocene part of the Netherlands. Krebs (1981, 1984) identified the adult males from many localities in the province of Zeeland.

IDENTIFICATION

The larvae are mainly distinguished from those of *C. tentans* by their smaller size. Because the size is greatly affected by the environmental conditions and there is a considerable overlap in the sizes of adults and exuviae (Palmén & Aho, 1966), some care is necessary. Identification of exuviae using Langton (1991) is not always possible. Hirvenoja (1998) stated that specimens with a broken shagreen band on sternite 3 have to be identified by studying the median armament of the sternites.

LIFE CYCLE

Palmén & Aho (1966) reported two generations a year and supposed a third generation is possible, depending on water temperature. The species clearly develops more rapidly than *C. tentans*. According to Shilova (1976) the cohorts are not very synchronised. The adults fly from May to September (Kruseman, 1933; Shilova, 1976); in Finland they occur only from the latter half of June (Palmén & Aho, 1966). In the Camargue they emerge the whole year round, with maxima in spring and autumn (Tourenq, 1975).

MICROHABITAT

Palmén & Aho (1966) found the species (together with *C. tentans*) on bottoms with rough plant material, such as decomposing *Fucus*, and hardly in places where the bottom material consisted of fine mud practically devoid of large particles. Moldován (1987) found the species in comparable places, but only when also submerged macrophytes were present. Krebs (1981, 1984) also reared the species from water bodies with muddy bottoms.

FEEDING

Vala et al. (2000: 299) noted that the larvae feed from grains of rice. The microhabitat mentioned above indicates that they also feed on decomposing plant material. See, however, under *C. tentans*.

WATER TYPE

According to Shilova (1976) the larvae live in stagnant water: pools, ditches and the littoral zone of lakes. Krebs (1981, 1984) reared the species mainly from wide (sometimes flowing) ditches, but also from pools in the province of Zeeland. Limnodata.nl also contains a number of records from stagnant water and only a few from flowing water. Orendt (2002a) noted the occurrence of exuviae in one fast-flowing stream in Bavaria, but most probably such specimens originated from more lentic stretches. Moldován (1987) found the larvae sometimes numerously (up to 3300 larvae/m^2) in a sewage treatment plant in Hungary.

Permanence

Driver (1977) collected the larvae in one case (scarcely) from a temporary pond. Krebs (1981, 1984) sometimes reared the species from temporary water. The larvae are not rare in rice fields (Tourenq, 1975; Vala et al., 2000).

pH

There are no records from strongly acid water and only a few records from water with a pH of 6 (Koskenniemi & Paasivirta, 1987; van Kleef, unpublished). The species seems to be most common in alkaline water. Steenbergen (1993) collected the whole subgenus *Camptochironomus* (143 localities in the province of Noord-Holland) rarely from water with pH values below 7.5.

TROPHIC CONDITIONS AND SAPROBITY

All Dutch records are from eutrophic to hypertrophic water bodies. Tourenq (1975: 176) stated that oxygen contents of 3–4 mg/l do not injure development. We collected the species in low numbers in ditches where the oxygen content every summer night fell to zero.

SALINITY

According to Kruseman (1933: 126) the species is euryhaline. Steenbergen (1993) stated that the subgenus *Camptochironomus* has no preference for fresh or brackish water. However, the larvae have not been found in the Netherlands in water with chloride contents above 2600 mg/l (Krebs & Moller Pillot, in prep.). Palmén & Aho (1966) found the species scarcely in brackish water and only in specific localities. Tourenq (1975) recorded an upper limit of 6000 mg chloride/l in the Camargue.

Chironomus parathummi Keyl, 1961

SYSTEMATICS AND IDENTIFICATION

Adult males and exuviae cannot be identified. The larva has been described and illustrated by Kiknadze et al. (1991) and is morphologically hardly different from *C. luridus*.

DISTRIBUTION IN EUROPE

C. parathummi has been collected in a few European countries, mainly in Central Europe (Saether & Spies, 2004). In the Netherlands its presence has been recorded on the basis of cytological investigation.

WATER TYPE

The larvae live in shallow stagnant water such as ditches and the littoral zone of lakes (Krieger-Wolff & Wülker, 1971; Matěna & Frouz, 2000). H. Vallenduuk (pers. comm.) collected the larvae in the seepage zone along the Oostvaardersdijk in the Netherlands.

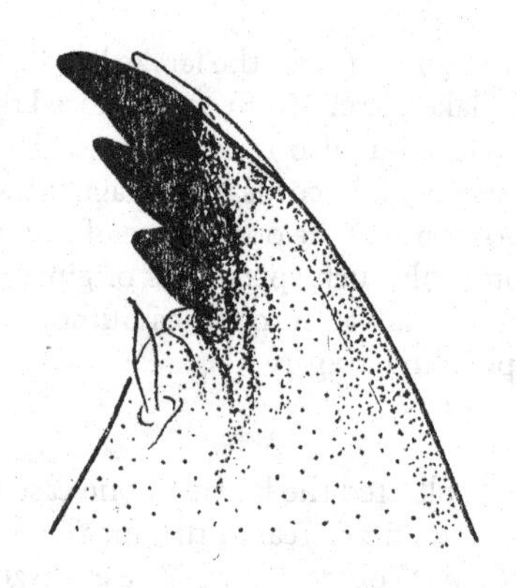

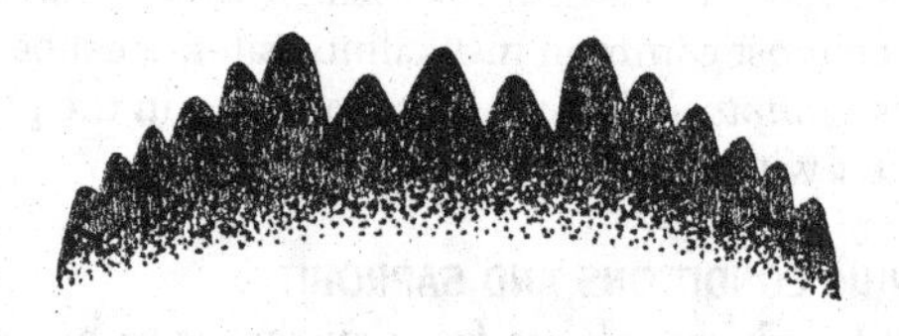

4 *Chironomus parathummi: mentum and mandible (from Kiknadze e.a., 1991)*

Matěna & Frouz (2000) reported them once from running water. Michailova (1989) found the larvae in Bulgaria in the muddy substrata of temporary waters.

TROPHIC CONDITIONS

Several of the above-mentioned records indicate that the species lives in more or less eutrophic water. Krieger-Wolff & Wülker (1971) stated that the species is an inhabitant of the Rhine floodplain and not of the Black Forest.

Chironomus piger Strenzke, 1956

Chironomus thummi piger Strenzke, 1959: 29 et seq., fig. 19; 1960: 115 et seq.

SYSTEMATICS AND IDENTIFICATION

C. piger has often been regarded as a subspecies of *C. riparius* (Strenzke, 1959). In the laboratory both species can hybridise, but hybrids have not been found in nature (Strenzke, 1960). Pinder (1978) did not distinguish between *C. riparius* and *C. piger*. Identification is possible using Strenzke (1959) and Langton & Pinder (2007). However, Dettinger-Klemm (2003: 315) stated that these two species cannot be distinguished from each other, not even by the combined use of the morphological characters of the larva and of the adult male, without cytological investigation. In this context it is remarkable that Russian workers (e.g. Sokolova et al., 1992) found mass development of *C. piger* in polluted running water, whereas in Western Europe *C. riparius* is the characteristic species in such circumstances. According to Spies & Saether (2004) the year of publication by Strenzke has to be 1956.

DISTRIBUTION IN EUROPE AND THE NETHERLANDS

The species is probably present in all parts of Europe, although this has not been stated in many countries (Saether & Spies, 2004). In the Netherlands the species is common and probably widespread.

LIFE CYCLE

According to Dettinger-Klemm (2000) the species needs about the same time for development as *C. dorsalis*: culture vessels at 20 °C for about 24 days. Sokolova et al. (1992) found 96,000–348,000 larvae/m^2 in a polluted river in Russia. In optimal conditions, 5 to 6 generations a year were possible. Near Moscow the first generation started to emerge as early as the beginning of April. Most larvae hibernated in fourth instar. Tourenq (1975) observed emergence in the Camargue from January to August. Strenzke (1959: 39) stated that rearing and copulation is possible in the laboratory.

OVIPOSITION

Strenzke (1959) noted that the egg masses are deposited on a firm substrate at the water's edge or on floating plant material (rarely directly on the water surface). He found 190 to 1320 eggs per egg mass. Sokolova et al. (1992) found 400 to 850 eggs per egg mass.

MICROHABITAT

Krebs (1981, 1984) collected the larvae in many pools and watercourses on a bottom of fine or course organic detritus on clay. However, the species can be found on every substratum; in urban waters they are common on firm bottoms.

WATER TYPE

Current

C. piger is mainly a stagnant water species; in running water the species is much less common than *C. riparius*. However, the larvae are often reported from brooks, streams and rivers (e.g. Sokolova, 1992; Izvekova, 1996; Matěna & Frouz , 2000). Krebs (1981, 1984) collected the species regularly in slow-flowing watercourses in the Dutch province of Zeeland.

Dimensions

C. piger is often collected in ponds and puddles (Strenzke, 1959; Scharf, 1973; Matěna & Frouz, 2000; own data). The larvae are also rather common in cattle pools and in watercourses 1 to 7 m wide (Krebs, 1981, 1984) and they are not rare in carp ponds (Matěna & Frouz, 2000).

Permanence

In the Camargue the larvae survived desiccation in summer by digging themselves in (Tourenq, 1975). However, they very often settle in small temporary water bodies, urban waters, rain pools, etc., which dry out completely (Matěna, 1986; Matěna & Frouz, 2000; own data).

pH

Krebs (1981, 1984) collected the species commonly in the province of Zeeland in alkaline water with a pH between 7 and 9 . We found the species also in an acid environment, but we have no quantitative data from such water bodies.

TROPHIC CONDITIONS AND SAPROBITY

According to Scharf (1973) the larvae are not as common in polluted water as the larvae of *C. riparius*. This author stated that in oxygen-poor conditions (approx. 2 mg O_2/l) the development of *C. riparius* larvae was obviously better. Compared with *C. riparius*, *C. piger* is more a species of eutrophic, not very oxygen-poor ponds and puddles; the larvae of *C. piger* are more resistant to cold and thus better adapted to freezing of small puddles. However, Sokolova et al. (1992) found the species numerously in a polluted river in Russia. Izvekova et al. (1996) found *C. piger* in heavily polluted (polysaprobic) streams, accompanied by very few other species. (See also under Identification above).

SALINITY

C. piger females deposit their eggs only in fresh water (Strenzke, 1960). However, Tourenq (1975) stated that the larvae survive a chloride content of 8000 mg/l. Krebs & Moller Pillot (in prep.) found the species in the province of Zeeland commonly in water with less than 1000 mg Cl/l and gradually less at higher chloride contents up to a single record at about 6000 mg/l.

Chironomus pilicornis Fabricius, 1787

IDENTIFICATION

Adult males identified as *C. pilicornis* can belong to two species: *C. pilicornis* and *C. heteropilicornis* (see Wülker, 1996). Identification of the larvae has not yet been possible. The larvae of *C. pilicornis* have lateral tubules at segment VII, those of *C. heteropilicornis* are without lateral tubules. The gula has a dark spot (Wülker, 1996). At least *C. pilicornis* larvae have notably large eyes (H. Vallenduuk, pers. comm.).

DISTRIBUTION IN EUROPE AND THE NETHERLANDS

C. pilicornis has been recorded in Western, Northern and Eastern Europe, but seems to be absent from the southern countries (Saether & Spies, 2004). In the Netherlands there is only one cytologically verified record (Deurnse Peel, April 2005). Records by Kruseman (1933) from Amsterdam and exuviae from Pleistocene moorland pools identified as *? pilicornis* have to be verified.

LIFE CYCLE

In northern Sweden *Chironomus pilicornis* is a relatively early flying univoltine species, emerging in May (Wiederholm et al., 1977; Wülker, 1996). Kalugina (1972) and Shilova & Zelentsov (1972) also call *C. pilicornis* univoltine, with a long diapause in winter in fourth instar. Smith & Young (1973) collected adults in an English pond at the end of April. Kruseman (1933) caught the species in four different years in April near Amsterdam. We saw a swarm in the Deurnse Peel nature reserve in early April 2005. The type material of *C. heteropilicornis* was caught in May in Finland (Wülker, 1996).

OVIPOSITION

Rempel (1936, cited by Nolte, 1993) found 650 eggs per egg mass.

FEEDING

According to Shilova (1955, fide Monakov 1972) the larvae feed on algae in different ways.

MICROHABITAT

Paasivirta & Koskenniemi (1980) reported the species as an inhabitant of the moss-covered littoral zone of a polyhumic reservoir in Finland. We collected fourth instar larvae in spring on a thick layer of partly decayed organic detritus.

WATER TYPE

Wülker (1996) recorded the occurrence of this species in shallow littoral parts of lakes in Sweden and Finland. Paasivirta & Koskenniemi (1980) collected emergent males in a polyhumic reservoir in Finland. We found the species in the Netherlands in an excavated peat pool. The specimens of Kruseman (1933) were collected near Amsterdam, possibly in a peaty area. Van Kleef (unpublished) collected exuviae probably belonging to this species in a number of moorland pools in the Pleistocene area of the Netherlands.

pH

In a Finnish reservoir the pH was 5.0–5.4 (Paasivirta & Koskenniemi, 1980). At least the Dutch larvae from the Deurnse Peel also lived in very acid water. In the majority of cases the exuviae collected in Dutch moorland pools that probably belong to *C. pilicornis* came from water with a pH value around 4 (3.5–6.5). Without doubt the specimens collected by Kruseman near Amsterdam had emerged from more or less alkaline water.

Chironomus plumosus (Linnaeus, 1758)

SYSTEMATICS AND IDENTIFICATION

As explained in the introduction to the genus, identification to species level within the genus *Chironomus* is extremely difficult and often not possible from morphological characters alone. Over time more and more species have been discovered within *C. plumosus*, although some other species appeared to be no more than colour varieties. Identification of larvae has presented yet more problems. Lenz (1954–62) classified the larvae into a number of groups, most without systematic significance, but easy for practical purposes. He distinguished a group *plumosus* s.l. and a group s.s. Independently of this, Konstantinov (1957) made another key to the species in which the name *C. plumosus* has a different meaning; old Russian literature cannot therefore be compared with Western literature.

At this moment gr. *plumosus* is a true systematic unit. Shobanov (1989a) assigned nine species to this group, belonging to Lenz's gr. *plumosus* s.s., gr. *semireductus* and gr. *reductus*. Within the *plumosus* group three species with ventral tubules are common in Western Europe: *C. plumosus*, *C. balatonicus* and *C. muratensis*. In the key by Geiger et al. (1978) these three species are taken together as *C. plumosus*; in the key by Webb & Scholl (1985) *C. muratensis* was split off, mainly on the basis of tubuli length. As demonstrated by Kiknadze et al. (1991), tubuli length is not characteristic for the species. The key by Vallenduuk et al. (1995) is based on new and better characters and *C. plumosus* and *C. balatonicus* are taken together as *C. plumosus* agg. The differences between these species in Shobanov (1989a) have to be verified with more material. Material identified from adult male or exuviae is not reliable at the species level. *C. balatonicus* and/or *C. muratensis* are absent from Pinder (1978), Langton (1991) and Langton & Pinder (2007).

The result of all these problems is that the name *C. plumosus* has had a different meaning in the past. *C. plumosus*, *C. balatonicus* and *C. muratensis* in particular are not well separated. In the descriptive text below we indicate where we think the data may apply to another species than the true *C. plumosus*.

DISTRIBUTION IN EUROPE AND THE NETHERLANDS

C. plumosus is widespread in Europe (Saether & Spies, 2004). In the Netherlands no material has been cytologically identified as *C. plumosus*. Nevertheless, the species will be common in all parts of the country.

LIFE CYCLE

The species usually has two (sometimes three) generations a year, but in cold water only one (Mundie, 1957; Shilova, 1958; Hilsenhoff, 1966; Janković, 1971; Kalugina, 1972; Sokolova, 1983), depending mainly on the water temperature. Johnson & Pejler (1987) stated that the species was univoltine in Lake Erken in Sweden, but that there were two cohorts. Pedersen (1988) reported the same difference between two populations inhabiting shallow and deeper lake areas. The total development of one generation in summer (in the temperate zone) lasts from 40 to 60 days. Kruseman (1933) collected adult males in the Netherlands from April to September. In the Camargue the adults emerge the whole year round, with maximum numbers in spring (Tourenq, 1975).
Predation by fish and/or invertebrate predators is probably the main factor determining whether the first generation is much more numerous than the second (see e.g. Matĕna, 1989: 605, 607).
In Russia the first and second instar can be found only from June until September; in winter nearly all larvae are in fourth instar in diapause; a small number is in third instar in winter (Sokolova, 1983). In Serbia second instar larvae were also found in small numbers in winter (Janković, 1971).

OVIPOSITION

The egg masses are thrown off or deposited on the open water surface and sink slowly to the bottom (Hilsenhoff, 1966). Sometimes the eggs are also deposited on water vegetation or on a firm substrate (Sokolova, 1983). The egg deposition is accurately described by Sokolova (1983: 158–159). The eggs have no diapause and hatch within 3 to 14 days, depending on temperature. Strenzke (1959) found 1100 to 2000 eggs per egg mass. Nolte (1993) found a maximum of 2300 eggs per egg mass.

DENSITIES

Natural densities of *C. plumosus* larvae are very different, depending on local circumstances. In the mesotrophic Lake Erken in Sweden densities ranged from 15 larvae/m^2 at a depth of 12 m to 104 larvae/m^2 at 20 m (Johnson, 1986). Kajak (1963) reported a natural density of 200–300 larvae/m^2 in Lake niardwy in Poland. Entz (1965) found 1500 larvae/m^2 in Lake Balaton in November, gradually decreasing during the winter months to 600/m^2. Janković (1971) found variable densities in different years in carp ponds in Serbia. One year in May she found more than 3000 second instar larvae/m^2 and in summer more than 2000 fourth instar larvae/m^2. Drawing on the literature, Kajak (1963) reported much higher densities up to about 60,000–70,000 larvae/m^2, and even higher in polluted water. As a rule, the individual condition of the larvae, and also their growth and development, decrease with increasing densities; however, this depends on the circumstances and sometimes the larvae exhibit a rapid rate of growth and a low degree of mortality even at very high densities (see, however, Reist & Fischer, 1987).

FEEDING

The larvae have three methods of feeding (Sokolova, 1983): filtration of water through a net, gathering food from the inner side of the tube supplied by undulating movements, and gathering food from the surroundings of the tube. The food consists of planktonic and benthic algae, detritus and, to a much lesser extent, animal remains (Walshe, 1951; Sokolova, 1983; Johnson, 1986). Diatoms are often more important and better digested than other algae (Johnson & Pejler, 1987; Kajak, 1987). However,

Matěna (1990) observed an increase in the number of young larvae after an increase of the amount of decaying filamentous algae. Johnson suggested that the larvae are able to coexist with *C. anthracinus* because of the differences in their dominant foraging strategies: filter feeding and collecting particles from the tube wall (*C. plumosus*) against scraping from the sediment surface (*C. anthracinus*). However, concurrence or hindrance can play a role (Kajak, 1987: 305).

Besides U-shaped feeding tubes the larvae can form J-shaped tubes sealed at the distal end, and open-ended horizontal tubes (McLachlan & Cantrell, 1976; McLachlan, 1977; Hodkinson & Williams, 1980). All these types of tubes can be formed in the sediment; horizontal tubes are also built on the undersurface of dead leaves. Larvae in J-shaped and horizontal tubes in particular are deposit feeders (although the latter are also capable of filter feeding). They are only able to feed on small particulate material. For instance, on places with leaf accumulations in eutrophic water, faeces of shredding organisms are rapidly colonised by microorganisms, providing a high quality diet (Hynes & Kaushik, 1969; Hynes, 1970: 432 et seq.; Hodkinson & Williams, 1980). The presence of coarse material can impair the effectiveness of filter feeding and lead to more deposit feeding. Wang (2000) reported that the species is an important rice pest, damaging the roots and leaves of rice seedlings. Ferrarese (1992) found the species in rice fields in Italy, but less commonly. He supposed that the larvae are mainly root growth disturbers. According to Hilsenhoff (1966) the larvae do not feed at temperatures of 5° C or less.

MICROHABITAT

Chernovskij (1938) found no other chironomids than *C. plumosus* (s.l.) in muddy bottoms without oxygen at a depth of more than 2 cm in the sediment. At a depth of 6.5 and 7 m in Lake Chainoe, in soft detritus, the larvae were most numerous at a depth of 2–8 cm and some larvae were found at a depth of 14–18 cm.

In Lake Erken the larvae (*C. plumosus* s.l.) live in muddy sediments. They are very dependent on sediment composition and still positively correlated with particle size less than 2 µm (Johnson, 1984). However, Palmén & Aho (1966) noted that the abundance of larvae (tentatively called *C. plumosus*) in two bays in southern Finland was much higher on bottoms containing coarse particles of organic detritus than in places where the particle size was small. Like other *Chironomus* species, the larvae are rarely collected on stones (Mol et al., 1982; cf. Brodersen et al., 1998).

WATER TYPE

Current

The larvae of *C. plumosus* are common inhabitants of reservoirs, lakes and ponds, but also live in flowing water. According to Sokolova (1983: 11) the larvae avoid places with much current in streams. Moog (1995) called the species a normal inhabitant of the littoral and profundal zone of lakes, but scarce in the potamal zone and very scarce in the rhithral zone of streams. In lowland brooks in the Netherlands the larvae can be numerous, but they are absent from the majority of brooks. Orendt (2002a) noted their occurrence in six rather fast-flowing streams in Bavaria. In the Dutch stretches of the rivers Rhine and Meuse they have been collected exclusively in the stagnant or nearly stagnant lower reaches (Klink, pers. comm.) and these larvae can belong also to *C. balatonicus*.

Dimensions

C. plumosus has been mentioned often as a lake species, occurring more scarcely in small water bodies such as small ponds and ditches (e.g. Langton, 1991). However, the larvae have been collected in ditches not more than 1 or 2 m wide and in ponds and pools of different dimensions (Krebs, 1981; Limnodata.nl). In running water they are not rare in brooks and streams, but in large rivers they seem to be common mainly in reservoirs and stretches before dams, etc. (see under Current). In lakes the larvae live in the littoral and profundal zone (Saether, 1979). Keyl (1962) and Krieger-Wolff & Wülker (1971), found a karyological difference between larvae from ponds and from the profundal zone of lakes. Strenzke (1960: 120, 124) did not find such karyological difference.

Permanence
The species has rarely been recorded from temporary pools (e.g. Krebs, 1981); we also rarely collected the larvae in temporary water. However, the species is a well-known pest in rice fields (Ferrarese, 1992; Wang, 2000).

pH
C. plumosus lives mainly in alkaline water (Limnodata.nl), but larvae and exuviae identified as *C. plumosus* are sometimes collected in moorland pools with a pH around 5 or 6 (e.g. van Kleef, unpublished). None of these records has been cytologically verified. The scarcity in acid water may be related to food shortage or, locally, to sensitivity to H_2S (see below).

TROPHIC CONDITIONS AND SAPROBITY
The larvae live in the littoral zone of mesotrophic and eutrophic lakes (Saether, 1979). In the profundal zone the species is more typical for very eutrophic lakes, where anoxia for a long period is possible; this in contrast to *C. anthracinus* (Saether, 1979; Wiederholm, 1980). The larvae of *C. plumosus* are a little larger and this difference is therefore consistent with Heinis (1993: 74), who stated that larger larvae are more resistant to anoxia than smaller ones. For further comments see under *C. anthracinus*.

On several occasions *C. plumosus* appeared to be absent in stagnant or slowly flowing water, whereas *C. riparius* was dominant and extremely numerous (e.g. Bazerque, 1989; Groenendijk et al., 1998; own data). In some cases, *C. plumosus* was only present after some improvement in the water (or silt) quality. Although the larvae of *C. plumosus* are more tolerant of anoxia, in many cases they appear to endure less pollution, probably because:
they are more sensitive to sulphide and/or nitrite;
and/or they are more sensitive to metal pollution (in the case of Groenendijk et al.);
and/or they need more oxygen for detoxification of toxicants.
An example can be that Rasmussen & Lindegaard (1988) found that larvae of gr. *plumosus* were more susceptible to high concentrations of Fe^{2+}-ions than larvae of gr. *thummi*. According to Neumann et al. (2001) the presence of nitrite can be the cause of reduced abundance of *Chironomus* in polluted waters. In less toxic environments, such as the profundal zone of some lakes, *C. plumosus* is the species which best survives anoxic periods for a long time. It remains to be investigated whether the differences in the presence of this species between different heavily polluted lowland brooks (e.g. stated by GWL at Boxtel) is caused by differences in toxic pollutants.
Sládeček (1973) and Moog (1995) supposed that *C. plumosus* is characteristic for polysaprobic water and is hardly present in β-mesosaprobic water. This is not correct. The larvae are sometimes common in the littoral zone of mesotrophic lakes (Saether, 1979), in pioneer habitats (see below) and in many non-polluted mesotrophic or eutrophic lakes, ponds and ditches with an organic layer of silt on the bottom (Limnodata.nl; van Kleef, unpublished; own data). Limnodata.nl gives a mean BOD5 of 5.5 and a mean oxygen content (during the day) of 7.5 mg/l.

C. plumosus is a typical pioneer species during the first years after the filling of a reservoir (Morduchai-Boltovskoi, 1961). Besides the high dispersal capacity of this species, an important reason for this is its high productivity in the initial phase, mainly as a result of decomposition of terrestrial vegetation (Brown & Oldham, 1984). The often reported decline in the population density after two or three years can be caused by lower productivity, the poor competitive ability of young larvae or predation and infection with parasites, which may have a relatively greater effect on this species.
In the Netherlands anoxia during ice cover in winter does not play an important role and is therefore not stated in our matrix table, which does not contain a column for very long-lasting anoxia. In Central European and northern countries ice cover may be more significant.

SALINITY

Tourenq (1975) stated the occurrence of *C. plumosus* (or related species) in the Camargue in water with a chloride content up to 10,000 mg/l. Krebs (1978) found *C. plumosus* (identified from adult male) in a brackish ditch (approx. 5000 mg Cl/l) only in periods of refreshment. Krebs (1981, 1984, 1985) collected the larvae very rarely in water with more than 3000 mg Cl/l (cf. Krebs & Moller Pillot, in prep.).

DISPERSAL

The relatively large adults of this species probably contribute to its colonisation success because large imagines have a greater dispersal capacity (Brown & Oldham, 1984). The larvae live planktonically in first instar (Lellák, 1968) and swim in later instars mainly under the influence of crowding, food shortage or changes in the environment (e.g. Hilsenhoff, 1966; Sokolova, 1983). Hilsenhoff stated that between November and May areas of Lake Winnebago having few larvae showed an increase in population and areas of high population showed a decrease. Sokolova (1983) noted that many larvae migrate into the deeper parts of a lake mainly in autumn. Dispersal takes place mainly at night.

Chironomus prasinus sensu Pinder, 1978

NOMENCLATURE

According to Spies & Saether (2004) the name *C. prasinus* Meigen, as used by Pinder (1978), is incorrect and should be replaced, possibly by *C. horni* Kieffer.

IDENTIFICATION

According to Langton & Pinder (2007: 2, 160) the adult male is hardly different from *C. entis*. The exuviae can be identified using Langton (1991) and Langton & Visser (2003).

DISTRIBUTION IN EUROPE

Saether & Spies (2004) called all records of *C. prasinus* doubtful, including the Dutch records of Kruseman (1933: 162).

Chironomus pseudothummi Strenzke, 1959

nec *Chironomus pseudothummi* Smith & Young, 1973

SYSTEMATICS AND IDENTIFICATION

The *pseudotummi* complex (sensu Keyl, 1962) is based on karyological characteristics and comprises species with different larval morphology, such as *C. luridus*, *C. aprilinus*, etc. (see also Michailova, 1989). It has not yet been possible to distinguish between the larvae of *C. luridus*, *C. pseudothummi* and *C. uliginosus* using morphological characteristics and they are treated together by Vallenduuk et al. (1995) as *C. luridus* agg. The larva has been described and illustrated by Kiknadze et al. (1991).

The adult male and the exuviae have been keyed by Langton & Pinder (2007) and Langton (1991) respectively, but *C. uliginosus* is missing from these keys. Strenzke (1959) pointed out that adult males of some other species (with morphologically and cytologically different larvae) cannot be distinguished from *C. pseudothummi*. In some cases such specimens have been collected in quite different environments. For instance, the species named *C. pseudothummi* by Smith & Young (1973) is not this species because it is cytologically different and the larva differs morphologically hardly at all from *C. annularius*.

DISTRIBUTION IN EUROPE AND THE NETHERLANDS

C. pseudothummi is widespread in Europe, but is possibly absent from the Mediterranean area (Saether & Spies, 2004). In the Netherlands only one reported identification (acid upper course of the Roodloop, see Moller Pillot, 2003) has been cytologically verified. Adults and larvae identified as cf. *pseudothummi* have been collected at several localities in the Pleistocene part of the country. An important part of these populations will belong to *C. 'uliginosus'* (see under this species).

LIFE CYCLE

Schleuter (1985: 127–129) stated three emergence maxima in a pool in Germany. Emergence began in early April and lasted until early July. Moller Pillot (unpublished) found emergence in the Roodloop as late as August. A significant proportion of the larvae of *C. luridus* agg. in the Roodloop were still in third instar in winter.

OVIPOSITION

Strenzke (1959) observed that the egg masses are deposited on a firm substrate. He found about 580 eggs in one egg mass.

WATER TYPE

Current and dimensions

The larvae (cytologically verified) were locally numerous in a narrow, slow-flowing acid upper course of the Roodloop in woodland (Moller Pillot, 2003). Matěna & Frouz (2000) found the species once in running water. All other records are from stagnant pools or ditches (Strenzke, 1959, 1960; Krieger-Wolff & Wülker, 1971; Schleuter, 1985, 1986; Matěna & Frouz, 2000).

Permanence

Schleuter (1985, 1986) called *C. pseudothummi* a characteristic inhabitant of permanent and temporary pools with fallen leaves. This seems to apply also to the Dutch records.

pH

Strenzke (1960: 120) called *C. pseudothummi* a stenotope inhabitant of acid bogs and woodland pools. Schleuter (1985, 1986) collected the species (identified from adult male) in nearly every temporary woodland pool at pH 4.2–6.5. Matěna & Frouz (2000) collected cytologically identified *C. pseudothummi* nearly exclusively in natural pools in peatland. In the Netherlands *pseudothummi/uliginosus* larvae are common in peat cuttings and acid woodland (pH 3.5–6), but there are few records from acid moorland pools in heathland. Krebs (1981, 1984) sometimes identified the species (as adult male) from non-acid pools, ditches and a creek in the province of Zeeland; this may be a different species.

TROPHIC CONDITIONS AND SAPROBITY

Schleuter (1985, 1986) collected the larvae very often in acid pools with fallen leaves, also with oxygen deficit. This also applies to the Dutch records. Larvae (of this species?) were collected even in peat cuttings where oxygen was nearly absent during summer nights (Werkgroep Hydrobiologie, 1993). The slow rate of decomposition in acid environments may be important in view of the oxygen regime and because decomposition products become available only gradually.

SALINITY

According to Tourenq (1975) the larvae survive a chloride content of 2000 mg/l. It is very probable that this applies to a related species.

Chironomus riparius Meigen, 1804

Chironomus thummi thummi Strenzke, 1959: 28–29

SYSTEMATICS AND IDENTIFICATION

Strenzke (1959) considered *C. piger* to be a subspecies of *C. riparius*. As a consequence of this many keys (e.g. Pinder, 1978) group both species together as *C. riparius*. Langton (1991) named exuviae of this combination *Chironomus* Pe 7a. Because identifying the larvae is difficult, both species are still often named *riparius* agg. (Vallenduuk et al., 1995). In some cases the addition 'agg.' has been omitted (e.g. in Limnodata.nl). Many authors are not convinced that both species can be distinguished as adults (e.g. Dettinger-Klemm, 2003: 315). See under *C. piger*.
In this book we have used publications on cytologically verified material as well as data based on identified adult males. Other data (*C. riparius* agg., gr. *thummi*, etc.) have been used with caution.

DISTRIBUTION IN EUROPE AND THE NETHERLANDS

C. riparius has been found nearly everywhere in Europe. In the Netherlands the species is widely distributed, but it can be locally scarce, for example in the province of Zeeland. In general, *C. riparius* agg. larvae are much more common in Pleistocene than in Holocene regions; moreover, the majority of records from the Holocene may apply to *C. piger*, as has been stated by Krebs (1981, 1984, 1990) for the province of Zeeland.

LIFE CYCLE AND DEVELOPMENT

C. riparius develops rapidly, with often 4–7 generations a year, emerging from March to November (Goddeeris et al., 2001). In laboratory studies Scharf (1972, 1973) discovered that one generation of *C. riparius* had a length of 34.8 days at 15 °C, 20.6 days at 20 °C and 15.5 days at 25 °C. Pinder (1986) supposed that the potentially rapid rate of development enables the species to exploit situations where there is little competition from other insects. In the Camargue, though, the species is monocyclic (Tourenq, 1975).
Strenzke (1959: 31, 39) noted that *C. riparius* is one of the few species which can be reared for several generations in the laboratory. In laboratory cultures the adults flew and mated in a box (height 30–40 cm) under a normal photoperiod (Credland, 1973). Most larvae have a diapause in winter in fourth instar, a smaller number also in third instar; sometimes non-diapausing first and second instar larvae can be collected in winter (Goddeeris et al., 2001).

OVIPOSITION

Strenzke (1959) noted that the egg masses are deposited on a firm substrate. He found 265–722 eggs per egg mass. However, Nolte (1993) found a maximum of 1800 eggs per egg mass. Strenzke (1960: 121) observed egg deposition especially near sewage outlets.

FEEDING

The larvae of *C. riparius* feed by ingesting fine particles at the sediment surface and do not filter particles from the water column (Walshe, 1951; Rasmussen, 1985). In the guts of the larvae we found, besides fine detritus particles, some diatoms and unicellular and filamentous algae, spores of fungi, small animal remains and a few decayed parts of plant tissue. However, the larvae are known to be injurious to rice seedlings, damaging especially the roots of the seedlings (Ferrarese, 1992). These larvae were probably mainly root growth disturbers.
We observed incidentally larvae eating young specimens of Tubificidae, but such behaviour is probably exceptional. In laboratory cultures Credland (1973) and Goddeeris et al. (2001) fed the larvae successfully with commercially prepared fish food (Tetramin) and filamentous algae. Vos (2001) tested different food items and found different results depending on the amount of food available.

MICROHABITAT

In a substratum-selection experiment, Credland (1973) observed that most larvae selected mud substrate in preference to sand or gravel. With the same foodstuffs added, the provision of a mud substratum resulted in a much larger yield of adults and many more eggs. Consistent with these findings, the species has been collected by many authors on organic mud. However, the species can be also numerous on stones or concrete bottoms with a thin layer of mud (Moldován, 1987; own data).

DENSITIES

Davies & Hawkes (1981) recorded densities up to 20,000 larvae/m^2 in an organically polluted stream in August and nearly 23,000/m^2 in October. Tourenq (1975) found densities above 50,000 larvae/m^2 . Koehn & Frank (1980) stated a maximum density of 132,715 larvae/m^2 in a polluted slow-flowing channel in Berlin. Rasmussen (1985) observed that the development and the number of generations can be influenced by density effects. In enclosures he found this effect using densities of about 20,000 larvae/m^2. Several authors stated a decline during winter (e.g. Matěna & Frouz, 2000; Moller Pillot, 2003).

WATER TYPE

Current

C. riparius is often stated to be a characteristic inhabitant of polluted flowing water (e.g. Thienemann, 1954; Krieger-Wolff & Wülker, 1971). Lehmann (1971) collected the species not only in the potamal zone, but also in the lower part of the rhithral zone. Braukmann (1984) collected *Chironomus* gr. *thummi* in large numbers in a polluted mountain brook. In the Netherlands the species lives nearly everywhere in more or less polluted lowland brooks and also in faster-flowing polluted brooks and streams in South Limburg. We also found them numerously in stream rapids. The main reasons the larvae are often absent in fast-flowing upper courses is because these are often less polluted or that the bottom is not as rich in organic silt. The species is also common in stagnant water like rain pools, ditches and ponds (e.g. Krieger-Wolff & Wülker, 1971; Matěna & Frouz, 2000; own data). However, the species is relatively scarce in pools and ditches in the Holocene part of the Netherlands compared with these water bodies in the Pleistocene regions (Limnodata.nl).

Dimensions

In stagnant water *C. riparius* is more or less characteristic of small water bodies like rain pools, ditches and ponds, but the species has sometimes been collected in the littoral zone of lakes (e.g. Geiger et al., 1978) and even in the profundal zone, although rarely (Thienemann, 1954). In running water the larvae are also especially common in brooks and small streams; they are scarce and often absent in large rivers and in stretches with much organic silt (e.g. Klink & Moller Pillot, 1982; Klink, 1991, 1994). Becker (1995) could rear the species only very locally from the littoral zone of the Lower Rhine in Germany.

Permanence

The larvae are common in astatic waters and temporary pools (e.g. Driver, 1977; own data). The species is more common in urban waters than in natural temporary pools (e.g. not found in woodland pools by Schleuter, 1985), notwithstanding the fact that these urban waters are usually not polluted. The species is more common in temporary pools than in comparable permanent pools, probably because of an absence of competition (see Driver, 1977). In polluted water this will play no role.

pH

Although larvae of *C. riparius* are found in water with a pH of 4 and lower (Moller Pillot, 2003), development deteriorates at pH < 5, especially < 4.5 (Palawski et al., 1989). Jernelov et al. (1981) stated that haemoglobin functions as an internal buffer, which enables the larvae to survive in water with a low pH. The larvae show no clear preference between pH 6 and 8, but we have no data about more alkaline water.

TROPHIC CONDITIONS AND SAPROBITY

Many authors mention the occurrence of *C. riparius* in polluted water. The species can be very abundant under organically polluted conditions (Davies & Hawkes, 1981; Pinder & Farr, 1987; Bazerque et al., 1989) and the larvae can present problems in sewage plants. According to Sládeček (1973) *C. riparius* (as *C. thummi*) is a little less typical for polysaprobic conditions than *C. plumosus*. Although this author means only the forma larvalis and not the species, this difference is consistent with Heinis (1993: 74), who stated that the smaller *C. riparius* is less resistant to prolonged anoxia than the larger *C. plumosus*. However, in extremely polluted brooks and ditches we found only *C. riparius*. In less toxic environments with long periods of anoxia (for instance several weeks) only *C. plumosus* survives. It must be stressed that Sládeček does not distinguish between the influence of pollutants and of oxygen. A further complication is that *C. riparius* is a species of small water bodies, while *C. plumosus* lives only in somewhat larger canals and lakes. Although in the Netherlands most problems with pollution occur in summer, we found mass mortality of *C. riparius* in winter after a long period of ice cover with snow.

C. riparius occurs often in hypertrophic brooks and ditches, but the species is here rarely dominant (Krebs, 1981; own data). The larvae are rare in less hypertrophic water with a better oxygen regime (cf. Moller Pillot, 1971). Vos (2001: 53 et seq.) found low survival and low growth rates of the larvae in poor sediments. Why the species is not rare in acid water, where the amount of valuable food seems to be low, is not fully understood.

Pinder (1986) stresses that the species is not confined to polluted waters and may also be dominant in temporary or newly created water bodies. The amount of food in such water bodies can be rather high (see section 2.16), but these conditions are quite aberrant. We therefore restrict the figures for saprobity and oxygen in the matrix table in Chapter 4 (only for *C. riparius*) to non-acid brooks and streams.

HIGH TEMPERATURE

Koehn & Frank (1980) stated that larvae of *C. riparius* survived well at temperatures between 25 and 27 °C.; at these temperatures they found no more deformities of the ventromental plates.

TRACE METALS

Several trace metals such as cadmium, zinc and lead are accumulated in larvae of *C. riparius* (Timmermans et al., 1992). Even low concentrations of these metals result in high mortality in first instar larvae and retardation of development. Older larvae are less sensitive (Williams et al., 1986; Timmermans et al., 1992). Redford (unpublished) stated that *C. riparius* was scarce at stations with much iron. In the laboratory he found high mortality at high iron concentrations and a strong retardation of development. However, Rasmussen and Lindegaard (1988) found larvae of *C.* gr. *thummi* to be still fairly numerous in water with 17 mg Fe^{2+}/l.

SALINITY

C. riparius seems to be scarce in brackish water. According to Strenzke (1960) the species is absent from brackish water in nature, but can be reared in oligohaline water. Krebs (1981, 1984) collected the species in the Dutch province of Zeeland scarcely (also in freshwater) and in only one case the larvae lived in mesohaline water (3400 mg Cl/l). Many data from other workers cannot be used because no distinction was made between *C. riparius* and *C. piger* (see under *C. piger*). Strenzke (1960) stated that in the Alps in water with a very low chloride content the larvae develop very large anal tubules. In extreme cases the larvae could not develop to the adult stage.

DISPERSAL

Groenendijk et al. (1998) stated mass drift of larvae of *C. riparius* up to 15,000 larvae/hour in the Dommel, a Dutch lowland brook. Young larvae were dominant in summer. In winter mass drift occurred only during high water discharge and most of these larvae were fourth instars.

PARASITISM

Tourenq (1975) stated that sometimes many larvae were parasitised by *Hydromermis contorta*, as a result of which only a small percentage emerged.

Chironomus salinarius Kieffer, 1915

IDENTIFICATION AND MORPHOLOGY

Identification of adult males, exuviae and larvae present no problems when using any of the usual resources. However, related species which can be identified only by cytological investigation are found in northern countries and in Bulgaria (Michailova, 1989). The females have been described and illustrated by Strenzke (1959). A comparison of measurements of Dutch, German and Norwegian populations is given by Parma & Krebs (1977). In the Netherlands the larvae are unique in having no lateral or ventral tubules.

DISTRIBUTION IN EUROPE AND THE NETHERLANDS

C. salinarius occurs in all European regions bordering on the sea and also in some countries in Central Europe (Saether & Spies, 2004). Palmén & Aho (1966) mention the absence of *C. salinarius* from the Northern Baltic, most probably because the species requires fairly high temperatures for development. In the Netherlands the species is very common in the delta region and scarce in the outermost northern part of the country (Krebs, 1981, 1984, 1985, 1990; Limnodata.nl).

LIFE CYCLE

Krebs (1978) found probably two generations a year in a ditch in the Dutch province of Zeeland; in winter the larvae were in third and fourth instar. In western Norway the species has only one generation a year (Koskinen, 1969). Strenzke (1960: 123) observed that for normal development the larvae need a temperature of more than 20 °C. At a mean temperature of 25 °C total development lasted only about 20 days. In the Netherlands the species appears to be adapted to lower temperatures: Krebs (1978) stated development and emergence at about 15 °C. It is not known whether three or four generations a year are possible at higher temperatures. In the Camargue emergence is possible from March to November (Tourenq, 1975).

FEEDING

In one case we found the gut to contain only detritus.

MICROHABITAT

C. salinarius can be found on fine organic silt or fine sand, sometimes also on a hard substratum (Michailova, 1989; Krebs & Moller Pillot, in prep.).

OVIPOSITION

Strenzke (1959: note p. 36) stated that the egg masses were deposited on the water surface. He counted 430 to 670 eggs per egg mass.

WATER TYPE

The larvae have been collected in salt or brackish creeks, lagoons, canals, ditches and ponds (Lenz, 1954–62; Krebs, 1981, 1984, 1985, 1990; Michailova, 1989). Other records from boreal regions (e.g. by Lenz) may apply to related species. Tourenq (1975) stated survival of the larvae in more or less desiccating bottoms. Krebs (1981) sometimes found the larvae in temporary pools.

TROPHIC CONDITIONS AND SAPROBITY

Krebs & Moller Pillot (in prep.) reported, on the basis of their own data, that *C. salinarius* endures poor oxygen conditions much better than *C. aprilinus.*

D

SALINITY

Krebs (1981, 1984, 1990) reared the larvae from brackish water with chloride contents of 750 to 20,000 mg Cl/l; in one case egg deposition and developing larvae were found at 28,700 mg/l (Krebs & Moller Pillot, in prep.). Tourenq (1975) recorded a maximum of 40,000 mg Cl/l. See also Parma & Krebs (1977).

Chironomus sororius Wülker, 1973

IDENTIFICATION

The adult males cannot be reliably distinguished from those of *C. aberratus* (Wülker, 1973). It is very difficult to distinguish between the larvae of *C. aberratus* and *C. sororius.* Geiger et al. (1978) illustrated a difference in the colouration of the frontal apotome, which was not copied by Webb & Scholl (1985) and Vallenduuk et al. (1995).

DISTRIBUTION IN EUROPE AND THE NETHERLANDS

The species has been collected in Scandinavia and Central and Eastern Europe (Wülker, 1999). In the Netherlands there are four records which still have to be verified.

WATER TYPE

The larvae have been collected in pools in the Black Forest, the Alps, Scandinavia and the Czech Republic, in several cases together with *C. aberratus* (Wülker, 1973, 1999; Ryser et al., 1980; Matěna & Frouz, 2000). The last authors also found the larvae once in running water. The Dutch records are from four (not very acid) moorland pools in the Pleistocene part of the country (Limnodata.nl). Most probably the species can easily colonise again after drying out of a pool (see Wülker, 1973: 368).

Chironomus storai Goetghebuer, 1954

SYSTEMATICS

C. storai belongs to the subgenus *Lobochironomus* (Webb & Scholl, 1987).

DISTRIBUTION IN EUROPE

C. storai has been collected in Scandinavia and the Alps. The species will not be treated here.

Chironomus striatus Strenzke, 1959

IDENTIFICATION

The adult males can be identified using Strenzke (1959), the exuviae using Langton (1991) and the larvae using Vallenduuk et al. (1995). Cytological verification is necessary.

DISTRIBUTION IN EUROPE AND THE NETHERLANDS

The species has been collected in few European countries, including some in Western Europe (Saether & Spies, 2004). There are several records from the Netherlands, only identified from larvae by morphological characters.

WATER TYPE

The larvae have been collected in many pools in the Alps (Ryser et al., 1980) and in some pools in the Black Forest (Krieger-Wolff & Wülker, 1971). The Dutch records are almost exclusively from moorland pools.

pH

Strenzke (1960: 120) called the species a stenotope inhabitant of mineral acid pools. The Dutch records are from more or less acid moorland pools (Limnodata.nl; H. Cuppen, pers. comm.; H. Vallenduuk, pers. comm.); karyological verification of these populations is necessary.

Chironomus tentans Fabricius, 1805

Camptochironomus tentans Strenzke, 1960: 116; Pinder, 1978: 102, 111; Kiknadze et al., 1991: 19, 90
nec *Camptochironomus tentans* Sadler, 1935: 1–25; Danks, 1971a: 1879 et seq.; Driver, 1977: 125 et seq.; Baker & Ball, 1995: 101–106 (= *C. dilutus* Shobanov et al., 1999)

SYSTEMATICS

Currently most authors regard *Camptochironomus* as a subgenus of *Chironomus*. For the discussion about the status and the use of the name *Camptochironomus* see Ashe (1983: 14–15) and Spies & Saether (2004). The American material belongs to another species: *Chironomus dilutus* Shobanov, Kiknadze & Butler, 1999.

IDENTIFICATION

The larvae are somewhat larger than those of *C. pallidivittatus*. However, there is most probably some overlap (see under *C. pallidivittatus*). Identification of exuviae using Langton (1991) is not always possible. Hirvenoja (1998) stated that specimens with a broken shagreen band on sternite 3 have to be identified by studying the median armament of the sternites.

DISTRIBUTION IN EUROPE AND THE NETHERLANDS

The species lives in nearly all European countries (Saether & Spies, 2004). In the Netherlands the species appears to be widespread; in comparison with *C. pallidivittatus* larvae and exuviae are reported more from the Holocene and Pleistocene regions (Limnodata.nl; van Kleef, unpublished). However, the species is scarce in the province of Zeeland (Krebs, 1981, 1984).

LIFE CYCLE

According to Palmén & Aho (1966), Shilova & Zelentsov (1972) and Shilova (1976) the species has two generations a year. Kruseman (1993) caught adult males in the Netherlands from April to September. In the Camargue the adults emerge from April to November (Tourenq, 1975). The larvae overwinter in different instars; development stops when day length becomes shorter (Shilova & Zelentsov, 1972; Shilova, 1976).

FEEDING

According to Mason & Bryant (1975) the guts of the larvae of *C. tentans* (identified from adult males using Coe) contained only detritus from decaying *Typha* stems and no algae, whereas the guts of other species on these stems contained mainly algae. The most detailed information about the feeding of *C. tentans* larvae actually concerns the related American species *C. dilutus*. The larvae of *C. dilutus* live in tubes constructed of algae or sediment particles, preferably on bare bottoms (Sadler, 1935; Baker & Ball, 1995). They scrape up filamentous and unicellular algae and detritus (rarely also animal material) around both ends of their tube. Some of the food is supplied by a current of water caused by undulating movements of the body. If no food is found, the larvae can leave their tubes entirely or extend their tubes up to 10 cm or (rarely) 30 cm long. The difference between both observations are not based on specific differences. Shilova (1976) observed in *C. tentans* the same behaviour as described above for *C. dilutus*: filtering detritus particles, bacteria and algae in addition to gathering food from the bottom around the tubes.

MICROHABITAT

Mason & Bryant (1975) collected the larvae from *Typha* stems. Palmén & Aho (1966) found the species on bottoms with rough plant material, such as decomposing *Fucus*, and hardly in places where the bottom material consisted of fine mud practically devoid of large particles. Moldován (1987) found the species in comparable places, but only when also submerged macrophytes were present. Marlier (1951) found the larvae abundantly at a strongly polluted place in a small lowland brook, the Smohain. The substrate was composed of fine organic silt with many plant fragments. Shilova (1976) collected the larvae mainly from organic silt and sandy bottoms.

MATING AND OVIPOSITION

Sadler (1935) stated that females of the related *C. dilutus* usually enter the swarm of males and that mating happens before reaching the ground. However, *Camptochironomus* is able to mate sitting on the substrate, not introduced by flight (Strenzke, 1960: 115). Sadler observed oviposition in the early morning and late afternoon, at the water's edge or on floating objects. The females never deposited more than one egg mass.

DENSITIES

Marlier (1951) recorded a density of 22,000 larvae/m^2 in a polluted upper course. Shilova (1976) found a maximal density of 7000 larvae/m^2.

WATER TYPE

The larvae live mainly in small stagnant waters and the littoral zone of lakes (Lenz, 1954–62; Shilova, 1976; Saether, 1979; Michailova, 1989). There are relatively few records from flowing water (Krebs, 1984; Orendt, 2002a; Limnodata.nl). The related North American species *C. dilutus* also lives in the profundal zone of lakes (Palmén & Aho, 1966); such records are not reported for *C. tentans*. Tourenq (1975) found the species regularly in summerdry marshes, where they survive for a long time during the drying up of the bottom.

pH

Like *C. pallidivittatus* the species seems to be most common in alkaline water. Steenbergen (1993) collected the whole subgenus *Camptochironomus* (143 localities in the Dutch province of Noord-Holland) rarely at a pH below 7.5. However, there are several records of larvae and exuviae of *C. tentans* from acid water with pH values between 4.5 and 6 (Limnodata.nl; van Kleef, unpublished).

TROPHIC CONDITIONS AND SAPROBITY

The larvae are characteristic for eutrophic lakes (Saether, 1979). They have been found in heavily polluted water (Marlier 1951; Thienemann, 1954: 300). Most records in the Netherlands are also from eutrophic water, but the larvae and exuviae have also been collected in more or less mesotrophic pools (e.g. van Kleef, unpublished).

SALINITY

According to Kruseman (1933: 126) the species is euryhaline. However, Palmén & Aho (1966) found the species scarcely in brackish water and only in oligohaline and rarely β-mesohaline conditions. *C. tentans* is very scarce in the Dutch province of Zeeland; Krebs (1981, 1984) does not report the species from water with more than 350 mg Cl/l.

Chironomus tenuistylus Brundin, 1949

DISTRIBUTION IN EUROPE

C. tenuistylus has been found only in Scandinavia and Russia (Wülker, 1999; Saether & Spies, 2004).

WATER TYPE

Brundin (1949) collected the species from oligotrophic polyhumic lakes in Sweden. Saether (1979) called the species characteristic for the profundal of mesohumic and polyhumic lakes.

Chironomus uliginosus Keyl, 1960

IDENTIFICATION

The adult male has not been included in the keys by Strenzke (1959) or Langton & Pinder (2007). The exuviae cannot be identified. The larva resembles that of *C. pseudothummi* and cannot be identified with certainty (Webb & Scholl, 1985; Vallenduuk et al., 1995). A second species has recently been collected in the Netherlands that is morphologically and ecologically very near to *C. uliginosus* and provisionally named ULI 2 by Kiknadze (unpublished).

DISTRIBUTION IN EUROPE AND THE NETHERLANDS

C. uliginosus has been collected in few scattered European countries. In the Netherlands only ULI 2 has been found, in peat pools (leg. H. Vallenduuk). Other material has not been verified cytologically.

WATER TYPE

In Western Europe *C. uliginosus* has been collected only in bog and woodland pools (Strenzke, 1960: 120; Krieger-Wolff & Wülker, 1971; Ryser et al., 1980). This also applies to ULI 2. However, Matěna & Frouz (2000) collected a larva in a newly flooded carp pond in the Czech Republic (cytologically identified) and Wülker (1999) mentioned a record by Kiknadze from a river in Russia.

Chironomus venustus sensu Pinder, 1978

NOMENCLATURE

According to Spies & Saether (2004) the interpretation of *C. venustus* Staeger, 1839 by Pinder is not correct and the use of this name must be discontinued. However, the name is used in Langton & Visser (2003) and Langton & Pinder (2007).

IDENTIFICATION

Identification of adult males is possible using Langton & Pinder (2007). However, definitive species identification requires cytological investigation and revision of the species group.

Goetghebuer (1937–54) and Kruseman (1933) treat *C. venustus* as a colour variety of *C. cingulatus*. The exuviae are not known with certainty (Langton, unpublished).

D

DISTRIBUTION IN EUROPE AND THE NETHERLANDS

C. venustus has been collected in few scattered European countries. The species has been placed on the Dutch list on the basis of identification of adult males by Kruseman (see above).

Cladopelma Kieffer, 1921

Cryptocladopelma Lenz, 1960a: 165–184; Moller Pillot, 1984: 237–240

SYSTEMATICS AND NOMENCLATURE; IDENTIFICATION

Lenz (1954–62: figs 277–279b) distinguished two types of larvae and pupae based on material from three species. Moller Pillot (1984) named these type groups and these names were borrowed by Pinder & Reiss (1983) and Nocentini (1985). Although it is questionable whether these groups can be maintained for all larvae and although they probably do not have a systematic value, we use this division provisionally. Use of these groups simply involves a change in the names as follows: *C.* gr. *goetghebueri* = gr. *lateralis* and *C.* gr. *viridulum* = gr. *laccophila.*

In this genus many authors have made errors in combining the names of larvae, pupae and adults. According to Ashe & Cranston (1990) and verification of material in Munich by Dr. P. Langton (in litt.), *C. laccophila* is a nomen dubium. The material named *C. laccophila* belongs to *C. viridulum.* Also Lenz (1954–62: 212, 214) noted that the identification of this species was not certain. Therefore this larval type will be named *Cladopelma* gr. *viridulum. Cryptochironomus* gr. *viridulus* Chernovskij, a name used by many East European workers, refers to gr. *lateralis* and not to the species *C. viridulum.* Also, *Harnischia viridulus* in Beck & Beck (1969: 299) and *Cladopelma viridulus* in Biró (1988: 239) refer to another species: this larva also belongs to gr. *lateralis* and the pupa is not *C. viridulum* as described by Langton (1991). Kruseman (1933: 194) saw that *C. virescens* in Edwards (1929) was not *C. virescens* Meigen, 1818 and renamed the former *Tendipes edwardsi.* Beck & Beck (1969: 301) and Biró (1988: 239) overlooked this and *C. virescens* in Biró (1998) refers to *C. edwardsi* and not to *C. virescens* Meigen. These misidentifications have led to mistakes in Moller Pillot (1984) and Nocentini (1985).

A further problem with names used in the literature is that *C. bicarinatum* is absent from the key by Pinder (1978) and *C. krusemani* and *C. edwardsi* are absent from Langton's key; these species can be identified unjustly under other names. *C. subnigrum* is hitherto absent from all keys.

Langton & Pinder (2007) give six species in the genus *Cladopelma*; a seventh species is only known from Sweden (see Ashe & Cranston, 1990: 269):
Cladopelma bicarinatum (Brundin, 1947)
Cladopelma edwardsi (Kruseman, 1933)
Cladopelma goetghebueri Spies & Saether, 2004 (= *C. lateralis* Goetghebuer, 1934)
Cladopelma krusemani (Goetghebuer, 1935)
Cladopelma subnigrum (Brundin, 1947)
Cladopelma virescens (Meigen, 1818)
Cladopelma viridulum (Linnaeus, 1767).

The question of which species belong to *C.* gr. *viridulum* and which to *C.* gr. *goetghebueri* is (as far as is known) treated under these groups below. The biology and ecology of the larvae is treated only under the group names.

Cladopelma bicarinatum (Brundin, 1947)

Cryptocladopelma torbara Lenz, 1960a: 175

IDENTIFICATION

The adult male is not keyed in Pinder (1978), but can be identified using Langton & Pinder (2007). The exuviae are described by Langton (1991), but the difference from *C. krusemani* is unknown. The larva belongs to gr. *goetghebueri.*

DISTRIBUTION IN EUROPE AND THE NETHERLANDS

C. bicarinatum has been recorded in a few countries in Western Europe and Scandinavia and in North America (Saether & Spies, 2004; Langton & Pinder, 2007). In the Netherlands the species has been recorded mainly from moorland pools in the eastern and southern parts of the country (e.g. van Kleef, unpublished), but they are not confined to this water type.

Cladopelma edwardsi (Kruseman, 1933)

IDENTIFICATION

Identification of the adult male presents no problems using Pinder (1978) or Langton & Pinder (2007). The exuviae cannot be identified. According to Beck & Beck (1969) the larva belongs to gr. *viridulum*, but in Sublette (1964) the larva corresponds with *C. goetghebueri*. The exuviae present the same problem.

DISTRIBUTION IN EUROPE AND THE NETHERLANDS

The species has been recorded in a number of countries scattered over Europe and in North America (Saether & Spies, 2004). In the Netherlands the species is only known from two males collected by Kruseman (1933) near Valkenswaard.

Cladopelma goetghebueri Spies & Saether, 2004

Cryptocladopelma lateralis Pinder, 1978: 118, fig. 57A, 148D

DISTRIBUTION IN EUROPE AND THE NETHERLANDS

C. goetghebueri has been collected in many European countries, but seems to be absent from the Mediterranean area (Saether & Spies, 2004). From the Netherlands only (many) unverified records of exuviae are known.

Cladopelma krusemani (Goetghebuer, 1935)

DISTRIBUTION IN EUROPE AND THE NETHERLANDS

C. krusemani is only known from some West European countries and from the East Palaearctic (Saether & Spies, 2004). The species has not yet been found in the Netherlands.

Cladopelma subnigrum (Brundin, 1947)

DISTRIBUTION IN EUROPE

The species is only known from two oligotrophic lakes in Sweden (Brundin, 1949: 742; Saether & Spies, 2004).

Cladopelma virescens (Meigen, 1818)

nec *Cladopelma virescens* Biró, 1988: 239

IDENTIFICATION

See under the genus description for misidentifications of the larvae. The larvae seem to belong to gr. *viridulum*, because only these larvae are collected in large rivers and C.

virescens appears to be the only *Cladopelma* species in the river Rhine (Caspers, 1991; Becker, 1994; see also Klink, 1985a).

DISTRIBUTION IN EUROPE AND THE NETHERLANDS

C. virescens has been collected in many European countries (Saether & Spies, 2004). In the Netherlands Kruseman (1933) reported several records from different parts of the country.

Cladopelma viridulum (Linnaeus, 1767)

? *Cryptocladopelma lacustris* Lenz, 1960a: 175–176, fig. 11
nec *Cladopelma viridulus* Biró, 1988: 239

IDENTIFICATION

See under the genus description for misidentifications of the larvae. Identification of the exuviae is a problem because the exuviae of *C. edwardsi* may be hardly different (see under *C. edwardsi*).

DISTRIBUTION IN EUROPE AND THE NETHERLANDS

C. viridulum has been collected in nearly the whole of Europe and in North America. In the Netherlands adult males have been collected by Kruseman (1933) and there are many records of exuviae.

Cladopelma gr. goetghebueri

Cryptocladopelma gr. *lateralis* Moller Pillot, 1984: 239–240

SYSTEMATICS AND NOMENCLATURE

The name *C. lateralis* has been replaced by *C. goetghebueri* Spies & Saether, 2004 (p. 40). At least *C. goetghebueri* and *C. bicarinatum* belong to this group (see Lenz, 1954–62: 214). For mistakes in the past see the introduction under the genus.

DISTRIBUTION IN EUROPE AND THE NETHERLANDS

The distribution of the species of this group in Europe has been treated under the species names. Larvae of gr. *goetghebueri* are widespread in the Pleistocene part of the Netherlands and in Holocene fen peat areas; they are rarely found on clay. Exuviae of *C. goetghebueri* and *C. bicarinatum* have been found in the Pleistocene (eastern and southern) parts of the country.

LIFE CYCLE

The life cycle of the species in this group is most probably the same as in gr. *viridulum*: usually two generations a year and wintering mainly as juvenile larva. Larvae in third instar have been found in the Netherlands until the end of April. *Cladopelma edwardsi*, possibly belonging to gr. *goetghebueri*, was observed overwintering in Canada in second and (mainly) third instar, of which some larvae in cocoons (Danks & Jones, 1978).

FEEDING

We found many diatoms and few other algae in the guts of the larvae, which seem to be selective grazers. However, Lenz (1960a) supposed that all members of the genus are mainly detritus feeders (see also under gr. *viridulum*).

MICROHABITAT

The larvae are typical bottom dwellers on more or less silty substrates (Lenz, 1954–62; 1960a). We also collected larvae mainly on sandy or peaty bottoms with much organic silt.

SOIL

In the Netherlands the larvae are widely dispersed in sandy and peaty regions (see map in Moller Pillot & Buskens, 1990: 40). Steenbergen (1993) found the larvae rarely on clay.

WATER TYPE

Current

The European literature concerning running water rarely mentions the species of this group (e.g. Lehmann, 1971; Pinder, 1974; Fittkau & Reiss, 1978; Braukmann, 1984; Orendt, 2002a). However, the larvae are not very rare in Dutch lowland brooks (Peters et al., 1988; Verdonschot et al., 1992; own data). There are no records from large rivers (cf. gr. *viridulum*).

Dimensions in stagnant water

In European lakes the larvae can be common in the littoral zone up to a depth of 9 m (Brundin, 1949; Lenz, 1954–62, 1960a). In the Dutch reservoirs in the Biesbosch some larvae were found at depths of up to 24 m (Kuijpers, pers. comm.). Larvae of this group are rare in Dutch sand pits (Buskens, unpublished.; H. Cuppen, unpublished.). In the Netherlands most records are from pools and fen-peat lakes and relatively few from ditches (Steenbergen, 1993; van Kleef, unpublished; own data).

Permanence

In one case larvae were found rather abundantly in a temporary pool (Belarus; own data).

pH

Raddum & Saether (1981) collected the larvae regularly in Norwegian lakes with a pH around 4.5. In Dutch moorland pools the larvae are often collected at pH 3.5–6 (Vallenduuk, 1990; Duursema, 1996; van Kleef, unpublished data; own data), but also at pH > 7. Steenbergen (1993) collected the larvae even at pH values above 8.6.

D

TROPHIC CONDITIONS AND SAPROBITY

Brundin (1949: 597–99, 766) collected *C. bicarinatum* only in polyhumic oligotrophic lakes and the lagg zone of a peat moor in Finland and Sweden. Exuviae of this species have been found in the Netherlands in oligotrophic, mesotrophic and eutrophic pools. Steenbergen (1993) never collected the larvae in water with low oxygen content and mainly at low contents of orthophosphate and chlorophyll-A. However, in some cases the phosphate and ammonium contents were very high and Limnodata.nl sometimes mentions a high BOD. Lenz (1960a) mentioned the absence of the whole genus on bottoms without decomposing organic material. Consistent with this the larvae are very rare in Dutch sand pits (see under Water type).

SALINITY

The larvae have been collected very rarely in slightly brackish water (> 300 mg Cl/l) (Steenbergen, 1993). However, there are two records by Tõlp (1971) from brackish water bodies in Estonia that probably concern this group (as *Cryptochironomus* gr. *viridulus*).

Cladopelma gr. viridulum

Cryptocladopelma gr. *laccophila* Moller Pillot, 1984: 237–238
Harnischia viridulus Mundie, 1957: 193, fig. 23
nec *Cryptochironomus* gr. *viridulus* Chernovskij, 1949: 64, fig. 27
nec *Harnischia viridulus* Beck & Beck, 1969: 299, fig. IV-3
nec *Cladopelma viridulus* Biró, 1988: 239, fig. 80 J–M
? *Microchironomus laccophilus* Chernovskij, 1938: 1047

SYSTEMATICS AND NOMENCLATURE

For the division in groups and misidentifications see under the genus. The species *C. viridulum* and most probably *C. virescens* belong to the *viridulum* group. The larva and pupa of *C. edwardsi*, as described and illustrated by Beck & Beck (1969: 301, fig. IV-5), could be attributed to this group, although in Sublette (1964: figs. 76–82) this species corresponds with *C. goetghebueri* (the descriptions and figures by these authors are different in all characters).

LIFE CYCLE

Prat & Rieradevall (1992) reported two generations a year for *C. virescens* in Lake Banyoles in northeast Spain, emerging in spring (April to June) and autumn (September/October). In winter the larvae were in (first?), second and third instar. From a Canadian lake in the autumn Moore (1979a) regularly collected a few larvae of *C. viridula* in fourth instar and the number of fourth instar larvae increased during winter, especially in February. Elsewhere in Europe the emergence periods of *C. viridulum* and *C. virescens* are more or less the same (Kruseman, 1933; Reiss, 1968; Shilova, 1976), but in Austria in spring not earlier than the middle of May (Janecek, 1995) and in Sweden only from the end of May to early September (Brundin, 1949: 742).

FEEDING

Lenz (1960a: 178) supposed that the larvae feed on detritus and we found much detritus and few diatoms and other algae in the guts of the larvae. The larvae are possibly less dependent on the presence of algae in their environment than *C.* gr. *goetghebueri*. Also Moog (1995) called the species detritophagous. However, Prat & Rieradevall (1992) considered the larvae of *C. virescens* to be microcarnivores and mentioned Ostracoda, Cladocera, Nematoda and Oligochaeta as potential prey animals.

MICROHABITAT

Chernovskij (1938) found larvae most probably belonging to this group in superficial layers of the muddy bottom of lakes up to 6 cm deep. Other investigations (e.g. Lenz. 1960a; Palomäki, 1989; own data) also suggest that the larvae are true bottom dwellers, most commonly on detritus and only in low numbers on mineral bottoms. Verneaux & Aleya (1998) reared *C. viridulum* from an artificial substrate on the bottom of Lake Abbaye in France. Brundin (1949), however, stated that the larvae are also numerous on isoetid carpets.

DENSITIES

In most cases the recorded densities of the larvae are low. Lenz (1960a) found 0–78 larvae/m² in Lake Maggiore. However, Prat & Rieradevall (1992) sometimes found very high larval densities, up to more than 9000/m², of *C. virescens* (in third and fourth instar) in a Spanish lake in spring. The summer generation displayed much lower densities, probably due to high densities of predators. The mean annual densities varied from 14 to 2246 larvae/m².

SOIL

The larvae prefer more silty habitats than those of C. gr. *goetghebueri* and are more often found on clay (Steenbergen, 1993; own data).

WATER TYPE

Current

In comparison with C. gr. *goetghebueri*, the larvae are more rare in Dutch lowland brooks (Peters et al., 1988; Verdonschot et al., 1992; own data), but they are more often collected in large rivers (Klink & Moller Pillot, 1982; Smit, 1982; Klink, 1986a). According to Caspers (1991) and Becker (1994) *C. virescens* is the only *Cladopelma* species in the river Rhine. Mackey (1976) collected *C. virescens* irregularly, but sometimes abundantly in the river Thames.

Dimensions in stagnant water

The larvae can be common in the littoral and sublittoral zones of lakes, but are often absent from the profundal zone (Lenz, 1960a; Paasivirta, 1976; see below under Trophic conditions). They are common in smaller pools (Reiss, 1968; van Kleef, unpublished; own data). Steenbergen (1993) collected the larvae mainly in wider ditches and canals.

Permanence

Driver (1977: 125) found *Harnischia viridula* only once in a temporary prairie pond.

pH

Although the larvae are more often collected in eutrophic environments than those of gr. *goetghebueri* (see below), exuviae (unverified) of *C. viridulum* have been found often and sometimes abundantly in very acid water (pH 4–5.5). Schleuter (1985) reared the adult males of this species from a pond with a pH of 5.0. *C. viridulum* is also a common inhabitant of Finnish lakes with a pH of 6–7 (Paasivirta, 1976; Palomäki, 1989) and larvae of this group have been collected in the Netherlands at pH values of 8 and more (Steenbergen, 1993; Limnodata.nl).

TROPHIC CONDITIONS AND SAPROBITY

According to Saether (1979) the larvae of *C. viridulum* (and *C. edwardsi*) are characteristic for oligotrophic and mesotrophic and to mesohumic to polyhumic lakes. Lenz (1960a) noted that the larvae are absent from the oxygen-poor layers of eutrophic lakes and he supposes that in the profundal zone of oligotrophic lakes food shortage is the limiting factor. Real & Prat (1992) suggested that Cladopelma can live in deep lakes when algal production and therefore also decomposition is not very high, because in this case in deep waters oxygen tension remains relatively high. Rieradevall & Prat (1989) collected the larvae very numerously in an oligotrophic lake rich in calcium. The larvae still completed their cycle in places where in summer the dissolved oxygen content in the water dropped below 3 mg/l and below 1 mg/l close to the sediment. Rossaro et al. (2007) sometimes found *Cladopelma virescens* and *C. viridulum* in water with only 1 mg O_2/l. In the Netherlands the larvae are found on average in more eutrophic conditions than gr. *goetghebueri* (Moller Pillot & Buskens, 1990; Steenbergen, 1993). Limnodata.nl report a higher mean ammonium content.

SALINITY

The larvae have been collected rarely in slightly brackish water (> 300 mg Cl/l) (Steenbergen, 1993).

D

Cryptochironomus Kieffer, 1918

IDENTIFICATION

The identification of adult males, exuviae and larvae of some species still presents problems, and in many cases it is not possible to know whether species are identified correctly (see under *C. albofasciatus*, *C. defectus* and *C. supplicans*). Apart from these problems, adult males can be identified using Langton & Pinder (2007) and exuviae using Vallenduuk & Morozova (2005). The females of most species have been described and illustrated by Rodova (1978). Larvae could not be identified at all until the publication of the key by Vallenduuk & Morozova (2005). Therefore it is not possible to treat all aspects of the biology and ecology of each species. A general picture is given for the genus as a whole.
Vallenduuk & Morozova (2005) also give many measurements of larvae from Russia and the Netherlands.

LIFE CYCLE

All species seem to have a winter diapause, but in some cases in winter the larvae are in second and third instar and in other cases in fourth instar. It is unclear whether these differences are characteristic of different species.

FEEDING

According to Konstantinov (1961) the larvae are obligate predators and detritus in the gut originates from the prey. Titmus & Badcock (1981) found mainly oligochaetes in the gut of *Cryptochironomus* larvae. Later investigations by Russian workers (Shilova, 1965a; Izvekova, 1980), summarised and supplemented by Morozova (2005), revealed that small larvae are hardly or not predatory. Larvae in fourth instar, especially those of the larger species like *C. redekei* and *C. psittacinus* feed mainly on Oligochaeta, small Chironomidae and small Crustacea. Larvae of smaller species and younger larvae eat more detritus. *Cryptochironomus* larvae do not chase their prey and eat mainly small prey and animals which they happen to meet. It is not clear if larger prey is sometimes sucked out, but as a rule the animals or parts of them are swallowed.

MICROHABITAT

Larvae of *Cryptochironomus* are typical bottom dwellers, rarely creeping on plants or stones (e.g. Brodersen et al., 1998). Usually they creep around freely, but the larvae of some species, at least *C. defectus*, also live in a carelessly built tube (Shilova, 1965a). All species build a tube for the pupa (Lenz 1954–62: 227, fig. 330; own observations). Under laboratory conditions they disappear into the bottom sediment immediately after release. However, Klaren (1987: 25) found larvae of this genus only in the upper 5 cm of the sandy sediment in the Haringvliet estuary.

As a rule, sand with some organic material is preferred (e.g. Tolkamp, 1980; own data), but larvae can also be found in pure sand (e.g. van Urk & by de Vaate, 1990; Buskens, unpublished data) or on thick layers of organic material (own data). Kashirskaya (1989) found the larvae of *Cryptochironomus* in equal numbers on silty sand, silt with detritus and silty clay in the Volgograd water reservoir. Steenbergen (1993) also collected the larvae in similar numbers on clay as on sand. The larvae are usually absent between plants and Steenbergen (1993) collected the larvae significantly less in water bodies with much vegetation. However, Verdonschot & Lengkeek (2006) observed their presence between plants relatively often in lowland brooks. They found very few larvae between dead leaves or coarse organic material.

DENSITIES

In most cases only very low densities are recorded, usually less than 100 larvae/m^2, but the densities can be much higher. Lindegaard & Jónsson (1983) found up to more than 1000 *C. redekei* larvae/m^2. On bare sand at 0.5 m depth in Lake Maarsseveen, the Netherlands, ten Winkel (1987) found densities of from 111 to 1991 larvae/m^2 (probably *C. albofasciatus*).

WATER TYPE

The larvae live in flowing and stagnant water, with different species exhibiting different preferences. The larvae are never abundant in large rivers (e.g. Smit, 1982; Becker, 1994). They are rare in fast running streams or upper courses and scarce and often absent in very small pools or narrow ditches (less than 3 m wide), but have been found in large lakes (rarely) up to more than 10 km from the shore (IJsselmeer: Maenen, 1983). Their presence at greater depth is dependent on the food and oxygen content. Reiss (1968a) suggested that the genus is more or less thermophilous and for this reason absent from alpine lakes above 1000 m.

pH

The genus is scarce, but not very rare in acid water. Most probably all the species can endure very acid conditions (pH approx. 5), but additional factors cause differences in the occurrence of different species in acid water bodies.

TROPHIC CONDITIONS AND SAPROBITY

Larvae of the genus can be collected in water with very little production and decomposition and also in polluted water. They are therefore not good indicators for water quality. Probably all the species can tolerate organic pollution (cf. Bazerque, 1989; Wilson & Ruse, 2005) and can endure anoxia for some time. The larvae are fiery red and have a high haemoglobin content (see under *C. supplicans*). The effects of toxic chemicals is unknown. The larvae of all species have been found more in less polluted water with a rather high oxygen content. As the species prefer different water types there are differences in their occurrence in oxygen poor conditions. It cannot be excluded that there are also differences in tolerance between the species.

SALINITY

Remmert (1955) found large numbers of *Cryptochironomus* in water with 4‰ and 7‰ salt. Brodersen et al. (1998: 589) also observed their occurrence in slightly brackish lakes. In both cases it is not clear which species is involved. Tourenq (1975) stated that the larvae of *C. supplicans* survive a chloride content of 10,000 mg/l. However, Krebs (1981, 1984, 1985, 1990) collected the genus scarcely in pools, canals and ditches in the province of Zeeland, the Netherlands, and rarely in (slightly) brackish water.

Cryptochironomus albofasciatus Staeger, 1839

Chironomus albofasciatus Munsterhjelm, 1920: 124

SYSTEMATICS AND IDENTIFICATION

Kruseman (1933) distinguished *C. albofasciatus* from *C. defectus* by the ringed forelegs. According to Langton (1991) his *C. albofasciatus* is the same species as *C. defectus* of Shilova, but Vallenduuk & Morozova (2005) stated that the exuviae are nearly indistinguishable from those of *C. supplicans*.

DISTRIBUTION IN EUROPE AND THE NETHERLANDS

C. albofasciatus has been reported from most parts of Europe (Saether & Spies, 2004). In the Netherlands the species has been identified mainly by Kruseman (1933) and Kouwets & Davids (1984). The records are from the fen-peat region in Holland and from Denekamp.

LIFE CYCLE

Mundie (1957) trapped the emerging adults only in August and September. In fish ponds in the Belgian Ardennes the species is bivoltine (Goddeeris, 1983), emerging in June and July–August. Kouwets & Davids (1984) stated two emergence periods: June and August–September. *C. albofasciatus* has an overwintering diapause in second and third instar (Goddeeris, 1983, 1986).

OVIPOSITION

Munsterhjelm (1920: 124) stated that oviposition takes place in flight over the water surface up to 0.5 km from the shore. He recorded 240–360 eggs per egg mass.

ECOLOGY

See under *C. defectus*.

Cryptochironomus defectus Kieffer, 1913

? *Cryptochironomus albofasciatus* Langton, 1991; Langton & Pinder, 2007 (see under *C. albofasciatus* above)

SYSTEMATICS

In the ecological sections below we will treat *C. albofasciatus* and *C. defectus* together.

IDENTIFICATION

C. defectus is absent from most West European keys, such as Pinder (1978), Langton (1991) and Langton & Pinder (2007), and so the species has not been identified by most ecologists. In many cases the exuviae will have been identified as *C. supplicans* (see Langton, 1991: 280 and Langton & Visser, 2003; compare Water type). Exuviae and larvae can be identified using Vallenduuk & Morozova (2005). In Eastern Europe the name has been used by most workers.

DISTRIBUTION IN EUROPE AND THE NETHERLANDS

The species occurs in large parts of Europe, but it has not been reported from Scandinavia and the British Isles. The reason is without doubt that *C. defectus* is absent from most keys. In the Netherlands most records are from the eastern and southern parts of the country.

LIFE CYCLE

Near Saratov (Russia) *C. defectus* has two generations a year, emerging in shallow or small water bodies from the end of April until the middle of October and in larger and deeper water from the end of May until September (Morozova, 2005). In the Netherlands the emergence periods are probably broadly the same. However, Klink & Moller Pillot (1982) and Buskens (in sand pits, unpublished data; partly this species?) collected the exuviae (as *C. supplicans*) relatively abundantly in summer, but never in spring. In winter the larvae are in diapause in second and third instar (own data, cf. Goddeeris, 1983, *C. albofasciatus*). In winter we also sometimes collected larvae in fourth instar.

FEEDING

In fourth instar the larvae are mainly predators, eating Oligochaeta and small Chironomidae and Crustacea. Younger larvae eat more detritus (Morozova, 2005). See also under the genus.

OVIPOSITION

See under *C. albofasciatus.*

MICROHABITAT

The larvae are found almost exclusively on the bottom. According to Shilova (1965a) the larvae of this species live in a carelessly built tube, contrary to other species of the genus. In the laboratory the larvae dig themselves in immediately. Tolkamp (1980) collected the genus (most probably mainly this species) significantly more frequently on a bottom of sand with detritus. This corresponds with our experience with *C. defectus.* However, the larvae are also sometimes collected from silt. We found the larvae very rarely on stones or plants.

DENSITY

In Russian reservoirs and ponds Morozova (2005) found mean densities of 5–66 larvae/m^2 and maximum densities of 114–150 larvae/m^2 . Most probably much higher densities can also occur; see under the genus and compare with *C. redekei.* In large rivers the larvae are present only in low numbers (Becker, 1994; cf. Smit, 1982).

WATER TYPE

Current

The species lives in flowing and stagnant water (Vallenduuk & Morozova, 2005). *C. defectus* is the most common species of the genus in the upper and lower courses of Dutch lowland brooks. It is scarce or absent in fast-flowing streams: Lehmann (1971) did nor collect the species along the Fulda and in streams in Bavaria *C. albofasciatus* appeared to be much scarcer than *C. rostratus* (Orendt, 2002a). Moog (1995) also noted *C. defectus* as occurring in the 'epipotamal' zone, in contrast to *C. supplicans.* The larvae are locally abundant in rather fast-flowing streams in South Limburg, but in the fast-flowing upper courses in this hilly landscape the whole genus is very rare and data on *C. defectus* are absent (Waterschap Roer & Overmaas, unpublished; own data).

Dimensions

C. defectus (and/or *C. albofasciatus*) and *C. rostratus* are the most common species of the genus in large rivers. Because Becker (1994) collected relatively large numbers of *C. albofsciatus* and never *C. supplicans* along the river Rhine in Germany, most exuviae collected by Klink & Moller Pillot (1982) in the Rhine and the Meuse and identified as *C. supplicans* must belong to *C. defectus/albofasciatus* (see Identification; however, compare with *C. supplicans*). In contrast to *C. obreptans,* Morozova (2005a) found *C. defectus* more in ponds than in large reservoirs. However, as stated for the whole genus, the larvae are significantly scarcer in ditches and brooks less than 4 m wide (Moller Pillot & Buskens, 1990; Steenbergen, 1993; Moller Pillot, 2003). According to Moog (1995) the species is absent from the profundal zone of lakes.

Permanence

The larvae can occur in temporary water, but this is exceptional (Fittkau & Reiss, 1978; Moller Pillot, 2003; see also Schleuter, 1985).

pH

The genus *Cryptochironomus* is not rare in acid water and is collected in goodish numbers also at pH 4 to 5 (Raddum & Saether, 1981; Buskens, 1983; Verstegen, 1985; Duursema, 1996). This also applies to *C. defectus*: the species has been found in acid moorland pools and the upper courses of lowland brooks (Moller Pillot, 2003; van Kleef, unpublished). In acid conditions this species will have little difficulty in finding food, but it is usually scarce or absent in acid pools. Steenbergen (1993) collected significantly fewer larvae of the genus in water with a pH < 7.5 and most at pH 8.1–8.5.

TROPHIC CONDITIONS AND SAPROBITY

The larvae of the genus *Cryptochironomus* are mainly collected in eutrophic water without much organic pollution and a moderate or high oxygen content (e.g. Steenbergen, 1993; Orendt, 1993). In Dutch lowland brooks, where *C. defectus* is the most common species, the data are different: Moller Pillot (1971, 2003) collected the larvae mainly in water of better quality and Peters et al. (1988) more in polluted water. The data given by Bazerque et al. (1989), who collected *C. supplicans* in a polluted stretch of the Somme, may also apply to *C. defectus* (see under Identification). It therefore seems probable that the tolerance of *C. defectus* to organic pollution is not much better than *C. supplicans* (which tolerates pollution very well). However, as a species that lives mainly in flowing water, *C. defectus* larvae seem to need a little more oxygen than *C. supplicans*.

SALINITY

We found no reliable data about salinity. See under *C. supplicans*.

Cryptochironomus denticulatus (Goetghebuer, 1921)

DISTRIBUTION IN EUROPE

C. denticulatus has been reported mainly from the countries surrounding the Netherlands, but not from the Netherlands itself (Saether & Spies, 2004). Everywhere it seems to be a rather rare species.

WATER TYPE

C. denticulatus is a typical inhabitant of brooks and streams (Fittkau & Reiss, 1978). The species has been reported from brooks and streams with moderate to fast currents (Pinder, 1974; Orendt, 2002a; Michiels, 2004). Klink (1985) also collected the exuviae in the river Meuse near Hastière in France. Buskens (unpublished) mentioned a record of the exuviae of this species in a sand pit in the Netherlands. This has to be verified.

Cryptochironomus obreptans (Walker, 1856)

DISTRIBUTION IN EUROPE AND THE NETHERLANDS

The species has been collected in many countries scattered over Europe (Saether & Spies, 2004). In the Netherlands *C. obreptans* is one of the most common species in the genus in all parts of the country, with the possible exception of Zeeland (Nijboer & Verdonschot, unpublished data; van Kleef, unpublished; Limnodata.nl; own data).

LIFE CYCLE

Near Saratov (Russia) *C. obreptans* has two generations a year, emerging in shallow or small water bodies from the end of April until the middle of October and in larger and deeper water bodies from the end of May until September (Morozova, 2005). In the Netherlands the emergence periods are probably broadly the same.

FEEDING

In fourth instar the larvae are mainly predators, eating Oligochaeta and small Chironomidae and Crustacea. Younger larvae eat more detritus (Morozova, 2005). See also under the genus.

MICROHABITAT

Smit et al. (1996) collected the larvae on the sandy bottom with some organic silt of the Ventjagers flats in a Dutch estuary. See further under the genus.

DENSITIES

Morozova (2005) found maximum densities of 114–128 larvae/m^2 in Russian reservoirs and only 19 larvae/m^2 in ponds. Koehn & Frank (1980) recorded a maximum density of 2836 larvae/m^2 in a polluted slow-flowing channel in Berlin.

WATER TYPE

Current and dimensions

According to Fittkau & Reiss (1978) and Moog (1995) *C. obreptans* inhabits only stagnant water. However, there are quite a large number of records from slow-flowing lowland brooks (Brabantse Delta, unpublished; Limnodata.nl; own data) and a low number of exuviae have been collected from the rivers Rhine and Meuse (Klink & Moller Pillot, 1982). Morozova (2005a) found the larvae more numerously in reservoirs than in ponds. In the Netherlands the species is rather common in lakes, canals and pools. Most probably the species is rare in narrow ditches and small pools (see *C. defectus*). Moog (1995) called the species rare in the profundal zone of lakes.

Permanence

There is only one record from a temporary pool (Belarus, own data).

pH

In the Netherlands *C. obreptans* has been collected very often in acid as well as in basic water (van Kleef, unpublished; Limnodata.nl). It has been collected relatively more at pH > 7.5, but also more than other species at a pH of around 5.

SALINITY

The species has occasionally been found in slightly brackish water (Limnodata.nl). However, it is certainly not common in brackish pools, canals and ditches (see under the genus).

Cryptochironomus psittacinus (Meigen, 1830)

DISTRIBUTION IN EUROPE AND THE NETHERLANDS

C. psittacinus lives in nearly the whole European mainland (Saether & Spies, 2004). There are about 10 records from the Netherlands, from stagnant water mainly in the western half of the country (Kruseman, 1933; Kuijpers et al., 1992; Nijboer & Verdonschot, 2001; Limnodata.nl; own data).

LIFE CYCLE

Adults and exuviae have been collected from early May to September (Kruseman, 1933;

Shilova, 1976; Janecek, 1995; own results). Most probably there are two generations a year, as in other species of the genus.

FEEDING

At least the fourth instar larvae of *C. psittacinus* are predators, eating Oligochaeta and small Chironomidae (Shilova, 1976).

MICROHABITAT

The larvae live on silty sand soils (Reiss, 1968; Shilova, 1976; Smit et al., 1996) and on the sapropel (Janecek, 1995).

WATER TYPE

C. psittacinus is an inhabitant of stagnant water (Fittkau & Reiss, 1978), but there are a few records from flowing water. Some exuviae and adult males have been collected in and along the river Rhine (Wilson & Wilson, 1984; Caspers, 1991; Becker, 1994). The species has been found in some lowland brooks in the Netherlands (Limnodata.nl) and even in a faster flowing stream in Bavaria (Reiss, 1984 fide Orendt, 2002a). There is a number of records from lakes and storage reservoirs (e.g. Reiss, 1968; Shilova, 1976; Kuijpers et al., 1992), where the larvae live at a depth of 2 to 12 or 15 m. Smit et al. (1996) collected the larvae in a Dutch estuary also near the mean water level. There are also some records from smaller water bodies like dune lakes and moorland pools (Dutch unpublished data).

pH

There are no records from acid water. The lowest pH (6.3) was measured in a Dutch moorland pool (van Kleef, unpublished).

Cryptochironomus redekei (Kruseman, 1933)

DISTRIBUTION IN EUROPE AND THE NETHERLANDS

C. redekei has been reported from most parts of Central and Northern Europe, but seems to be absent from the Mediterranean area (Saether & Spies, 2004). In the Netherlands there are about 20 records, mainly from the Holocene part of the country and from the province of Noord-Brabant, often not far from the sea coast (Kruseman, 1933; Limnodata.nl; own data).

LIFE CYCLE

Mundie (1957) in England and (probably also) Lindegaard & Jónsson (1987) in Denmark stated two generations a year, emerging mainly in June and August. In the Netherlands exuviae and adults have been collected from the end of April until September (Kruseman, 1933; Buskens, 1989; own data). In Denmark all larvae wintered in fourth instar (Lindegaard & Jónsson, 1987).

FEEDING

At least the fourth instar larvae of *C. redekei* are predators, eating Oligochaeta and small Chironomidae (Morozova, 2005: 26, based on Shilova and Izvekova).

DENSITY

Lindegaard & Jónsson (1983) stated densities of less than 10 to more than 1000 larvae/m^2 in Hjarbæk Fjord, Denmark.

OVIPOSITION

Morozova (2005) noted that the eggs are deposited in a long thread, usually fastened at both sides to the stems of water plants or other substrates. On average there are about 500 eggs in one egg mass.

WATER TYPE

Current and dimensions

C. redekei has been recorded only from stagnant water. Many records are from relatively large and shallow lakes at less than 6 m depth (Lindegaard & Jónsson, 1987; Janecek, 1995; own data), but Shilova (1976) also recorded occurrence up to a depth of 30 m, although the highest densities were found at a depth of less than 10 m. Mundie (1957) reported most emergences from a depth of 4–5 m. In the Netherlands there are also some records from smaller water bodies like moorland pools (van Kleef, unpublished). Lindeberg (1958) collected the species in small rock pools.

pH

The larvae can survive at very high pHs (10.5–11), as stated by Lindegaard & Jónsson (1983), although the species has also been collected in some cases in water with a pH of around 6 (Koskenniemi & Paasivirta, 1987; Buskens, 1989; van Kleef, unpublished). Van Kleef collected one specimen in a Dutch moorland pool with a pH of 4.6. Because most data are from Holocene regions not far from the coast, the pH optimum seems to be rather high.

TROPHIC CONDITIONS AND SAPROBITY

The species was very numerous in hypertrophic lakes like Hjarbæk Fjord (Lindegaard & Jónsson, 1983, 1987), but has also been collected in mesotrophic to eutrophic water (Kuijpers, 1992; own data). As far as we know, the oxygen content of the water column was more or less stable or sometimes decreasing to less than 1 mg/l (e.g. Lindegaard & Jónsson, 1987).

SALINITY

C. redekei is known to inhabit (slightly) brackish water up to 3000 mg Cl/l (Lindeberg, 1958; Ringe, 1970). The larvae were present in low numbers in Hjarbæk Fjord, Denmark, in 1968, when the water was still slightly brackish (< 540 mg Cl/l). The numbers increased after further freshening (Lindegaard & Jónsson, 1983). The species was seen swarming at Lake Markiezaat in the Netherlands (approx. 1000 mg Cl/l). Most water bodies where the species has been collected contained fresh water.

Cryptochironomus rostratus Kieffer, 1921

DISTRIBUTION IN EUROPE AND THE NETHERLANDS

C. rostratus has been reported from all parts of Europe (Saether & Spies, 2004). In the Netherlands the species has been collected in the large rivers and locally elsewhere in the eastern and southeastern part of the country, especially in South Limburg.

LIFE CYCLE

The exuviae have been collected in the Dutch stretch of the river Rhine from May to August (Klink & Moller Pillot, 1982). In South Limburg prepupae were also collected in September (Waterschap Roer and Overmaas, unpublished).

MICROHABITAT

Caspers (1980) reared (few) adults from the sandy littoral zone of the Rhine in places with much detritus. We also found the larvae on such sites in the rivers Rhine and Waal. Becker (1994) also reared the species from stones and gravel. Unidentified larvae of the genus were scarcely found on stones in the river Meuse, but here the larvae are most common on sand (Smit, 1982; own data).

WATER TYPE

C. rostratus lives almost exclusively in flowing water, mainly in rivers, but also in smaller streams. Lehmann (1971) collected the adult males scarcely along the Fulda. Wilson & Wilson (1984) and Caspers (1991) reported the species from the whole German stretch of the Rhine. In most stretches, including the Dutch stretch of the river, it appears to be the most common species of the genus (Klink & Moller Pillot, 1982; Klink, 1985). The exuviae and adults have been collected often in fast-flowing streams in Bavaria and Baden-Württemberg (Orendt, 2002a, Michiels, 2004). There are some records from Dutch lowland brooks and streams (Kruseman, 1933: Denekamp; H. Cuppen, unpublished; Limnodata.nl). In the faster flowing streams in South Limburg the species seems to be the most common species of the genus (Waterschap Roer and Overmaas, unpublished).

Saprobity and oxygen

Moog (1995) gives the same saprobity figures for *C. rostratus* as for *C. defectus*. Because most published and unpublished records concern rather clear water, we think that the mean quality in the streams where *C. rostratus* occurs is much better than that given by Moog. As a characteristic inhabitant of flowing water the larvae probably need a rather high oxygen content.

SALINITY

The larvae are able to survive a chloride content of 3000 mg/l (Tourenq, 1975). Because they are confined to running water, their presence in brackish water will be exceptional.

Cryptochironomus supplicans (Meigen, 1830)

IDENTIFICATION

See the information under the genus. Reports of *C. supplicans* exuviae often apply to *C. defectus*; see under this species. Brundin (1949) did not separate *C. supplicans* and *C. obreptans*. He also stated that the long bearded setae on the fore tarsi of *C. supplicans* often fall off. This character is hardly used in the keys by Pinder (1978) and Langton & Pinder (2007).

DISTRIBUTION IN EUROPE AND THE NETHERLANDS

C. supplicans has been reported from nearly the whole European mainland (Saether & Spies, 2004). In the Netherlands it seems to be one of the most common species of the genus throughout the whole country.

LIFE CYCLE

The species has two, possibly sometimes three generations a year (Orendt, 1993: 104). Sometimes only a (late) summer generation has been observed (Orendt, 1993; Buskens, unpublished).

MICROHABITAT

See under the genus.

WATER TYPE

C. supplicans has been found mainly in lakes (Brundin, 1949; Mundie, 1957; Shilova, 1976; Orendt, 1993). Rossaro (2007) collected the larvae in the Po River in Italy. The species seems to be absent from most European streams and rivers because it is not mentioned by Lehmann (1971), Caspers (1991) and Becker (1994). The reports by Orendt (2002a) from streams in Bavaria were based on identification of exuviae and probably refer to *C. defectus*. In the Netherlands the species has been collected mainly in lakes, but also in pools and ditches and in some cases in rivers (Krebs, 1981; Schmale, 1999; Vallenduuk & Morozova, 2005; van Kleef, unpublished; Vallenduuk, pers. comm.; own data). According to Lenz (1954–62: 227, 229) *Cryptochironomus* collected in the profundal zone of the Plön Lake did not belong to *C. supplicans* (pupal tergites without reticulation), but the species was fairly common in the littoral zone.

pH

C. supplicans appears to be less common in acid moorland pools than *C. obreptans*. However van Kleef (unpublished) collected the exuviae of one specimen in a pool with a pH of 4.3 and two exuviae in a pool with a pH of 5.2.

TROPHIC CONDITIONS AND SAPROBITY

In ditches in Bergambacht (the Netherlands) we collected larvae and exuviae in organically polluted water in which the oxygen content was less than 0.5 mg/l for the whole night. However, the larvae have been found in mesotrophic lakes (Orendt, 1993; Schmale, 1999) and at least some of the populations found by Brundin in oligotrophic lakes in Sweden will belong to this species.

Rossaro et al. (2007) sometimes found *C. supplicans* in water with only 1 mg O_2/l. These authors found larvae from the Po River to have a very high haemoglobin content (16.5 µg Hb/mg). The relatively large ring organ of the thoracic horn of the pupa corresponds with the ability of the species to respond to oxygen shortage.

SALINITY

Tourenq (1975) collected the species in water with 600–980 mg Cl/l and reported that the larvae can endure 10,000 mg Cl/l (for some time?). Krebs (1981) reported a record from a pool with 600 mg Cl/l. The identifications by both these authors are not fully reliable (see Identification). The species is certainly scarce in brackish pools, canals and ditches in the Netherlands (see under the genus).

'Cryptochironomus' macropodus Ljachov, 1941

SYSTEMATICS AND IDENTIFICATION

This species belongs to an unknown genus of which the adult and pupa are unknown. The larva has been described and illustrated by Chernovskij (1949) and Pankratova (1983).

DISTRIBUTION IN EUROPE AND THE NETHERLANDS

The species was known only from Russia, but Klink (2002) collected a larva in the river Waal in the Netherlands.

MICROHABITAT

The larva lives in shifting sand and gravel in fast-flowing rivers (Chernovskij, 1949).

Cryptotendipes Lenz, 1941

IDENTIFICATION

Identification of the species is only possible as adult male and as pupa/exuviae. The key by Pinder (1978) for adult males was still incomplete, but all five European species can be identified using Langton & Pinder (2007). The exuviae can be identified using Langton (1991), but the exuviae of *C. nigronitens* are still unknown.

DISTRIBUTION IN EUROPE AND THE NETHERLANDS

All five European species have been recorded from a limited number of countries scattered over nearly the whole of Europe (Saether & Spies, 2004). In the Netherlands *C. holsatus* has been collected regularly, mainly in the Pleistocene areas of the country. There are some records of *C. usmaensis*. *C. pseudotener* has been collected once, as an adult male, in 1921 by De Meijere (Kruseman, 1933). The reported presence of *C. nigronitens* by Kruseman (1933) is based on a misidentification. However, this species will live in the Netherlands because it is not rare along the river Rhine in Germany (Becker, 1994).

Larvae of the genus have been found more often (about 50 localities) in different parts of the country.

LIFE CYCLE

Goddeeris (1983) observed two generations a year in fish ponds in the Belgian Ardennes, emerging mainly in June and August. The species could not be identified. The larvae had an overwintering diapause from late summer in second instar; the first fourth instar larvae appeared in May (Goddeeris, 1983; 1986). A larva in third instar was collected in Lake Maarsseveen in early February. Elsewhere we found many larvae in fourth instar before the end of April. These differences may be due to temperature or to innate differences between species. The adult male of *C. pseudotener* was collected by De Meijere in May.

FEEDING

Detritus, diatoms and other small particles can be found in the gut of the larvae (own observations). The species is probably a selective collector. Wilson & Ruse (2005) called the larvae detritivores.

MICROHABITAT

The larvae are mentioned only as bottom dwellers, in lakes as well in brooks. They inhabit sandy bottoms, silt and coarse decaying plant material (Lenz, 1959a; 1954–62; Mol et al., 1982). Mol et al. (1982) collected some larvae from roof tiles laid out on the bottom of Lake Maarsseveen. The pupa has also been found in silt, where its long thoracic horn branches protrude above the silt layer (Lenz, 1959a).

DENSITIES

Different authors have recorded strikingly low densities of the larvae (Brundin, 1949; Lenz, 1959a; Pankratova, 1964; Mol et al., 1982). Most data from the Netherlands concern less than 5 larvae or exuviae. It is likely that the larvae have been missed at many sampling localities.

WATER TYPE

Most species of *Cryptotendipes* appear to live in flowing water and lakes. In lakes the larvae have been collected down to a depth of 11 m (Lenz, 1954–62; 1959a).
C. usmaensis and *C. pflugfelderi* have been collected in lakes (Lenz, 1959a; Fittkau & Reiss, 1978). In the Netherlands there is also one record of a pupa of *C. usmaensis* in a very slow-flowing stream, the Kromme Rijn (T. van Haaren, unpublished). Most records of *C. holsatus* are from very slow-flowing brooks (Limnodata.nl). *C. pseudotener* has also been collected in brooks (Lenz, 1959a; Orendt, 2002a; Michiels, 2004), but the Dutch specimen probably lived in a lake (collected at Linschoten). This is the only species that seems to occur in faster-flowing streams.
C. nigronitens has been reared from the silty bottom of the river Rhine and has been caught as an adult male in several stretches of this river (Becker, 1994). According to Becker, Fritz (1982) also stated the presence of this species in dead river arms. However, the genus is absent from most publications on brooks and streams (e.g. Lehmann, 1971; Braukmann, 1984) and exuviae were not collected in the river Rhine in the Netherlands by Klink & Moller Pillot (1982).

pH

There are no records from acid water.

TROPHIC CONDITIONS AND SAPROBITY

Brundin (1949) collected *C. usmaensis* in small numbers in oligotrophic lakes. Lenz (1959a) inferred a certain oxygen need because the larvae live in flowing water and lakes. Most Dutch records are from mesotrophic or eutrophic lakes or eutrophic slow-flowing brooks. We think that *C. usmaensis* is less tolerant of some organic load than *C. holsatus*. The American species of the genus are not restricted to oligotrophic or mesotrophic conditions (Saether, 1979). Wilson & Ruse (2005) called the genus rather tolerant (group C).
The scarcity of the larvae is probably not caused by very high demands on water quality.

SALINITY

Fittkau & Reiss (1978) mentioned the presence of *C. nigronitens* in brackish water, but this was most probably based on a misidentification by Kruseman (1933).

D

Cyphomella cornea Saether, 1977

IDENTIFICATION

Cyphomella cornea is the only species of the genus known from Western and Central Europe. The larva of *C. cornea* is unknown. Saether (1977) described the larva of a probably related species from Canada, which is hardly different from *Paracladopelma* larvae (see also Pinder & Reiss, 1983). Langton (1991) keyed the exuviae of *C. cornea* and a second species from Greece.

DISTRIBUTION IN EUROPE

C. cornea has been collected in Austria and Bavaria (Saether, 1977; Michiels, 1999).

WATER TYPE

C. cornea has been found in fast-flowing rivers in Bavaria and in lakes in North America (Saether, 1977; Michiels, 1999).

Demeijerea rufipes (Linnaeus, 1778)

DISTRIBUTION IN EUROPE AND IN THE NETHERLANDS

Demeijerea rufipes has been recorded from most countries in Western, Central, Northern and Eastern Europe, but in the Mediterranean region only from Italy (Saether & Spies, 2004). According to Limnodata.nl and Nijboer & Verdonschot (unpublished) the larvae have been found in many water bodies in the provinces of Noord-Holland, Zuid-Holland and Flevoland. Elsewhere, especially in the Pleistocene areas, the species seems to be rare.

LIFE CYCLE

Shilova (1976) collected adults at the end of July and in August in Russia and supposed that the species has only one generation there. However, Kruseman (1933) caught the adult males from May until October in the Netherlands; the species will therefore have at least two generations a year.

FEEDING

According to Wundsch (1943a) the larvae feed on sponges.

MICROHABITAT

Because the larvae live only in sponges, they can be found only on firm substrates like stones and wood (Smit 1982; Peeters 1988; own data).

WATER TYPE

Shilova (1976) mentioned only records from lakes. In the Netherlands the larvae have also been found in slow-flowing canals and, rarely, in rivers (Steenbergen, 1993; Limnodata.nl; own data).

Current

In rivers *Demeijerea* is much less common than *Xenochironomus*, the larvae of which are also inhabitants of sponges. In the Fulda region of Germany Lehmann (1971) observed only a swarm of this species along the lower part of the river Fulda. In the Rhine the species is only known from the upper stretch (between Basle and Bingen: Caspers, 1991). Becker (1994) considered the rare occurrence along this river to be remarkable, because sponges are common in this river. There are several scattered records from the Meuse, most concerning one or a very few specimens. However, Steenbergen (1993) collected the larvae somewhat more often in slow-flowing than in stagnant canals. Limnodata.nl gives only very few records in slow-flowing water bodies in the Netherlands and a record from a moderately flowing brook in Limburg.

Dimensions

The larvae live mainly in the larger water bodies such as canals and lakes (Moller Pillot & Buskens, 1990; Verdonschot et al., 1992). Steenbergen (1993) did not find the species in ditches or canals less than 4 m wide. Braukmann (1984) did not mention the species at all as an inhabitant of small rivers in Germany.

Permanence

The occurrence of the larvae is confined to permanent water.

SOIL

The species has been found in the Dutch province of Noord-Holland about equally often on sand, clay and peat (Steenbergen, 1993).

pH

There are no records of the species from acid water. The species is confined to water rich in calcium because sponges cannot live in calcium-poor conditions.

TROPHIC CONDITIONS AND SAPROBITY

A precondition for this species is the presence of sponges; these live mainly in β-mesosaprobic water (Sládeček, 1973). Larvae of *Demeijerea* have been found in the Netherlands only in eutrophic water with variable or high oxygen content (Steenbergen, 1993; Limnodata.nl). Phosphate and nitrogen contents are sometimes rather high.

SALINITY

We found no published records from water with a chloride content above 1000 mg/l.

Demicryptochironomus (Irmakia) neglectus Reiss, 1988

IDENTIFICATION

The adult male can be identified using Langton & Pinder (2007) and the exuviae using Langton (1991). The exuviae of the subgenus have been described and illustrated by Pinder & Reiss (1986) as Chironomini Genus D. The larva is unknown.

DISTRIBUTION AND ECOLOGY

Adults and exuviae have been collected in and along streams and rivers in the alpine region (Caspers, 1991; Saether & Spies, 2004), the Black Forest (Michiels, 2004) and the British Isles (Langton, 1991, as Pe 1).

Demicryptochironomus vulneratus (Zetterstedt, 1838)

SYSTEMATICS

According to Pinder & Reiss (1983) *D. vulneratus* is the only representative of the genus in Europe and *D. ploenensis* Lenz is regarded as a synonym. However, Langton (1991) described the exuviae of a second species as *Demicryptochironomus* Pe 1. Both species will be taken together here as *D. vulneratus* because this is common practice in most of Europe. A new species, *D. neglectus*, has been described by Reiss (see above).

DISTRIBUTION IN EUROPE AND THE NETHERLANDS

D. vulneratus has been recorded from nearly the whole European mainland (Saether & Spies, 2004). In the Netherlands the species has been recorded from about 250 localities, but not from the Holocene part of the country, except from the fenland and dune regions (Nijboer & Verdonschot, unpublished data; Limnodata.nl; own data).

LIFE CYCLE

Lehmann (1971) and Shilova (1976) called *D. vulneratus* a univoltine summer species, flying from June to September. If this is true there is no strong synchronisation. In Austria (Janecek, 1995) and in the Netherlands pupae and adults have been recorded from May until September; two generations cannot be excluded.
In winter larvae have been collected in third and fourth instar.

FEEDING

Most authors, for instance Lenz (1960) and Pinder & Reiss (1983), mention that the larvae are predatory, especially on Oligochaeta.

MICROHABITAT

The larvae live mainly on sandy bottoms, but they are often collected from (organic) silt and rarely from stones (Lenz, 1960; Shilova, 1976; Srokosz, 1980; Pinder, 1980; Mol et al., 1982; Brodersen et al., 1998; own data). Chernovskij (1938) found them mainly superficially, rarely down to 10 cm deep in the substrate. Mol et al. (1982) sometimes collected them from artificial plants and Brundin (1949) reported the occurrence of this species in isoetid vegetation. The larvae probably rarely creep on plants.

DENSITIES

The larvae appear never to live at high densities. Lenz (1960) recorded a density of 21 larvae/m² in the littoral zone of the Plön Lake. Buskens (unpublished data) found the larvae widely distributed in Dutch sand pits, but scarce everywhere else.

WATER TYPE

Current

The larvae are mentioned in the literature as inhabitants of the potamal zone of rivers and streams (e.g. Lehmann, 1971; Pankratova, 1983) and rarely from brooks or streams with a faster current (Orendt, 2002a; Michiels, 2004). Caspers (1980, 1991) does not mention the species from the river Rhine. In the Netherlands the species is rather common in lowland brooks and less common in lakes, canals and moorland pools, but absent from narrow ditches, fast-flowing streams and large rivers.
Langton (1991) stated that his species Pe 1 (see above under Systematics) lives in streams while *D. vulneratus* is an inhabitant of lakes.

Dimensions

As stated above, the larvae appear to live mainly in large stagnant water bodies and in small streams. They can also be found in the profundal zone of lakes, sometimes up to 50 m deep (Brundin, 1949; Lenz, 1960; Reiss, 1968; Shilova, 1976).

Temperature

Reiss (1968a) called the species thermophilic and argued that this is why the species is not found in alpine lakes above 1000 m. However, the species is also common in subarctic lakes (Brundin, 1949).

Permanence

The larvae have never been reported from temporary water bodies.

pH

The larvae are scarce in acid conditions, but have been collected several times in low numbers in moorland pools and sand pits at pH 4.1–6 (Raddum & Saether, 1981; Vallenduuk, 1990; Buskens, unpublished; van Kleef, unpublished). The specific prey (Oligochaeta) may be too scarce in water with a low pH. In most localities a pH > 7 has been measured.

TROPHIC CONDITIONS AND SAPROBITY

According to Saether (1979) the species is absent from strongly eutrophic lakes. Lenz (1960) supposed that the larvae are eurythermic, but more or less stenoxybiontic. The larvae are scarce in oligotrophic environments due to a shortage of food. In the

Netherlands *D. vulneratus* has been collected mainly at high oxygen contents and low BOD, phosphate and chlorophyll-A contents (Limnodata.nl). However, Brundin (1949) recorded the presence of the species also in the profundal zone of Swedish lakes at lower oxygen contents and Peters et al. (1988) collected the larvae most often in moderately polluted small streams in the Netherlands. Wilson & Ruse (2005) supposed that the species tolerates organic pollution.

SALINITY

Demicryptochironomus larvae are not recorded from brackish water.

Dicrotendipes Kieffer, 1913

Limnochironomus Chernovskij, 1938: 1047; 1949: 72–74, fig. 38–39

SYSTEMATICS AND IDENTIFICATION

A revision of the genus in Europe has been made by Contreras-Lichtenberg (1986). She gives descriptions and figures of adult males, pupae and larvae. For a discussion about the type species see Spies & Saether (2004).

Dicrotendipes lobiger (Kieffer, 1921)

Limnochironomus lobiger Pinder, 1978: 126, fig. 61 J-K, 158C
Dicrotendipes gr. *lobiger* Moller Pillot, 1984: 194
Limnochironomus pulsus Higler, 1977: 37–78 (pro parte) (nec aliis)

IDENTIFICATION

Identification of the male using Langton & Pinder (2007) presents no serious problems. However, Kruseman (1933: 175) possibly interchanged *D. lobiger* and *D. tritomus*. The female has been described by Contreras-Lichtenberg (1989).

DISTRIBUTION IN EUROPE AND IN THE NETHERLANDS

Dicrotendipes lobiger inhabits nearly the whole European mainland except for the Iberian peninsula (Saether & Spies, 2004). A few hundred records are available from the Netherlands. The species is especially common in the seepage regions: the inland fringe of the coastal dunes, the Vecht region and northwest Overijssel (Steenbergen, 1993; Nijboer & Verdonschot, unpublished data; Limnodata.nl). Within the dune fringe region the larvae are characteristic of the water bodies receiving seepage water from the dunes (van der Hammen, 1992). In other parts of the country the species is scarce and probably absent from some areas of the provinces of Zeeland and Flevoland (Nijboer & Verdonschot, unpublished data; Limnodata.nl).

LIFE CYCLE

Most probably there are two emergence periods within the period May–September (Macan, 1949; Mundie, 1957; Kouwets & Davids, 1984; Janecek, 1995; van Kleef, unpublished). Mundie saw swarming from 1 June to 12 July. At least a part of the larvae overwinter in third instar.

MICROHABITAT

The larvae are common, but rarely abundant on *Stratiotes* and other plants (Müller-Liebenau, 1956; Mol et al., 1982; Higler, unpublished data; own observations). Janecek

(1995) reared them from colonies of Bryozoa. Paasivirta & Koskenniemi (1980) found *D. lobiger* rather common in moss-covered littoral sites on the bottom of two Finnish lakes, which dried out and froze in late winter. The larvae most probably also live in small numbers on the bottom of lakes and pools.

WATER TYPE

Current

The larvae have been scarcely recorded in brooks or streams (e.g. Verdonschot, 1990; Orendt, 2002a), but live regularly in canals with slow currents and emergent and submerged vegetation (Steenbergen, 1993). Most records, however, are from stagnant water in lakes, pools, canals and ditches (Limnodata.nl).

Dimensions

The records from the Netherlands show the species to be present in water bodies ranging in size from small ditches and pools to large lakes. Steenbergen (1993) stated no clear preference for small or large water bodies. They most probably also live in the profundal zone of lakes.

Permanence

There are few records of the species from temporary water (e.g. Schleuter, 1985; ten Cate & Schmidt, 1986).

pH

Dicrotendipes lobiger has been rarely found in acid water, but appears to endure it because there are several records of larvae and exuviae in acid Dutch moorland pools (minimum pH 4.3: van Kleef, unpublished). Relatively few larvae are found in water at pH 8 and higher (Steenbergen, 1993; Limnodata.nl).

TROPHIC CONDITIONS AND SAPROBITY

The preference for regions and water bodies with seepage water (see Distribution) is an indication of sensitivity to water quality. In the fenland region in Overijssel the larvae are found mainly in the cleanest ditches (Verdonschot, 1990). Steenbergen (1993) stated a significant preference for water with a low phosphate content, but also found the species sometimes in water with a low oxygen content (saturation during daytime less than 40%). Elsewhere, the presence of the larvae has been reported in ditches where the oxygen content sometimes drops to zero in summer (Prov. Waterstaat Zuid-Holland, unpublished; Grontmij | Aqua Sense, unpublished). Although there are no records from very polluted water, it is still not clear what levels of pollution the larvae can survive. The figures in the matrix table have to be used with care.

SALINITY

Dicrotendipes lobiger seems to be a typical freshwater species. At 61 sites investigated by Steenbergen (1993) in the province of Noord-Holland the chloride content was below 300 mg/l. There are only a few records from slightly brackish water (Limnodata.nl, as conductivity).

Dicrotendipes nervosus (Staeger, 1839)

Limnochironomus gr. *nervosus* Chernovskij, 1949: 72–74, fig. 38
Dicrotendipes gr. *nervosus* Moller Pillot, 1984: 195–196, fig. IV.17.e

IDENTIFICATION

The identification of males, pupae and larvae presents no problems. The female has been described by Rodova (1978) and Contreras-Lichtenberg (1989).

DISTRIBUTION IN EUROPE AND IN THE NETHERLANDS

Dicrotendipes nervosus has been recorded in nearly the whole of Europe (Saether & Spies, 2004). The species also lives in all regions of the Netherlands.

LIFE CYCLE

Mundie (1957) inferred three generations a year based on emergences in English reservoirs from May to the end of September (cf. Otto, 1991: 80). Orendt (1993) found two generations in mesotrophic lakes in Bavaria and three in eutrophic lakes. In Germany (Lehmann, 1971) and in the Netherlands the first pupae have been found in April. In some lakes the first generation is small (Orendt, 1993). Macan (1949) found only one generation in May/June in an English tarn. Sokolova (1966) found that larvae wintering under ice in Russia were in cocoons on the bottom and attached to *Elodea* plants. She suggested that oxygen deficit in winter may play a role in the formation of cocoons by this species. In the Netherlands the larvae winter in second, third and fourth instar. In April many larvae are still in third instar.

MICROHABITAT

Urban (1975) found the larvae in *Potamogeton* stems in mines. In most cases the larvae live in tubes on plants and stones, sometimes in littoral muds and algal mats (Chernovskij, 1938; Mundie, 1957). They are rather scarce on mineral bottoms (Smit, 1982; Peeters, 1988; Buskens, unpublished). In rivers they live mainly on stones and only locally on plants (Lehmann, 1971; Peeters, 1988; Becker, 1994); Smit (1982) found them as plentiful on the underside as on the upper side of stones.

OVIPOSITION

Balushkina (1987) and Nolte (1993) reported 350 eggs per egg mass for *D. nervosus*.

DENSITIES

Brodersen et al. (1998) sometimes found very high densities of *Dicrotendipes* larvae (most probably *D. nervosus*) on stones in the littoral zone of Danish lakes: up to more than 40 larvae per 1 dm^2 stone surface. Smit (1982) found 0.1–21 larvae/dm^2 on stones in the river Meuse. In lowland streams the larvae are likely to occur irregularly in patches on plants because of the dynamic nature of these systems (Tokeshi & Townsend, 1987).

FEEDING

Lenz (1954) described the different feeding methods of the larvae of this genus. The larvae spin a pouch-shaped net within a short tube. The surrounding water is usually filtered by undulating movements. In addition, the larvae gather food around their tubes and eat from dead animals like the pupae of other chironomids. The food therefore consists of algae, detritus, bacteria and animal remains. Izvekova (1980) confirmed the observations of Lenz, but stated that *D. nervosus* larvae do not spin a net but attach many threads lengthwise to the inner side of the tube, from which they collect their food. The gut contents consist mainly of detritus of plant origin (Moore, 1980). Nevertheless, the larvae may possibly select algae, as stated by Izvekova for *D. pulsus*. The scarcity of planktonic and benthic algae in many brooks in the Netherlands is possibly the reason for the absence of *D. nervosus* in these brooks. The species is absent or rare in totally shaded brooks (see e.g. Tolkamp, 1980) and was also missing in the Balgoijse Wetering, a turbid lowland brook containing suspended clay particles (unpublished own investigations).

WATER TYPE

Current

In brooks and small streams the larvae live in small numbers on sites with a moderate or slow current (Cuijpers & Damoiseaux, 1981; Braukmann, 1984; own data). The species is a little more abundant in regulated books in the Netherlands (Peters et al., 1988). Records from fast-flowing streams (e.g. Orendt, 2002) are probably occurrences at sites with a slow current. In large rivers the larvae live in large numbers (mainly on stones) in the potamal zone (Lehmann, 1971; Peeters, 1988). They are found at sites with moderate or slow current (Ertlová, 1970; Lehmann, 1971: 530), but endure a little more current than *P. arcuatus* (Bazerque et al., 1989; see also under Microhabitat). In stagnant water the larvae can be numerous, especially on stones in larger lakes and on plants in small lakes.

Shade

The larvae seem to be rare or absent in brooks and ditches totally overshadowed by trees (e.g. Tolkamp, 1980). The scarcity of algae is probably the most important reason for this (see under Feeding). The absence of plants as a substrate or the narrow width of some watercourses can also play a role.

Dimensions

Dicrotendipes nervosus is rarely recorded from narrow ditches and narrow brooks. Steenbergen (1993) found the number of larvae to gradually increase as the width of the water body increased from less than 4 m to more than 100 m. Verneaux & Aleya (1998) did not find the species in the profundal zone of lakes. Humphries (1936) collected the larvae only in the sublittoral zone of Lake Windermere down to a depth of 6 m. However, Tõlp (1971) found the species in Estonian lakes down to a depth of 16.8 m. In the Netherlands many larvae of this genus have not been identified to species level, but *D. nervosus* most probably also lives in the profundal zone (Mol et al., 1982; Kuijpers, pers. comm.).

Permanence

D. nervosus is rarely recorded as an inhabitant of temporary stagnant water (Driver, 1977; Fritz, 1981; Schleuter, 1985) or temporary flowing waters (Verdonschot et al., 1992). Most temporary water bodies are not attractive to this species because of their small dimensions. The figures in the matrix table in Chapter 4 do not refer to very small water bodies, but are nevertheless influenced by the scarcity of the species in narrow brooks and ditches.

pH

Roback (1974) reported the occurrence of the larvae only in water with a pH of 6.3–8.3. In Norwegian lakes the larvae live in water with a pH of 4.5 and more (Raddum & Saether, 1981). In Dutch lakes and pools the larvae have very rarely been found at pH < 6 (Leuven et al., 1987; Duursema, 1996; Buskens, unpublished; van Kleef, unpublished). In lakes and canals in the province of Noord-Holland the species was significantly more common at pH > 8.

TROPHIC CONDITIONS AND SAPROBITY

The larvae live in more or less oligotrophic to strongly eutrophic lakes (Humphries, 1936; Brundin, 1949; Saether, 1979; Brodersen et al., 1998; Langdon et al., 2006). In Bavaria the species was absent in oligotrophic lakes and is characteristic of eutrophic lakes (Orendt, 1993). The larvae are rare in oligotrophic to mesotrophic pools, lakes, sand pits, etc. in the Netherlands, whereas they are very common and abundant in more or less eutrophic canals and lakes (Buskens, 1983; Buskens & Verwijmeren, 1989; Steenbergen, 1993; van Kleef, unpublished).
The larvae are very resistant to organic pollution: 'tolerance group D' (Peters et al., 1988; Bazerque et al., 1989; Wilson & Ruse, 2005). In the river Meuse they were more abundant in the polluted stretch in South Limburg (Smit, 1982; Smit & Gardeniers, 1986). Compared with *Parachironomus arcuatus* the species seems to be more characteristic of a little heavier pollution (Smit & Gardeniers, 1986; Bazerque et al., 1989), but the latter can live in water with a much lower oxygen content (several own investigations; see also Steenbergen, 1993). *D. nervosus* can survive for a short time in water with an oxygen content lower than 1 mg/l (Rossaro et al., 2007), but most probably the larvae move up from the bottom to escape oxygen-poor conditions. A low oxygen content may often be the reason for the absence of this species, whereas *P. arcuatus* can be present.

SALINITY

Many authors have recorded the occurrence of *Dicrotendipes nervosus* in water with 1000–3000 mg Cl/l (Remmert, 1955; Tõlp, 1971; Krebs, 1984; Steenbergen, 1993). The species is rarely recorded in water with higher chloride contents (e.g. Tõlp, 1971, salinity 6.3‰).

Dicrotendipes notatus (Meigen, 1818)

Limnochironomus notatus Lehmann, 1969: 265

DISTRIBUTION IN EUROPE AND THE NETHERLANDS

Dicrotendipes notatus lives in all parts of Europe (Saether & Spies, 2004). In Scandinavia it is possibly a southern species (Brundin, 1949). In the Netherlands the species is present nearly everywhere, but possibly absent from some regions with clay bottom (Nijboer & Verdonschot, unpublished).

LIFE CYCLE

Adults emerge from early May until September (own observations). There are at least two generations a year. Hibernation is in third and fourth instar. A significant proportion of the larvae are still in third instar in March (own data).

FEEDING

Nothing has been published about larval feeding. The frequent occurrence of larvae in polluted water or in water with decaying leaves indicates detritus feeding. Bacteria may be the most important food source (see *D. pulsus*).

MICROHABITAT

Larvae have been found mostly on organic substrates, often on coarse material (leaves, etc.), but young larvae in particular (?) can be also observed on wood, plants and stones.

WATER TYPE

Current
Most authors do not mention *Dicrotendipes notatus* as an inhabitant of running water (Lehmann, 1971; Fittkau & Reiss, 1978; Braukmann, 1984). In the Netherlands, however, a relatively large number of records are from brooks and slow-flowing water in ditches and canals (see Verdonschot et al., 1992; Limnodata.nl; cf. Lehmann, 1969).

Shade
The larvae are often found in pools, brooks and ditches in woodland or relatively shaded water.

Dimensions
In the Dutch province of Noord-Holland Steenbergen (1993) found the larvae more in lakes and canals than in narrow ditches. In the more sandy regions in the Netherlands a larger proportion of records are from brooks and ditches between 1 and 4 m wide. The larvae are rare or absent in large rivers. It is not clear whether the specimens from the river Rhine mentioned by Caspers (1991) are exuviae, possibly brought down from stagnant water near the river. Becker (1994) did not collect or rear the species at all.

Permanence
The larvae are regularly collected in temporary non-acid pools and ditches.

SOIL

Steenbergen (1993) collected the species in the province of Noord-Holland mainly on sand and very rarely on clay.

pH

There are very few records from very acid water (cf. Buskens, 1983; Leuven et al., 1987; Duursema, 1996), but the species is also less common in water with a pH higher than 8 (Steenbergen, 1993; Limnodata.nl). The only records at low pH mentioned in the literature are for three specimens in an acid pond (Schleuter, 1985: pH 5.0) and two larvae in a slow-flowing ditch with a pH of 5.3 (Moller Pillot, 2003). See under Feeding.

TROPHIC CONDITIONS AND SAPROBITY

Brundin (1949) found the species (numerously) only in an eutrophicated lake in Sweden. Steenbergen (1993) collected the larvae significantly more often in water with a low phosphate and chlorophyll-a content. As mentioned under Feeding, the species seems to be specialised in feeding on bacteria.
Bazerque et al.(1989) attributed *D. notatus* to the pollution-resistant species in lentic areas. In the Netherlands the species has been collected relatively often in polluted ponds and brooks and there are (unpublished) records from heavily polluted brooks where the oxygen content was not very low. Rossaro et al. (2007) sometimes found *D. notatus* in water with only 1 mg O_2/l. The larvae probably endure at least the same pollution and oxygen deficit as those of *D. nervosus*.

OTHER ASPECTS OF WATER QUALITY

D. notatus larvae are without doubt very pollution-resistant and are particularly numerous between decaying leaves, etc. However, in comparison with *D. nervosus* the species appears to make special demands on water quality. In the Dutch province of Noord-Holland the larvae occur nearly exclusively in the coastal dunes, dune-fringe and Vecht regions (van der Hammen, 1992; Steenbergen, 1993) where groundwater seepage has a significant effect on abiotic conditions. At most localities the potassium and magnesium contents are very low. Elsewhere (e.g. on in the Pleistocene areas) the larvae live mainly in flowing water or pools and ditches fed by groundwater seepage. The differences are probably not only caused by the fact that *D. notatus* is to a greater extent a bottom dweller than *D. nervosus* and that it is more common in smaller water bodies; feeding on bacteria may be the key factor, but this is no more than a suggestion.

SALINITY

Tourenq (1975) found the species quite often in oligohaline water with a chloride content up to 3000 mg/l. Steenbergen (1993) found the larvae only scarcely in brackish water and not in water with a chloride content higher than 1000 mg/l.

Dicrotendipes pallidicornis (Goetghebuer, 1937)

IDENTIFICATION

The larva, pupa and adult male have been described by Contreras-Lichtenberg (1986). The female has been described and keyed by Contreras-Lichtenberg (1989).

DISTRIBUTION IN EUROPE AND THE NETHERLANDS

The species is known from the Mediterranean area, the British Isles and the Netherlands (Krebs, 1988; Saether & Spies, 2004). In the Netherlands there are only records from two localities in the province of Zeeland.

LIFE CYCLE

There are at least two generations a year (Krebs, 1988).

MICROHABITAT

The larvae live on plants and firm substrates (Moller Pillot & Buskens, 1990).

WATER TYPE

The European records are for brackish pools with a depth of 1 to 5 m (Krebs, 1988).

SALINITY

Larvae have been found in water with 920 to 8000 mg Cl/l (Krebs, 1988).

Dicrotendipes pulsus (Walker, 1856)

Dicrotendipes modestus Contreras-Lichtenberg, 1986: 682 et seq.; Langton, 1991: 258 nec *D. modestus* Danks & Jones, 1978: 667–669; Moore, 1979a: 302 et seq.
D. objectans Langton & Visser, 2003
Limnochironomus gr. *tritomus* Chernovskij, 1949: 74, fig. 39 (pro parte)
Dicrotendipes gr. *tritomus* Moller Pillot, 1984: 197–198 (pro parte)

NOMENCLATURE

Contreras-Lichtenberg (1986) treated *D. pulsus* as a synonym of the American species *D. modestus* (Say, 1823). Langton & Visser (2003) stated that these two are different species and used the name *D. objectans* (Walker, 1856). Spies & Saether (2004) argued that the latter has no priority and stated that the valid name is *D. pulsus*. The larvae of *D. pulsus* and *D. tritomus* are taken together in Moller Pillot (1984) as gr. *tritomus*.

IDENTIFICATION

The larva, pupa and male have been described and keyed by Contreras-Lichtenberg (1986), the female by Contreras-Lichtenberg (1989).

DISTRIBUTION IN EUROPE AND THE NETHERLANDS

Dicrotendipes pulsus lives in nearly the whole European mainland (Saether & Spies, 2004). In the Netherlands the species is especially common in the fenland regions and in the sandy Pleistocene areas; there are a few records from dune lakes. Elsewhere only some records of 'gr. *tritomus*' are known.

LIFE CYCLE

According to Sokolova (1968) this species needs relatively few degree-days to develop (770–1110 after reaching the critical temperature of 9 degrees in spring), so that often two or three generations a year are possible. Wotton et al. (1992) stated that this species needs less time for development than most other Chironomini. They found the first peak of emergence in sand filter beds after 25–38 days.

In fish ponds in the Belgian Ardennes the species is bivoltine, emerging in April–May and July–August (Goddeeris, 1983). In mountain areas the species has only one generation (Laville, 1971); in the Netherlands two generations will be the rule, with probably a third generation in late summer at somewhat higher temperatures (cf. Mundie, 1957). In this latter situation a majority of the larvae may hibernate in second and third instar as stated for the related American species *D. modestus* (Dank & Jones, 1978; Moore, 1979a). Based on his results in Mirwart (Belgian Ardennes), Goodderis (1983, 1986) argued that the larvae of *D. pulsus* have an overwintering diapause, mainly in fourth instar, starting from the middle of August. He found only few larvae wintering in second and third instar. Armitage (1968) observed in a Finnish lake that in winter the larvae built cocoons or sealed off their tubes.

FEEDING

Mundie (1957) supposed that *D. pulsus* is the least sensitive species of the genus in its food requirements. Armitage (1968) found the guts of larvae from a Finnish woodland lake to contain mainly detritus and occasionally diatoms. Mackey (1979) also stated that the gut of the larva contains much fine detritus and some diatoms.
Izvekova (1980) found 20% algae in the gut, mainly in autumn, and especially the blue-green alga *Microcysytis aeruginosa*. The larvae do not spin a true filtration net (see under *D. nervosus*); they appeared actively to select algae more than detritus. However, Moore (1979a) stated that in the related species *D. modestus* algae were badly digested and were not an important energy source. This author supposed that bacteria may be the most important food source.

EGGS

Balushkina (1987) recorded 800 eggs per egg mass for *D.* gr. *tritomus*.

MICROHABITAT

Larvae of the *tritomus* group have been found on plants, but all published or verified own records of *D. pulsus* are from stones, from artificial substrate or from organic (and rarely sandy) bottoms (Lenz, 1954–62; Bijlmakers, 1983; Buskens et al., 1986; Palomäki, 1989; Verneaux & Aleya, 1998). The related American species *D. modestus* has been found in summer mainly on plants, but after November when water temperatures had decreased below 5 °C only in the sediment (e.g. Menzie, 1980).

DENSITIES

Charles et al. (1974) collected in summer, autumn and winter on mud in Loch Leven in Scotland 465 larvae/m^2 . Menzie (1980) found mean densities of the related American species *D. modestus* of 6392 larvae/m^2 (on aquatic plants, *Myriophyllum spicatum*, and sediments together).

WATER TYPE

Current

There are only a few records of the species from lowland brooks and the larvae appear to be absent from fast-flowing streams (Limnodata.nl).

Dimensions

D. pulsus has been collected in pools and lakes, also at a great depth (Contreras-Lichtenberg, 1986). However, in the Biesbosch reservoirs in the Netherlands, which have a high oxygen content also at great depth, the larvae have not been collected at the deepest point (27 m) (cf. Ketelaars et al., 1992; Heinis, 1993). There are few records from ditches and these are all for larvae of gr. *tritomus*, and possibly not *D. pulsus*.

Permanence

In one case we collected the species in a temporary pool. Wotton et al. (1992) found the species regularly in sand filter beds which were filled with water for periods of 16 to 77 days.

SOIL

Steenbergen (1993) collected the larvae of *Dicrotendipes* gr. *tritomus* in the Dutch province of Noord-Holland mainly on peat and only rarely on sand and clay. The sandy regions in this province have very few acid water bodies, so it cannot be compared with other regions (see under Distribution and pH). Elsewhere in the Netherlands the species is mainly recorded on sand.

pH

The occurrence of *Dicrotendipes pulsus* seems to be more or less independent of pH, although there are a relatively large number of records from acid water. Van Kleef (unpublished) found the exuviae in many Dutch moorland pools with pH values from 4.15 to 7.5. The species was also rather common in Lake Maarsseveen with a pH of around 8 (Mol et al., 1982; Kouwets & Davids, 1984) and in some dune lakes with pH values from 8 to 8.7 (Schmale, 1999). Brodin & Gransberg (1993) found an increase in numbers of larvae during acidification of a Scottish lake, when the pH fell from 5.5 to 4.8. Larvae of gr. *tritomus* were found numerously at pH 3.5 by Buskens et al. (1986), while in the Dutch province of Noord-Holland Steenbergen (1993) found relatively few larvae at pH <7.5 and still found larvae at pH >8.6. In this province acid waters are very scarce and the species is only locally common (see under Soil).

TROPHIC CONDITIONS AND SAPROBITY

Brundin (1949) collected the species in many lakes in southern Sweden, in oligotrophic oligohumic and polyhumic as well as eutrophic conditions. The exuviae are collected abundantly in the storage reservoirs in the Biesbosch, where the productivity is low: chlorophyll-a 11 and 20 (Ketelaars et al., 1992; Kuijpers et al., 1992). Elsewhere in the Netherlands the larvae of gr. *tritomus* and/or exuviae of *D. pulsus* are mainly found in clear, well oxygenated water with a rather low phosphate content, although in some cases with a high chlorophyll-a content (up to more than 200 µg/l) (Verdonschot, 1990; Steenbergen, 1993; Schmale, 1999; van Kleef, unpublished). In one ditch the

D

oxygen content could sometimes drop to zero during the night (Prov. Waterstaat Zuid-Holland, unpublished). There are no records from severely polluted water.

SALINITY

Steenbergen (1993) collected the larvae of gr. *tritomus* rarely at chloride contents > 300 mg/l. Tõlp (1971) reported gr. *tritomus* from brackish water.

Dicrotendipes tritomus (Kieffer, 1916)

IDENTIFICATION

Although at present identification presents hardly any problems, the older literature contains several mistakes, which makes the available information less reliable. Specifically this concerns the identification of male adults by Kruseman (1933), Brundin (1949) and Kouwets & Davids (1984). The larvae of *D. tritomus* and *D. pulsus* are taken together in Moller Pillot (1984) as gr. *tritomus*. At present adult males can be identified using Langton & Pinder (2007), exuviae using Langton (1991) and larvae using Contreras-Lichtenberg (1986). The female has been described by Contreras-Lichtenberg (1989).

DISTRIBUTION IN EUROPE AND THE NETHERLANDS

Dicrotendipes tritomus lives in Western, Central and Eastern Europe, but is absent from the Mediterranean area and North Scandinavia (Contreras-Lichtenberg, 1986). In the Netherlands the species is probably common and widespread, at least in the Pleistocene areas.

MICROHABITAT

The larvae live on plants and between the overgrowth of stems and stones (Lenz, 1954–62: 195; Reiss, 1968).

WATER TYPE

Lenz (1954–62) mentioned many records from lakes. According to Contreras-Lichtenberg (1986) the larvae also live at greater depth. The latter supposed that the species is less common than *D. pulsus*. This is not the case in Dutch moorland pools. Van Kleef (unpublished) collected exuviae of *D. tritomus* in 18 moorland pools scattered throughout the southern and eastern part of the country. In many cases the two species were found together. There are no records from running water or from temporary water.

pH

Van Kleef (unpublished) found exuviae of *D. tritomus* in Dutch moorland pools with pH values from 4.0 to 7.8 (over the whole range of the investigated pools).

TROPHIC CONDITIONS AND SAPROBITY

The chemical data on the pools where *D. tritomus* has been found are not very different from those of *D. pulsus*: in most cases the water bodies had rather clear water, sometimes with a high phosphate content and more often with a high ammonium content (van Kleef, unpublished).

As stated for *D. pulsus*, the larvae of gr. *tritomus* have not been found in severely polluted water.

Einfeldia Kieffer, 1924

SYSTEMATICS

For many years it has been recognised that the genus *Einfeldia* in the usual sense is not monophyletic. In the older literature even *Chironomus longipes* was placed in *Einfeldia*. Cranston et al. (1989) also placed *E. carbonaria* and *E. dissidens* in *Chironomus (Lobochironomus)*. As the larvae of these species are quite different from *Chironomus* the next step will be to place them in a separate genus. Martin et al. (2007) proposed the genus name *Benthalia* Lipina. It is also possible that both species will be placed in *Fleuria* (M. Spies, in litt.). While waiting for a definite solution we prefer to use the name *Einfeldia* for these two species. Moreover, both these species have often been mistaken: see under the species names.

Einfeldia s.s. is represented in Europe by two species: *E. pagana* and *E. palaearctica*. The larva and ecology of the last species is unknown.

Einfeldia carbonaria Meigen

Chironomus (Lobochironomus) carbonarius Saether & Spies, 2004
Einfeldia gr. *carbonaria* f.l. *reducta* Kolosova & Lyachov 1957: 1101–1104
Einfeldia gr. *insolita* f.l. *reducta* Moller Pillot, 1984: 220–221, fig. IV.23.a/b
Einfeldia gr. *insolita* Steenbergen, 1993: 511 (pro parte)
nec *Einfeldia carbonaria* Shilova, 1976: 111; Shilova, 1980: 166 et seq.; ? Rodova, 1978: 55, 60, fig. 17; Pinder & Reiss, 1983: 312, fig. 10.22

SYSTEMATICS AND IDENTIFICATION

The use of the generic name *Einfeldia* has been treated above. A second problem is the confusion between *E. carbonaria* and *E. dissidens*. Only the new key by Langton & Pinder (2007 vol. 2: 158) gives the correct differences between these species in the adult stage. All older keys for males are unreliable. The exuviae can be identified using Langton (1991). According to Friedrich Reiss (pers. comm.) the identification of *E. carbonaria* larvae in Pinder & Reiss (1983) was based on a mistake; *E. carbonaria* is the species without tubuli, as pointed out by Moller Pillot & Wiersma (1997). Larvae of *Fleuria lacustris* are sometimes incorrectly identified as *Einfeldia carbonaria*.

DISTRIBUTION IN EUROPE AND IN THE NETHERLANDS

According to Saether & Spies (2004) *E. carbonaria* has been collected throughout Western and Central Europe and Russia, but is absent from the British Isles, the Mediterranean area and Scandinavia. However, not all records are reliable because of the confusion about the species name. In Russia and other countries the larvae are often identified using Shilova (1980), in which larvae of *E. dissidens* are called *E. carbonaria*. *E. carbonaria* has been recorded in Russia as *Einfeldia* gr. *carbonaria* f.l. *reducta* by Kolosova & Lyachov (1957).
In the Netherlands most records are from the western and northern parts of the country and from the Rhine-Meuse floodplain region. Elsewhere the species has been rarely found. A complete and reliable distribution has not been published.

LIFE CYCLE

In the Netherlands Smit et al. (1996) collected large larvae (8–10 mm) in spring (May–June) and in August, and after both periods on average smaller larvae (5–7 mm), which without doubt indicates two generations. Elsewhere, however, we found mass emergence early July. In Russia Kolosova & Lyachov (1957) stated emergence in May, but not in late summer. In autumn and winter these authors collected mainly larvae in third instar and some in fourth instar; the same has been found in the Netherlands (Kuijpers, pers. comm.).

MICROHABITAT

The larvae are found on the bottom of lakes outside the vegetation zone (Kolosova & Lyachov, 1957; Smit et al., 1994). The latter authors found the larvae (as *E. dissidens*) mainly on more silty bottoms (grain size for the greater part < 63 μm). The larvae are very rarely found on stones and never on plants. However, Verneaux & Aleya (1998) reared the species from an artificial substrate in Lake Abbaye in France.

FEEDING

The larvae are thought to be selective microphytobenthos grazers in the Haringvliet and Lake Volkerak-Zoommeer (van der Velden et al., 1995).

OVIPOSITION

On mud flats in the Haringvliet small larvae of *E. carbonaria* appeared in late summer at the highest site and appeared gradually at lower sites in the course of time. This fact suggests that oviposition had taken place at one of the exposed sites on the flats and that the larvae then migrated to other areas (Smit et al., 1996).

DENSITY

Smit et al. (1994) found the larvae (as *E. dissidens*) on a silty bottom in densities up to 2600 larvae/m^2. Van der Velden et al. (1995) stated mean densities in Lake Volkerak-Zoommeer of 239 larvae/m^2 in the optimal year 1988.

WATER TYPE

The larvae have been found in the Netherlands especially in large lakes and freshwater estuaries (Smit et al., 1994, as *E. dissidens*; own data). They have been collected rarely in (the very large) Lake IJssel (Maenen, 1983), but even more rarely in ditches less than 6 m wide. There are very few records from streams (Orendt, 2002a) and the larvae have been found in very few lowland brooks in the Netherlands (Limnodata.nl). Verneaux & Aleya (1998) collected the larvae in the littoral zone of Lake Abbaye in the Jura, in contrast to *E. dissidens* which lived more in the sublittoral zone.

pH

There are no records of this species at pH values lower than 7.0 and the larvae appear to live usually in water with somewhat higher pH values than *E. dissidens* (Limnodata.nl; own data).

TROPHIC CONDITIONS AND SAPROBITY

The larvae are common in eutrophic and somewhat polluted lakes, but there are no records from severely polluted water. In comparison with *E. dissidens* the species prefers larger water bodies, which have more oxygen in the water column.

SALINITY

The records from slightly brackish water mentioned by Steenbergen (1993) may apply to this species.

Einfeldia dissidens (Walker, 1856)

Chironomus (Lobochironomus) dissidens Saether & Spies, 2004
Einfeldia dissidens agg. Moller Pillot & Buskens, 1990: 11, 22, 43
Einfeldia gr. *insolita* Moller Pillot, 1984: 220–221, fig. IV.23.a/b (pro parte)
Einfeldia carbonaria Shilova 1976: 111; Shilova 1980: 166 et seq.
nec *Einfeldia dissidens* Pinder & Reiss, 1983: 312, fig. 10.21

IDENTIFICATION

For a discussion of the confusion between *E. carbonaria* and *E. dissidens* see under the former species. The adult male can be identified only using Langton & Pinder (2007). The description of the female by Rodova (1978) most probably refers to this species. Exuviae can be identified using Langton (1991). Because of the problems of identification not all published and unpublished data could be used for the description of the ecology of the species.

DISTRIBUTION IN EUROPE AND THE NETHERLANDS.

E. dissidens has been recorded from Western, Central and Northern Europe and from Italy (Saether & Spies, 2004). Records from Russia are not mentioned by these authors because they have been published as *E. carbonaria* (Shilova 1976, 1980). The species has been found throughout the Netherlands at more than 200 localities, but it appears to be rare in the delta region (Nijboer & Verdonschot, unpublished data; Limnodata.nl). The scarce records from the delta region have not been verified.

LIFE CYCLE

According to Shilova (1976; sub *E. carbonaria*), who caught adults from June to August, the species probably has two generations a year, but it is possible that in many cases only one generation emerges (compare with *E. carbonaria*). Janecek (1995) collected exuviae from the end of May until early September in a carp pond in Austria. Our data indicate wintering mainly in third instar, like *E. carbonaria*. Third instar larvae have been found until end of April.

FEEDING

We found the guts of four larvae (from one locality) to contain detritus, many filamentous blue-green algae and a few diatoms.

MICROHABITAT

Bijlmakers (1983) collected the larvae (in low numbers) mainly on organic bottoms without vegetation in Belversven (a eutrophic moorland pool), sometimes also on sand. Otto (1991) collected them on organic bottoms with H_2S build-up. However, we also found the larvae on clay-silt without very much organic material and not solely on bottoms without any vegetation. Verneaux & Aleya (1998) collected the larvae in large numbers on an artificial substrate in Lake Abbaye in France.

DENSITIES

Janecek (1995) found average abundances during the year of 518 and 790 larvae/m^2 in a carp pond in Austria. Bijlmakers (1983) usually found about 100 larvae/m^2 .

WATER TYPE

The species lives mainly in stagnant water, but there are some records from slow-flowing brooks and streams. The larvae are most commonly collected in rather small lakes (e.g. fen-peat lakes), water courses and large ponds and only rarely in narrow ditches.

D

However, the larvae were numerous in Lake Abbaye in the French Jura on an artificial substrate at a depth of 5 m (Verneaux & Aleya , 1998).

pH

Koskenniemi & Paasivirta (1987) found the larvae commonly at pH 5.5–6 in a Finnish reservoir. However, the species seems to be rare in acid water and has not been collected often at pH values between 6.5 and 7.0 (Buskens, 1989; van Kleef, unpublished, Limnodata.nl), but it is rather common at pH values from 7 to 8.5.

TROPHIC CONDITIONS AND SAPROBITY

Brundin (1949) collected the species nearly exclusively in eutrophic lakes. Langdon et al. (2006) reported the occurrence of larvae of the '*Einfeldia dissidens* type' in eutrophic turbid lakes with very high numbers of waterfowl. Saether (1979) stated that the species is characteristic of eutrophic lakes. However, the Dutch data do not indicate a very hypertrophic environment because the orthophosphate content and chlorophyll-A content are usually low (Limnodata.nl). Decaying phytoplankton appears not to be the food base for the larvae (see also Feeding). The larvae are probably not very sensitive to organic pollution, but they do not have a preference for severely polluted water. The larvae can live on bottoms almost devoid of oxygen (see Microhabitat), but may need some oxygen in the water column.

Einfeldia pagana (Meigen, 1838)

Einfeldia synchrona Danks, 1971: 1597 et seq.; 1971a: 1878 et seq.

DISTRIBUTION IN EUROPE AND THE NETHERLANDS

E. pagana has been reported from many countries in Europe and from North America (Saether & Spies, 2004). The species lives throughout the Netherlands, but only scarcely and it has not been collected in the south of the province of Limburg (Nijboer & Verdonschot, unpublished data; Limnodata.nl). It is possible that larvae and/or exuviae identified as *E. pagana* belong to *E. palaearctica* because this species is only known as an adult male.

LIFE CYCLE

Danks (1971) stated only one generation a year near Ottawa (Canada), emerging mainly (highly synchronised) at the end of May. Shilova (1976) supposed the species has two generations a year; she collected adult males from the end of May until the middle of August. According to Macan (1949) and Otto (1991) there is only one generation a year, emerging in June–July (northern Germany) or July–August (English Lake District). Schmale (1999) collected mainly exuviae in Dutch dune lakes in May and few from June to August.
The larvae winter in third and fourth instar (Danks, 1971a; own data). In Canada Danks found the larvae in great numbers wintering in cocoons, which were formed well after ice cover. The percentages of third and fourth instar larvae in different winters were different in the same pond.

FEEDING

Danks (1971) observed that larvae feed by surface scraping and observed no filter feeding mechanism. Dissection revealed that small soil fragments, scraps of vegetation and diatoms had been ingested.

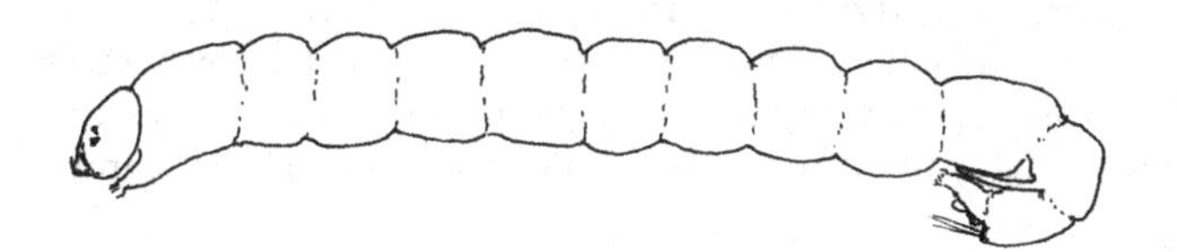

5 *Einfeldia pagana: third instar larva in overwintering position (after Danks, 1971a)*

MICROHABITAT

The larvae live in tubes of soil and plant fragments, usually in the top few centimetres of mud, but in late summer and in winter they move deeper into the mud (Danks, 1971).

OVIPOSITION

Danks (1971) observed that the females visit the pond surface together at dusk to lay eggs. Many females produced a second egg mass after 4 days. The first egg masses contained 183–808 eggs (mean 516), the second 226–431 (mean 351). The author also describes the eggs and first instar larvae.

DENSITY

Danks (1971) found a density of 10,000 to 20,000 larvae/m² in the central area of a pond. After the peripheral areas dried up the density in the central area was still much higher. The density decreased sharply during winter.

WATER TYPE

Current

The species is only rarely recorded from brooks or streams, but the exuviae are quite often collected in fast-flowing rivers in Bavaria (Michiels, 1999; Orendt, 2002a). It seems probable that not all these exuviae came from stagnant water bodies near the river (cf. Fittkau & Reiss, 1978). In the Netherlands there are some records from slowly flowing water, but the larvae are very rare in lowland brooks (Steenbergen, 1993; Limnodata.nl).

Stagnant water

The larvae are collected in ditches, pools and lakes. Danks (1971) stated a preference for the deepest parts of the pond studied (at that moment about 40–60 cm deep). A very high proportion of Dutch records are from water bodies with groundwater seepage and dune lakes (H. Cuppen, pers. comm.; Steenbergen, 1993; Schmale, 1999; own data).

Permanence

The larvae have been reported from temporary water only twice (Driver, 1977: 125; Schleuter, 1985: 92). Because there is as a rule (?) only one generation a year, completing the life cycle in temporary water is nearly impossible.

pH

E. pagana seems to be most common at pH values between 7 and 8 (Steenbergen, 1993; Limnodata.nl). However, Cannings & Scudder (1978) found numerous larvae in lakes with a pH of 9. There is only one record from acid water: the larvae occurred scarcely at pH 5.5–6 in a Finnish reservoir (Koskenniemi & Paasivirta, 1987).

TROPHIC CONDITIONS AND SAPROBITY

The species has been collected relatively often in oligotrophic/mesotrophic conditions with a low phosphate and chlorophyll-A content (Macan, 1949; Steenbergen, 1993; Limnodata.nl; H. Cuppen, pers. comm.). The larvae are also found in very eutrophic water, for example in highly productive lakes rich in blue-green algae (Cannings & Scudder, 1978).
Walshe (1948) stated that the larvae are extremely tolerant of anoxia: the larvae survived anaerobic conditions longer than any other chironomid tested (more than 50 hours). Consistent with these findings, the larvae are often found on bottoms rich in organic material and sometimes also in water bodies with little oxygen in the water column (e.g. Steenbergen, 1993; Schmale, 1999: pool Wittenberg), but in most cases the oxygen conditions in the water body as a whole are relatively good (e.g. Limnodata.nl).

The biology of the larvae is still not completely understood, but in any case the species appears to live in an environment conducive to rapid decomposition (see under pH and Water type). The larvae do not seem to be interested in phosphate or phytoplankton, but as detritus feeders they need decomposing plant material.

SALINITY

In Estonia Tõlp (1971) reported the occurrence of this species in water containing more than 2000 mg Cl/l. Cannings & Scudder (1978) often found *E. pagana* to be a dominant chironomid species in Canadian lakes containing up to 1600 mg Cl/l and less common in more saline lakes. In the Netherlands the species is only known from fresh water.

Einfeldia palaearctica Ashe, 1990

Tendipes (Einfeldia) dilatatus Goetghebuer, 1954: 30, fig. VII-98

SYSTEMATICS AND IDENTIFICATION

E. palaearctica belongs to *Einfeldia* s.s. and is related to *E. pagana*. The species can be identified only as adult male: Langton & Pinder, 2007: 170. Its biology and ecology are unknown.

DISTRIBUTION IN EUROPE AND THE NETHERLANDS

E. palaearctica has been collected in few countries, including Belgium, Germany and the British Isles. It is very probably present in the Netherlands.

Endochironomus Kieffer, 1918

SYSTEMATICS

As Spies & Saether (2004) state, the division of the genus *Endochironomus* sensu lato into two genera *Synendotendipes* and *Endochironomus* has been based on Nearctic material and for the European species it is still somewhat tentative. In the Netherlands it has been decided to use the original genus name *Endochironomus* sensu lato pending further investigations.
Endochironomus abranchius Lenz is still an accepted name and Michailova (1989) described the cytology of a species as *Endochironomus* sp. In the older literature *Tribelos intextum* was also placed under *Endochironomus* as *E. intextus*. *E. donatoris* Shilova also belongs to *Tribelos*.

IDENTIFICATION

For many years there have been problems with identifying the adults (and therefore also the pupae and larvae) of the genus. Brundin (1949: 746) questioned whether *E. albipennis* and *E. tendens* were different species. Chernovskij (1949), Lenz (1954–62) and many other authors have used the name *E. tendens* for *E. albipennis*. In this book we follow Kalugina (1961), but in some publications we could not determine which species was meant.
Spies & Saether (2004) asserted that the true number of West Palaearctic species and the diagnostic characters for identifying them had still not been established and that all published records under any name are uncertain. However, the larvae of the West European species seem to be well known, although the proof of the identity of *E. lepidus* is not convincing and no difference is known between the larvae of *E. impar* and *E. dispar*.
Kalugina (1960) warned that the mental teeth can be worn, especially after hibernation, leaving hardly any difference between the larvae of *E. impar* and *E. tendens*. The larvae cannot be identified to species level in first instar and not always in second instar. The median mental tooth of the first instar larva is trifid, as in many other Chironomini (Kalugina, 1959).

EGGS

The eggs and egg masses of the genus have been described by Kalugina (1961). The egg masses are cylindrical and attached to the substrate by both adhesive ends. The eggs are yellowish to brown. Nolte (1993) recorded 300 to 1000 eggs per egg mass.

LIFE CYCLE

As far as is known, the larvae of *Endochironomus* hibernate in third and fourth instar, but some larvae hibernate in second instar.

FEEDING, MICROHABITAT AND WATER TYPE

All species of *Endochironomus* (with the exception of *E. lepidus*) are generally filter feeders, in summer living mainly on plants in stagnant water. However, there are some differences between the species, as can be seen in the matrix tables in Chapter 4.

Endochironomus albipennis (Meigen, 1830)

D

Endochironomus gr. nymphoides Lenz, 1954–62: 182–187, fig. 161–167, 169, 171–173
Endochironomus tendens Chernovskij, 1949: 81, 83, fig. 54; Lenz, 1954–62: 182–187, fig. 161–167, 171-173 nec aliis

IDENTIFICATION

For a discussion of identification problems in the past see under the genus. At present all keys give reliable differences between the species. The female of *E. albipennis* has been described and keyed by Rodova (1978). Living larvae can often be distinguished from other Chironomini because of their more or less orange (sometimes partly greenish) colour. In second instar the larvae cannot be distinguished from those of *E.* gr. *dispar* (see there).

DISTRIBUTION IN EUROPE AND THE NETHERLANDS

E. albipennis lives throughout Europe (Saether & Spies, 2004). In the Netherlands the species is one of the most common chironomids, recorded by Nijboer & Verdonschot (2001) and Limnodata.nl in about 5000 localities, mainly in the Holocene part of the country. In the Pleistocene areas the larvae are common in non-acid stagnant water bodies.

LIFE CYCLE

Different authors have reported two generations a year without strong synchronisation, emerging during a long period from May (sometimes the end of April) to September or even October (Mundie, 1957; Kalugina, 1961; Shilova, 1976; Otto, 1991; Janecek, 1995). Three generations may occur in Bavaria (Orendt, 1993). Whether the spring or primarily the summer generation is better developed depends on circumstances not fully understood (cf. Mundie, 1957; Brown & Oldham, 1984; Otto, 1991; and see the section 2.5 on life cycle in this book). The larvae hibernate in third and fourth (rarely in second) instar; especially in a cold climate often in cocoons (see figure).

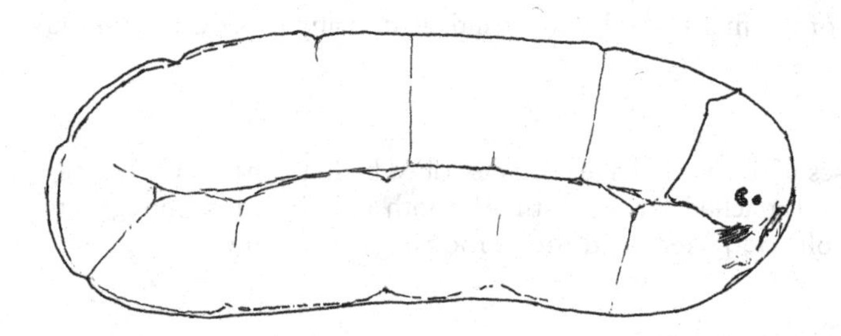

6 *Endochironomus albipennis larva in cocoon*

FEEDING

Walshe (1951) described the filter-feeding mechanism of three species of *Endochironomus*, including *E. albipennis*. The larvae are usually leaf miners and spin a net within their tube. Walshe also frequently found larvae spinning their nets on the surface of the leaves and observed that the food of all the larvae consisted of algae. However, the larvae have been observed to be also detriticollectors and grazers (Izvekova, 1980; Dvořák, 1996). Detritus enriched with bacteria appears to be a very nutritious food for them (Izvekova & Lvova-Katchanova, 1972; see also Brock, 1984; Izvekova, 2000).

MICROHABITAT

The larvae are found on nearly all types of plants (Dvořák & Best, 1982; Dvořák, 1996; own observations). They are also common between filamentous algae (Meuche, 1939) and on decomposing leaves on the bottom (Brock, 1984). Higler (unpublished data) found them abundantly on artificial plants and they can also be numerous on stones, although the numbers there are often lower (Mol et al., 1982; Peeters, 1988; Buskens, unpublished data; cf. Verneaux & Aleya, 1998). In summer they are scarce on the bottom, but Olafsson (1992) observed in a lake in southwest England that third and fourth instars of *E. albipennis* showed vertical migrations in autumn and built cocoons 6–8 cm deep in the bottom from September to March. In winter some larvae can be found still in their mines in the decaying leaves (Thienemann, 1954:173). In the Netherlands we also found many larvae on the bottom in winter, but most larvae did not build cocoons.

DENSITIES

In May Izvekova (2000) found more than 4000 larvae/m^2 on dead terrestrial plants. Orendt (1993) found from 2 to 120 emergent adults/0.5 m^2 in eutrophic lakes in Bavaria.

WATER TYPE

Current

E. albipennis is a typical stagnant water species, but the larvae are found regularly in small numbers in lowland brooks with very slow current and much more commonly in the littoral vegetation of small rivers (Brabantse Delta, unpublished; own data). The species is not often collected in large rivers, but there are several records from the Rhine and Meuse (Klink & Moller Pillot, 1982; Peeters, 1988).

Dimensions

Steenbergen (1993) collected significantly fewer *E. albipennis* in water bodies less than 4 m wide or less than 50 cm deep. The larvae are more common in large sand pits than other species of the genus (Buskens, unpublished data). Humphries (1936) collected the larvae in low numbers down to 60 m deep in the profundal zone of Lake Windermere. They are also collected, though rarely, in Lake IJssel up to more than 1 km from the coast (Maenen, 1983).

Permanence

E. albipennis has been reared regularly from flooded grassland (Steinhart, 1998; H.v.d.Hammen, unpublished), but there are no records from temporary water, except from water bodies flooded by streams.

SOIL

In the Dutch province of Noord-Holland *E. albipennis* is significantly less present on sand and most common on peat (Steenbergen, 1993). In the whole country the species is relatively scarce on sand, but is the most common species of the genus on clay.

pH

E. albipennis has been collected regularly in acid moorland pools in the Netherlands and sometimes at pH < 4 (Duursema, 1996; van Kleef, unpublished). In sand pits in the Netherlands Buskens (unpublished) found the larvae rarely in acid water with a pH below 6.0. The absence of the larvae may be caused by the absence of suitable plant species (Bijlmakers, 1983). In contrast to *E. tendens*, the species appears to be absent from peat pools, even when plants such as *Juncus effusus* are present (Werkgroep Hydrobiologie, 1993).

TROPHIC CONDITIONS AND SAPROBITY

Saether (1979) reported the occurrence of *E. albipennis* in oligotrophic to strongly eutrophic lakes. Duursema (1996) even collected some larvae in acid moorland pools with less than 0.03 mg orthophosphate/l. However, Ruse (2002) did not find the species in water with low conductivity and most authors stated an obvious preference for eutrophic conditions (e.g. Mol, 1982a; Orendt, 1993; Langdon et al., 2006). Steenbergen (1993) recorded a mean chlorophyll-A content of 123 µg/l.

Bazerque et al. (1989) found the larvae abundantly in rather seriously polluted water in the river Somme in France. The authors (in agreement with Wilson & Ruse, 2005) attribute the species to the moderately tolerant species (group D). We found the larvae still present (in low numbers) in ditches with an oxygen saturation percentage of less than 5% in the water column every night in summer. Without doubt this is possible because in summer the larvae do not live on the bottom. Steenbergen (1993) stated that the species is more common under conditions with a better oxygen content.

SALINITY

Steenbergen (1993) stated that *E. albipennis* is more tolerant of brackish water than other species of the genus. The larvae were less common only in water bodies with a chlorinity > 1000 mg/l.

DISPERSAL OF LARVAE

Brown et al. (1980) stated that *E. albipennis* was consumed by trout more than other Chironomini species, especially in early autumn. In agreement with Kalugina (1959) they observed that at the least disturbance the larvae leave their tubes and migrate through the water column. Mundie (1965) found many larvae in the water mass of lakes from May to October and supposed that the larvae make vertical migrations to escape deoxygenation of the bottom water.

Endochironomus gr. dispar

Endochironomus dispar Kruseman, 1933: 140–142
Synendotendipes dispar Saether & Spies, 2004
Endochironomus impar Kruseman, 1933: 142; Kalugina, 1961: 916–917
Synendotendipes impar Saether & Spies, 2004

IDENTIFICATION

No method has yet been found to distinguish the larvae in second instar from those of *E. albipennis*. The teeth of the mentum look quite similar, as in fourth instar of *E. albipennis* (Kalugina, 1959: fig. 1r).

DISTRIBUTION IN EUROPE AND THE NETHERLANDS

Endochironomus dispar and *E. impar* have been recorded from most parts of the European mainland. However, *E. impar* in particular may be absent from large parts of the Mediterranean area (Saether & Spies, 2004). Adult males of both species from the Netherlands have been identified by Kruseman (1933). The larvae of gr. *dispar* live everywhere in the Netherlands, but they are scarce in brackish water regions and not very common on clay (Nijboer & Verdonschot, unpublished).

LIFE CYCLE

Kalugina (1961) reported two generations a year near Moscow, but Shilova (1976) supposed that near Borok the second generation could be incomplete. As in *E. albipennis*, she found no strong synchronisation in *E. impar*. Kruseman (1933) collected *E. impar* from April to September and *E. dispar* mainly in summer. The larvae hibernate in third and fourth instar.

FEEDING

Walshe (1951) observed that the larvae are leaf miners, filtering planktonic algae with a net constructed in their tubes. Brock (1984) found many larvae of *E.* gr. *dispar* in litter bags with decomposing leaf blades of *Nymphoides peltata*. (especially in shallow water, 20 cm deep, between helophytes). Soszka (1975a) found mainly periphytic algae as well as detritus in the guts of the larvae. Relatively more detritus was eaten in winter.

MICROHABITAT

Kalugina (1961) found the larvae of E. impar mining in vascular plants and among decaying plant material on the bottom. However, many larvae do not mine, but live on the surface of plants or on bottoms with coarse organic material (Bijlmakers, 1983). Bijlmakers found the larvae rarely on silt or sand. The larvae are very rare on stones in rivers (Smit 1982, Peeters 1988).

WATER TYPE

Current

The larvae are scarcely recorded from lowland brooks and more regularly in ditches and canals with very slow currents (Brabantse Delta, unpublished; Limnodata.nl). The larvae are rare in the river Meuse (Smit 1982, Peeters 1988).

Dimensions

The larvae are often found in small pools, but also in lakes (e.g. Brundin, 1949). Buskens (unpublished) collected the larvae very rarely in sand pits in the Netherlands, possibly because of a lack of sheltered habitats. Steenbergen (1993) found the larvae in the province of Noord-Holland no more often in ditches and canals than in lakes.

Permanence

Schleuter (1985) collected *E. impar* more often in temporary pools than any other *Endochironomus* species.

SOIL

The species group is scarce on clay and significantly more common on sand (Steenbergen, 1993). This also appears from the distribution of records across the different regions of the Netherlands (Nijboer & Verdonschot, unpublished data).

pH

E. gr. *dispar* is more common in acid moorland pools than *E. albipennis* (Leuven et al., 1987; Vallenduuk, 1990), but Leuven at al. collected the larvae only scarcely at pH values lower than 5. Duursema (1996) never found the larvae at pH < 4, but in 60% of the pools with a pH > 4.5. Its absence in many acid moorland pools seems to be caused more by the scarcity of a suitable microhabitat than by pH; for instance Bijlmakers (1983) rarely found the larvae in the acid Staalbergven moorland pool, where plants like *Typha* were absent. Steenbergen often found the larvae, sometimes abundantly, in water with a pH up to 7.5, but sharply decreasing at higher pHs.

TROPHIC CONDITIONS AND SAPROBITY

The larvae live in oligotrophic and eutrophic water, but are more common when the phosphate content is not very low (Brundin, 1949; Duursema, 1996). Steenbergen (1993) collected the larvae in less eutrophic water than *E. albipennis.* They were more common when the orthophosphate content was less than 0.05 mg P/l and the chlorophyll-A content not very high.

There are few records from water with severe pollution, but the larvae also live in water bodies in which the oxygen content sometimes falls to zero in summer (Prov. Zuid-Holland, unpublished; Grontmij | Aqua Sense, unpublished). When living on plants they are not affected by a rotting silt layer on the bottom. The fourth instar larva have a deeper red colour than *E. albipennis*, indicating a higher haemoglobin content.

SALINITY

Steenbergen (1993) collected the larvae in the province of Noord-Holland scarcely in brackish water and very rarely in water with a chloride content > 1000 mg/l.

DISPERSAL

Kalugina (1959) observed even older larvae of this species leave their shelter at night and swim in the water column, as observed in many other Chironomini (see Vallenduuk & Moller Pillot, 2007: 17–18).

Endochironomus lepidus (Meigen, 1830)

Endochironomus spec. Ubbergen Moller Pillot, 1984: 125
Endochironomus II Pinder & Reiss, 1983: fig. 10.25
Synendotendipes lepidus Saether & Spies, 2004

IDENTIFICATION

As stated in the introduction to the genus, the identity of the larva has not been proved convincingly (only a male pupa with larval skin has been identified, see van der Velde & Hiddink, 1987). The exuviae cannot be identified using Langton (1991).

DISTRIBUTION IN EUROPE AND THE NETHERLANDS

E. lepidus has been recorded from many European countries except large parts of the Mediterranean area (Saether & Spies, 2004). In the Netherlands the larvae have been collected scarcely (at about 60 localities), but scattered over nearly the whole country (Limnodata.nl; own data). Kruseman (1933) identified adult males from different parts of the country.

LIFE CYCLE

Van der Velde & Hiddink (1987) stated two well separated generations near Nijmegen. The pupae were collected from the end of May until early July and from early August until the middle of September. Most larvae appear to hibernate in third instar because the majority of them were still in third instar in the first half of May. However, larvae were found in fourth instar as early as October. The relatively late development of the larvae in spring must be an adaptation to the development of the food plant in relation to the nutritive value of their stem tissues.

FEEDING

In contrast to the other species of the genus, the larvae are phytophagous, feeding on the petioles of floating leaves of *Nuphar lutea* in spring and mainly on peduncles of this plant in summer. The plant forms characteristic galls when inhabited by the larva (van der Velde & Hiddink, 1987).

MICROHABITAT

As mentioned above, the larvae are known to live as miners in the petioles and peduncles of *Nuphar lutea*.

WATER TYPE

E. lepidus has been recorded mainly from oligotrophic to eutrophic lakes and pools (Brundin, 1949, ident. adult males; van der Velde & Hiddink, 1987, ident. male pupa with larval skin; van Kleef, unpublished, ident. exuviae by P. Langton). In last case the pH was only 4.3. Schleuter (1985) collected the species (adult males) in a few temporary pools. There are also some records from slow-flowing regulated lowland brooks (Peters et al., 1988, ident. larvae).

Endochironomus tendens (Fabricius, 1775)

Tendipes signaticornis Gripekoven, 1913: 185–188, fig. 41–42

IDENTIFICATION

See the comments under the genus *Endochironomus*. As noted by Shilova (1976), data recorded by Reiss (1968) relate to *E. albipennis* because at that time Reiss did not know of the work by Kalugina (1961).

DISTRIBUTION IN EUROPE AND THE NETHERLANDS

E. tendens lives in the whole of the European mainland (Saether & Spies, 2004). The species occurs everywhere in the Netherlands, but is scarce in clay regions (Nijboer & Verdonschot, unpublished).

LIFE CYCLE

Kalugina (1961) reported two generations a year near Moscow. The synchronisation is probably not very strong (as in other *Endochironomus* species); see Shilova (1976). Kruseman (1933) and Schleuter (1985) collected adult males from May to September; Schleuter observed three emergence maxima in a pool in Germany. The larvae hibernate in third and fourth instar (own unpublished data).

FEEDING

According to Walshe (1951) the larvae are leaf miners or spin their tubes on the leaves of plants. They filter planktonic algae by spinning a net in their tubes. Gripekoven (1913) observed that they also eat leaf tissue. In all probability they also feed on detritus, as do the other *Endochironomus* species, and possibly more so than the other species because *E. tendens* lives abundantly in acid water with little phytoplankton. The claim by Gaevskaya (1969: 96) that the larvae are obligate phytophages feeding mainly on the live tissue of higher aquatic plants is incorrect.

MICROHABITAT

The larvae are leaf miners or spin their salivary tubes on the surface of leaves (Walshe, 1951). Our own observations suggest that the larvae only penetrate into soft plant tissues or damaged plants. According to Meuche (1939) the larvae are only found between filamentous algae by accident, in contrast to *E. albipennis*. They can be found on very slender plants (e.g. *Glyceria fluitans*) as well as on firm species (Gripekoven, 1913). Many authors have observed the larvae in or on decaying plant material on the bottom (e.g. Gripekoven, 1913; Kalugina, 1961; Shilova, 1976; Izvekova, 2000), but in summer the larvae are scarce on sandy and silty bottom and on stones (Reiss, 1968; Bijlmakers, 1983; Buskens, unpublished). Although the larvae are very common in acid moorland pools, they are found in peat cuttings only if firm plants such as *Juncus effusus* are present (Werkgroep Hydrobiologie, 1993; cf. Waajen, 1982).

D

WATER TYPE

Current

The larvae are scarce in flowing water, but recorded in lowland brooks more frequently than other *Endochironomus* species (Gripekoven, 1913; Peters et al., 1988; Limnodata.nl; own unpublished data). The larvae are more numerous in the littoral zone of small rivers (Brabantse Delta, unpublished). They are very rare in the river Meuse (Klink & Moller Pillot, 1982; Peeters, 1988).

Dimensions

In contrast to other species of the genus Steenbergen (1993) found the larvae more frequently in smaller water bodies (the larvae were recorded in 427 localities). Buskens (unpublished data) collected the species in sand pits less numerously then *E. albipennis.*

Permanence

The larvae are very rare in temporary water (Schleuter, 1985; Duursema, 1996; Moller Pillot, 2003).

SOIL

E. tendens is significantly more common on peaty soil and scarce on clay (Steenbergen, 1993).

pH

E. tendens is very common and often abundant in acid moorland pools and peat cuttings (Leuven et al., 1987; Werkgroep Hydrobiologie, 1993; Duursema, 1996; van Kleef, unpublished) and is also found in other acid water bodies like sand pits (Buskens & Verwijmeren, 1989). Duursema (1996) collected the larvae in nearly all the investigated permanent moorland pools, also when the pH was lower than 4. On the other hand, the species is still rather common in water bodies with a pH > 8.5 (Steenbergen, 1993).

TROPHIC CONDITIONS AND SAPROBITY

The larvae can be very numerous in oligotrophic and mesotrophic water (Brundin, 1949; Duursema, 1996), but also often live in eutrophic water (Steenbergen, 1993). In regulated streams the species is more or less characteristic of a rather low level of organic pollution (Peters et al., 1988). The larvae seem to be hardly affected by a lower oxygen content (Steenbergen, 1993) and in two cases they have been found in a ditch in which the oxygen content can drop to zero in summer (Prov. Zuid-Holland, unpublished; Grontmij | Aqua Sense, unpublished). There are no records from severely polluted water.

SALINITY

The larvae are rare in water with a chloride content higher than 1000 mg/l (Krebs, 1984; Steenbergen, 1993). In the Camargue Tourenq (1975) found a maximum content of 2000 mg/l.

Fleuria lacustris Kieffer, 1924

IDENTIFICATION

The adult male was absent from the keys in Pinder (1978), but has been illustrated in Langton & Pinder (2007 vol. 2: 163, fig. 271). The female has been keyed and figured by Rodova (1978). Exuviae can be identified using Langton (1991). The larva can be distinguished from *Einfeldia* by the small field of spinules between the labral lamella and the S I (see Pinder & Reiss, 1983; Moller Pillot, 1984). The additional toothlets on the surface of the pecten epipharyngis can be absent (Pinder & Reiss, 1983), but seem to be present in other populations (cf. Kiknadze et al., 1991: fig. 40).

DISTRIBUTION IN EUROPE AND THE NETHERLANDS

F. lacustris has been found in many countries scattered over the whole European mainland (Saether & Spies, 2004). In the Netherlands the species has been recorded from the Holocene parts of the country at about 50 localities (Moller Pillot & Buskens, 1990; Nijboer & Verdonschot, 2001 and unpublished data; Limnodata.nl). There is one record from the Pleistocene areas, in the province of Limburg (Limnodata.nl, det. O. Duijts).

LIFE CYCLE

According to Bakhtina (1980) and Otto (1991) *F. lacustris* has one generation a year. Adults emerged in Germany at the end of May and in June, and in Russia mainly in June, but a low number still in July or early August. Young larvae of the new generation appeared in July. Lindegaard & Jónsson (1987) reported two generations in Denmark, emerging from June until early August. These authors obtained similar results to Bakhtina (1980) but interpret them differently. In winter all larvae were in third instar, but we found larvae in fourth instar in early March. In the Netherlands a second generation seems to be rare. Bakhtina (1980) noted that about 70% of the larvae survived when the pond bed was dry during winter (in Russia!).

FEEDING

Lindegaard & Jónsson (1987) suggested that the larvae fed on sedimented phytoplankton. Schlee (1980) observed only the presence of filamentous algae and in Bergambacht (Netherlands) the main food seemed to be detritus.

MICROHABITAT

As far as is known, the larvae live mainly on organic silt (often thick layers).

DENSITIES

Bakhtina (1980) found about 10,000 to 30,000 young larvae/m^2 in carp ponds in July, and in spring a maximum of 4000 larvae/m^2. Lindegaard & Jónsson (1983) found densities up to more than 10,000 larvae/m^2 in Hjarbæk Fjord, Denmark. Schlee (1980) noted that large differences in colonisation by *Fleuria lacustris* can occur within and between ponds. Every year the species was numerous in one pond (or at one side of a pond), but entirely absent elsewhere. We found the same situation in ditches in Bergambacht (the Netherlands), where a large population was found only very locally; in adjacent, apparently similar, ditches very few or no larvae were present. Such mass development has never been found elsewhere in the Netherlands.

WATER TYPE

Fleuria lacustris is a stagnant water species which lives in permanent water bodies of very different dimensions, such as ponds, lakes, ditches and canals.

pH

Fleuria lacustris is able to live in water with a very high pH (10.5–11), such as Hjarbæk Fjord, Denmark (Lindegaard & Jónsson, 1983). Steenbergen (1993) collected the larvae only in water with a pH > 7.5.

TROPHIC CONDITIONS AND SAPROBITY

Schlee (1980) found the species in a very eutrophic pond. Lindegaard & Jónsson (1983, 1987) observed mass development in water with a very high phytoplankton production (net primary production of 17,724 kJ/m^2 /y). It is most likely that a second generation failed to emerge if a large population of *Daphnia hyalina* reduced the sedimentation of algae.

In a polluted ditch near Bergambacht (Netherlands) the larvae lived in water which contained almost no oxygen during summer nights (own data). Auch Schlee (1980) reported a very low oxygen content.

SALINITY

In Hjarbæk Fjord, Denmark, the larvae were absent in water with a salinity of 1‰ (= 540 mg Cl/l) and appeared only after almost complete freshening (Lindegaard & Jónsson, 1983). Steenbergen (1993) collected the species only in water with less than 1000 mg Cl/l. In the western part of the Dutch province of Noord-Brabant the larvae are not rare in slightly brackish water up to 1000 mg Cl/l (data from Brabantse Delta).

D

Glyptotendipes Kieffer, 1913

SYSTEMATICS

The genus *Glyptotendipes* has been divided into three subgenera by Heyn (1992). However, the names of these subgenera caused problems and were ultimately established by Spies & Saether (2004: 50) as follows: *Glyptotendipes* Kieffer, 1913 (syn. *Phytotendipes*), *Caulochironomus* Heyn, 1993 and *Heynotendipes* Spies & Saether, 2004 (syn. *Trichotendipes* Heyn, 1993). In Contreras-Lichtenberg (2001) the subgenus *Caulochironomus* is named *Glyptotendipes*.

IDENTIFICATION

Adult males can be identified using Contreras-Lichtenberg (1999, 2001) and Langton & Pinder (2007). Langton & Visser (2003) is the most complete key for exuviae. However, these publications do not treat all species and there are many problems with nomenclature, which are described below. Most females have been described and keyed by Rodova (1978). Larvae in third and fourth instar can be identified using Vallenduuk (1999) and Moller Pillot et al. (2000). Kalugina (1975) noted that the genus can be recognised as larvae because they become green after fixation in alcohol and the head is more or less brown.

ECOLOGY

Reiss (1978a) suggested that the genus is probably thermophilous and is therefore absent from alpine lakes above 1000 m.

Glyptotendipes aequalis Kieffer, 1922

IDENTIFICATION

The male, female, pupa and larva have been described by Hirvenoja & Michailova (1991). However, Spies & Saether (2004) are not sure if the species described by Hirvenoja & Michailova is indeed Kieffer's species.

DISTRIBUTION IN EUROPE

G. aequalis has been collected in Germany (Westfalen) and Finland (Hirvenoja & Michailova, 1991).

MICROHABITAT

The German larvae were mining in the stem of *Alisma*. The Finnish larvae had overwintered among leaves and in mud (Hirvenoja & Michailova, 1991).

Glyptotendipes barbipes (Staeger, 1839)

Glyptotendipes polytomus Chernovskij, 1949: 69

IDENTIFICATION

A key for male adults can be found in Langton & Pinder (2007); a description and key for females is given in Contreras-Lichtenberg (1996), for exuviae in Langton (1991) and for larvae in second, third and fourth instar in Vallenduuk (1999) and Moller Pillot et al. (2000).

DISTRIBUTION IN EUROPE AND THE NETHERLANDS

Apart from the Mediterranean area and Norway and Sweden, the species has been reported from the whole European mainland (Saether & Spies, 2004). The species is especially common in the western provinces of the Netherlands, but not only in the brackish regions (Steenbergen, 1993; Limnodata.nl). There are scattered records elsewhere throughout the country.

LIFE CYCLE

In the Netherlands pupae, exuviae and adults can be found from May (rarely April) to October (Kruseman, 1933; own data). In the Camargue (southern France) Tourenq (1975: 188, 250) sometimes observed three generations, emerging from April to October or November. The larvae overwinter in second, third and fourth instar (own data).

FEEDING

Monakov (2003) reported the species (as *G. polytomus*) to be a filter feeder of detritus, planktonic and epiphytic algae. The larvae are probably also partly grazers, as reported for *G. pallens*. Moog (1995) called the larvae mainly detritus feeders and only to a lesser extent filterers.

MICROHABITAT

G. barbipes is mainly a bottom dweller (own data). According to Tourenq (1975) the larvae live mainly on bottoms with coarse organic material, where they construct tubes. In severely polluted water, however, many more larvae can be found on stones or other hard substrates, where pollution is less problematic. Driver (1977: 130) found the larvae in the axils of the plant *Scolochloa festucacea* and on *Myriophyllum* and *Potamogeton*.

SOIL

In the Dutch province of Noord-Holland Steenbergen (1993) collected the larvae significantly more often on clay and rarely on sand.

WATER TYPE

Current

Glyptotendipes barbipes is practically absent from most European brooks, streams and rivers (e.g. Lehmann, 1971; Braukmann, 1984; Caspers, 1991; Becker, 1994). However, in the Pleistocene parts of the Netherlands the larvae have been found mainly in (very) slow-flowing streams and canals (Limnodata.nl; own data). In the Holocene areas it is a normal inhabitant of stagnant water. Becker (1994) reared one male from the littoral zone of the river Rhine.

Dimensions

In the Dutch province of Noord-Holland Steenbergen (1993) collected significantly fewer larvae in ditches or other water bodies less than 4 m wide and/or less than 30 cm deep. However, in the delta region Krebs (1981, 1984) found them both in small drinking pools and ditches as well in larger creeks, pools and watercourses, with no clear preference for one or the other. They also occur regularly in southern Finland in small rock pools, although not abundantly (Lindeberg, 1958).

Permanence

The species has been recorded rather often from temporary water, but more rarely from pools where the bottom dries out completely (Tourenq, 1975; Driver, 1977; Krebs & Moller Pillot, in prep.).

pH

G. barbipes has rarely been recorded in acid water (e.g. Koskenniemi & Paasivirta, 1987). Most records are from water with a pH above 7.5 (Steenbergen, 1993; Limnodata.nl).

TROPHIC CONDITIONS AND SAPROBITY

Cannings and Scudder (1978) collected the species in Canada in very eutrophic lakes rich in blue-green algae. Izvekova et al. (1996) reported *G. barbipes* to be a typical species accompanying *Chironomus piger* in heavily polluted (polysaprobic) streams which contained hardly any other chironomids. In severely polluted water we sometimes found the larvae only on stones and in other cases also on the anaerobic bottom. However, the larvae appeared to be less resistant to anoxia than *Chironomus* larvae; as bottom dwellers they have more problems with anoxia than *G. pallens*, which lives more on plants.

Ten per cent of the records mentioned by Limnodata.nl are from water bodies where less than 2 mg oxygen/l had been measured during the day. Steenbergen (1993) collected the larvae a little less often in water with stable oxygen conditions and significantly less in water with a chlorophyll-A content less than 20 µg/l.

SALINITY

Parma & Krebs (1977) and Krebs (1978, 1984) reported the occurrence of the species at chloride contents between 1000 and 5000 (rarely 9000) mg/l. In the Camargue Tourenq (1975) found a maximum chloride content of 15,000 mg/l. Cannings & Scudder (1978) found the larvae in large numbers in Canadian lakes with up to 3‰ salinity and in lower numbers in more saline lakes. The larvae are not rare in fresh water, especially in regions where brackish water is also present.

Glyptotendipes caulicola (Kieffer, 1913)

SYSTEMATICS AND IDENTIFICATION

The species belongs to subgenus *Caulochironomus* Heyn, 1993. Identification of male adult, exuviae and larva is possible using Contreras-Lichtenberg (2001). The male is absent from the keys by Pinder (1978) and Langton & Pinder (2007). The female has been described and keyed by Rodova (1978: 72, 81, fig. 24 A-E). Identification of the exuviae using Langton (1991) is not always possible because the differences between his *G. foliicola* and *G. caulicola* are not clear.

Larvae can be identified using Vallenduuk (1999) and Moller Pillot et al. (2000), as mentioned under the genus. In some Dutch publications all larvae of the subgenus *Caulochironomus* are named *Glyptotendipes* cf. *caulicola* because only this name is used in Moller Pillot (1984). Some workers have omitted the 'cf.' and have thus created some confusion.

DISTRIBUTION IN EUROPE AND THE NETHERLANDS

C. caulicola has been reported from only a few European countries, probably because of the problems with the identification of the species (Saether & Spies, 2004). The species may be absent from the Mediterranean area and Scandinavia. There are only a few verified records from the Netherlands; in addition to the fenland areas in Holland and Friesland-Overijssel, we only have data from the provinces of Gelderland and Noord-Brabant.

MICROHABITAT

The larvae are miners, mainly in *Stratiotes* and sometimes in other plants, such as

Alisma plantago-aquatica, *Nuphar lutea*, *Iris* and *Sparganium* (Contreras-Lichtenberg, 2001; own data).

pH

G. caulicola has been collected at least once in acid water at pH 5.4 (van Kleef, unpublished). All other records are from non-acid water bodies.

SOIL

Steenbergen (1993) collected the larvae of the subgenus *Caulochironomus* almost only on peat. This also applies to our own records of *G. caulicola* in the Netherlands.

Glyptotendipes cauliginellus (Kieffer, 1913)

Glyptotendipes gripekoveni Kalugina, 1975: 1830–1834; Pinder, 1978: 124, fig. 59I, 153D; Langton, 1991: 230, fig. 96i; Vallenduuk, 1999: 3 et seq.; Moller Pillot et al., 2000: 11 et seq.

NOMENCLATURE

Spies & Saether (2004: 60–62) stated that *G. cauliginellus* is the valid name for the species usually named *G. gripekoveni*, and since then this view has been generally accepted. Chernovskij (1949) used the name *Glyptotendipes* gr. *gripekoveni* for all species of the genus without tubuli. This name has been used in this sense in Eastern Europe for a long time. Owing to a mistake this group name was used in the Netherlands from 1975 to 1979 for larvae of different genera with one pair of tubuli.

DISTRIBUTION IN EUROPE AND THE NETHERLANDS

G. cauliginellus has been reported from the whole European mainland (Saether & Spies, 2004). In the Netherlands the species has been collected throughout the country, but mining chironomids have not been investigated in most regions. The species is probably particularly common in the Holocene part of the country.

LIFE CYCLE

The egg mass and first instar larva have been described by Kalugina (1963: 902, fig. 1r, 2). There are two generations a year (Kalugina, 1963a; Shilova, 1976). Janecek (1995) collected adults and exuviae from June until early September.

FEEDING

As a rule the larvae are leaf miners, filtering the water with the help of a net (Walshe, 1951; Izvekova, 1980). They feed on detritus, green algae and diatoms and only rarely eat a small animal by accident if it comes into the tube. The opinion of Gaevskaya (1969: 96) that the larvae are obligate phytophages feeding mainly on the live tissue of higher aquatic plants is incorrect.

MICROHABITAT

According to Gripekoven (1913), Kalugina (1963a) and Shilova (1976) the larvae are miners in decaying wood and plants. In wood they are less numerous than *G. pallens* agg., but in plants they can be more numerous. They have been found on *Stratiotes*, *Scirpus*, *Typha*, *Phragmites*, *Sagittaria*, *Sparganium* and *Alisma*. In first and second instar the larvae live in tubes on the surface of plants (Izvekova, 1980: 84). We also found larvae in fourth instar in tubes on *Phragmites* stems.

DENSITIES

Kalugina (1963a) found 660 larvae/m (stem length) on stems of *Scirpus* and 130 larvae/m on *Typha*.

WATER TYPE

We collected no larvae in flowing water. However, Moog (1995) called the species from the lower courses of streams as common as *G. pallens*. According to known data from the Netherlands, the larvae are rare or absent from temporary water.

pH

G. cauliginellus is rare in acid water. We found only one record from water with a pH lower than 6: 2 exuviae collected in the moorland pool De Banen near Nederweert, the Netherlands, at pH 4.85 (van Kleef, unpublished). The species has also been collected in other moorland pools with pH values fluctuating around 6 (e.g. Brabantse Delta, unpublished).

TROPHIC CONDITIONS AND SAPROBITY

The larvae are often collected in very eutrophic water, but there are no records from severely polluted water, although Gripekoven (1913: 162, 166) found the larvae mainly in 'natural polluted' water. The figures in the matrix table in Chapter 4 are tentative.

SALINITY

Although nearly all records in the Netherlands are from fresh water (Limnodata.nl), the species has also been collected in brackish water, even with chloride levels higher than 1000 mg/l (Ruse, 2002; own data). The figures in the matrix table are tentative.

DISPERSAL

Kalugina (1959) observed that older larvae of this species (as *G. gracilis*) also leave their shelter at night and swim in the water column. This behaviour has been observed in *G. pallens* and many other Chironomini (see Vallenduuk & Moller Pillot, 2007: 17–18).

Glyptotendipes foliicola sensu Kalugina, 1979

Glyptotendipes foliicola Rodova, 1978: 72, 81, fig. 24 Ж-Н (female)
nec *G. foliicola* Walshe, 1951 (larvae); Langton 1991: 226 (exuviae)

SYSTEMATICS

As Spies & Saether (2004) noted that *G. foliicola* Kieffer, 1918 is probably not the species described by Contreras-Lichtenberg (2001). We follow here Kalugina (1979), although these larvae probably belong to another species. We could not investigate the adult males collected by Kalugina. The description of the female by Rodova (1978) is of the species referred to here because Kalugina advised her on this genus (see Rodova, 1978: 4). We tried to rear these (very characteristic) larvae from Belarus but obtained only a prepupa with epaulettes. The larvae as described by Walshe (1951) and the exuviae of Langton (1991) belong to another species (possibly to *G. caulicola*?).

IDENTIFICATION

Identification is only possible for the female (using Rodova, 1978) and the larva (using Vallenduuk, 1999, and Moller Pillot et al., 2000).

DISTRIBUTION IN EUROPE

According to Contreras-Lichtenberg (2001), *G. foliicola* has been collected at least in Germany and France. Because it is not sure that the larvae described by Kalugina (1979) belong to the same species as the verified adults, only the records of Kalugina from Russia and our records from Belarus belong without doubt to this taxon.

LIFE CYCLE

The larvae most probably have only one generation a year; they emerge in spring from temporary pools. A summer diapause is probable.

MICROHABITAT

The larvae live on decaying leaves on the bottom of temporary pools (own observations).

WATER TYPE

Kalugina (1979) collected the larvae in Russia in pools with melted water. We also found the larvae locally very common in temporary woodland pools in Belarus.

Glyptotendipes glaucus (Meigen, 1818)

See under *Glyptotendipes pallens* agg.

Glyptotendipes imbecilis (Walker, 1856)

Glyptotendipes imbecillis Burtt, 1940: 118–119; Langton, 1991: 228

SYSTEMATICS AND NOMENCLATURE

Kalugina (1975) gives two differences between adult males of *G. imbecilis* and *G. viridis*: the foreleg of *G. viridis* has no beard and the hypopygia are slightly different. The larvae are also different (see also Pankratova, 1983; Vallenduuk, 1999). Contreras-Lichtenberg (2001) considers *G. imbecilis* to be a synonym for *G. viridis* (Macquart, 1834). However, Spies and Saether (2004) argue that her evidence is insufficient and that most probably two or more species do exist, one of which has to be named possibly *G. imbecilis*. For instance Walshe (1951) observed that the two species exhibited different behaviours. Until the type material has been investigated the name *G. imbecilis* should be retained. In this book *G. imbecilis* and *G. viridis* are treated separately, although identification by some authors can sometimes be wrong.

D

IDENTIFICATION

G. imbecilis is not keyed separately in the keys by Contreras-Lichtenberg (see above). For adult males the keys by Pinder (1978) or Langton & Pinder (2007) can be used. Langton (1991) has keyed the exuviae, but *G. viridis* is absent in his book. The larvae can be identified using Vallenduuk (1999) or Moller Pillot et al. (2000). These authors follow the interpretation by Kalugina (1975). The female has been described and keyed by Rodova (1978).

DISTRIBUTION IN EUROPE AND THE NETHERLANDS

G. imbecilis has been reported from many countries throughout European (Saether & Spies, 2004). There are only a very few records from the Netherlands (Klink & Moller Pillot, 1996; van Kleef, unpublished).

MICROHABITAT

The larvae are exclusively leaf miners in living plants like *Sagittaria*, *Sparganium* and *Alisma* (Burtt, 1940; Kalugina, 1963a, 1975).

FEEDING

According to Burtt (1940) and Monakov (1972, 2003), the larvae are filter feeders of diatoms, planktonic algae and bacteria. According to Burtt (1940) the larvae make a tube like *G. glaucus* and feed in the same manner, although their tubes are smaller and their movements are more rapid.

WATER TYPE

Shilova (1976) collected the species in ponds, in the littoral zone of lakes and in slow-flowing water.

pH

There is one record from a moorland pool with a pH of 5.4 (van Kleef, unpublished). The other Dutch records are not from acid water.

Glyptotendipes ospeli Contreras-Lichtenberg & Kiknadze, 1999

Glyptotendipes spec. Ospel Vallenduuk, 1999: 27–29

IDENTIFICATION

The adult male, female, pupa and larva are described by Contreras-Lichtenberg & Kiknadze (1999) and Contreras-Lichtenberg (1999). They cannot be identified using Langton & Pinder (2007) or Langton (1991). The larva has been keyed by Vallenduuk (1999) and Moller Pillot et al. (2000).

DISTRIBUTION IN EUROPE AND THE NETHERLANDS

G. ospeli has been found only in the Netherlands, despite an international appeal for information by Tempelman (2002). The occurrence in Belarus reported by this author is based on an incorrect identification. In the Netherlands the species seems to be widespread: Tempelman (2002) gives 36 localities dispersed over the country.

MICROHABITAT

The larvae often live on wood, mainly wooden timbering along canals in towns and sometimes (or usually?) in the presence of Bryozoa (Tempelman, 2002).

FEEDING

From observations of the larvae, Tempelman (2002) inferred that they were collecting coarse organic material.

WATER TYPE

Most of the water bodies in which *G. ospeli* has been found are canals in towns with little or no current (Tempelman, 2002). There are also records from pools and other mostly stagnant water bodies.

pH

The pH measured in 22 localities varied from 6.0 to 8.5 (Tempelman, 2002). At Ospel (Netherlands), where the species was first collected, the larvae were only present in a pool with a pH of 6.0 and not in other more acid pools and peat cuttings.

TROPHIC CONDITIONS AND SAPROBITY

A significant proportion of the water bodies inhabited by *G. ospeli* and mentioned by Tempelman (2002) were polluted to a greater or lesser degree (ß-mesosaprobe). The oxygen content measured at 16 sites varied from 0.4 to 12.1 mg/l.

SALINITY

Tempelman (2002) sometimes collected *G. ospeli* in water with approx. 500 mg chloride/l. No records are known from more brackish water.

Glyptotendipes pallens agg.

Glyptotendipes pallens agg. sensu Vallenduuk, 1999; Moller Pillot et al., 2000
Glyptotendipes pallens Pinder, 1978: 124, figs. 59J, 154A
Glyptotendipes glaucus Kalugina, 1958: 1045–1057; Izvekova, 1980: 83

SYSTEMATICS AND IDENTIFICATION

The larvae of *Glyptotendipes glaucus* and *G. pallens* are extremely difficult to distinguish one from another. The differences are described by Michailova & Contreras-Lichtenberg (1995) and Klink & Moller Pillot (2003). Most of the information about the biology and ecology of these species has been published under one of these names, but may apply to the other species or to both species. We therefore take both species together and use the species names only when reliable information exists. All life stages of both species are described and keyed in Contreras-Lichtenberg (1999). Adult males can be identified using Langton & Pinder (2007), the females using Contreras-Lichtenberg (1996). Langton & Visser (2003) give the differences between the exuviae of both species. The larvae of the aggregate in second, third and fourth instar have been described and keyed by Vallenduuk (1999) and Moller Pillot et al. (2000).

DISTRIBUTION IN EUROPE AND THE NETHERLANDS

Both species have been reported throughout Europe (Saether & Spies, 2004). The aggregate is common nearly everywhere in the Netherlands. Only Kruseman (1933) distinguished between the two species; he collected mainly 'var. *glaucus*'.

LIFE CYCLE

Orendt (1993) recorded locally three generations, but elsewhere probably only two. Beattie (1982) reported two generations in Lake Tjeuke (Tjeukermeer), the Netherlands, but at lower temperatures the second generation came only partly to emergence and overwintered in fourth instar. Van der Velde & Hiddink (1987) probably found only two generations. The adults emerge from the end of April to early October. We collected larvae in second, third and fourth instar in winter (third instar dominating); the second instar is still present as late as early April. According to Beattie (1982) growth stops from December to March. Considering that the larvae do not display a clear synchronisation in winter, one cannot expect a good synchronisation during the whole year. The bad synchronisation in this species contradicts the supposition by Verberk (2008) that large species have a better synchronisation and overwinter in the final stage. The egg masses and first instar larvae have been described by Kalugina (1963).

FEEDING

The larvae are filter feeders and grazers, feeding on all that comes into the net (detritus, bacteria, algae, diatoms) and also decaying plant material; small animals are only

eaten when they come into the net by accident (Burtt, 1940; Walshe, 1951; Kalugina, 1958; Izvekova, 1980). Kalugina found them also feeding on decaying wood in large numbers. Although there are many indications that the larvae prefer algae, they can complete their development feeding only on detritus (Izvekova, 1980).
The larva spins a net, pumps water through the net for 5–10 minutes and eats the ensnared particles (Heinis et al., 1990). If too little food is obtained, the larvae scrape food from the area around the tube (Kalugina, 1958).

MICROHABITAT

In summer most larvae can be found in silty tubes on stems and leaves of plants like *Stratiotes, Potamogeton, Typha, Phragmites, Nymphaea* (Burtt, 1940; Walshe, 1951; Kalugina, 1958), more scarcely on tender plants such as *Callitriche*. They mine only in the decaying parts of these plants or in mines abandoned by the original inhabitants (Kalugina, 1958; van der Velde & Hiddink, 1987). Second instar larvae are not found in mines (Izvekova, 1980). Besides plants, many larvae can be found on other firm substrates, such as wood and stones, where they live in tubes constructed on the surface or in self-made mines within the decaying wood (Kalugina, 1958). Higler (unpublished data) also found the larvae to be equally or more abundant on artificial (plastic) '*Stratiotes*' plants than on true plants. In large rivers the larvae are more numerous on stones (and especially on artificial substrates) than on plants (Peeters, 1988; Klink, 1991).
In summer the larvae are scarce on different types of water bottoms. Mason & Bryant (1975) observed that the larvae settled on *Typha* stems in spring and left them as the periphyton declined during summer. In winter they lived on the bottom of the broad. Beattie (1978, as *G. paripes*) and Beattie (1982) reported that in the autumn they found larvae coming from the reed beds in the adjacent sand, but rarely in the centre of the lake. In Hungary Koskenniemi & Sevola (1989) found the larvae in organic sediment in the winter, some 3–7 cm deep. The larvae survived in frozen sediment better than most other chironomid species. In very polluted ditches we found the larvae on living and decaying plants even in winter, and not on the anaerobic bottom.

DENSITIES

On the bottom of Lake Tjeuke the larvae were present in and around the reed beds in densities from 500–9000/m^2 (Beattie et al., 1978, as *G. paripes*, see Beattie, 1982). Izvekova (2000) found up to 13,220 larvae/m^2 on decaying immersed terrestrial vegetation. We found 0–72 larvae per leaf on *Nuphar lutea*.

WATER TYPE

Current

The larvae are absent or nearly absent from many brooks and small streams (e.g. Lehmann, 1971; Lindegaard-Petersen, 1972; Tolkamp, 1980; Braukmann, 1984). However, Orendt (2002a) reported the occurrence (of exuviae?) in several fast-flowing streams in Bavaria. They are found regularly in large rivers, mainly on stones (Peeters, 1988; Becker, 1994; Klink, 1991). In the Netherlands the larvae appear to be often present in streams and canals with a very slow current and more scarcely where the current flows faster than 10 cm/sec (Limnodata.nl).

Dimensions

G. pallens agg. can be found in small and large water bodies, but seems to be often absent from narrow ditches and small pools (e.g. Schleuter, 1985; Steenbergen, 1993; Krebs & Moller Pillot, in prep.). In deeper lakes the larvae inhabit the littoral zones (Heinis & Swain, 1986). However, Koskenniemi (1992) found the larvae in the profundal zone of Lake Kyrkösjärvi in Finland, probably because of the unusual quality of the lake bottom. In flowing water they are restricted almost entirely to large rivers.

Permanence
G. pallens agg. can be found sometimes in temporary water (Dettinger-Klemm, 2003; own data), but this is rather exceptional. The females may not be attracted to small water bodies (see under Dimensions). The larvae are probably not rare in somewhat larger temporary pools. They commonly settle in floodplains and temporary pools flooded by rivers (Fritz, 1981; Steinhart, 1999). The figures in the matrix tables in Chapter 4 do not apply to very small water bodies.

pH

G. pallens agg. is rare in acid water and records from water with a pH below 5 are always for very few specimens (van Kleef, unpublished; own data).

TROPHIC CONDITIONS AND SAPROBITY

Heinis & Swain (1986) noted that the larvae of *G. pallens* agg. stopped filter feeding at oxygen concentrations below 3 mg/l. They are therefore not able to live in oxygen poor water for a long time. Nevertheless, we found the larvae sometimes numerously in ditches where the oxygen content dropped below 0.5 mg/l every night. In such cases the larvae lived only on plants and stones and not on the anaerobic silt.
In severely polluted water the larvae are usually less numerous than *G. barbipes*. However, in such cases the latter species requires the presence of stones or other hard substrates, whereas *G. pallens* agg. also lives on plants (own unpublished data). Bazerque et al. (1989) found fair numbers of the larvae in a polluted stretch of the river Somme in France; they attribute *G. pallens* agg. to the pollution-resistant species in lentic areas.

SALINITY

Krebs (1981, 1984) collected *G. pallens* agg. rarely at chloride contents above 1000 mg/l. Other sources also indicate that the aggregate appears to be nearly restricted to fresh and slightly brackish water (Steenbergen, 1993; Limnodata.nl; own data). The species have not been recorded in the Camargue (Tourenq, 1975).
Ruse (2002) found the species more often in water with a conductivity of about 1000 µS/cm.

DISPERSAL

Kalugina (1959) observed that even older larvae of this species leave their shelter at night and swim in the water column, as observed in many other Chironomini (see Vallenduuk & Moller Pillot, 2007: 17–18).

Glyptotendipes paripes (Edwards, 1929)

nec *G. paripes* Beattie, 1978; Beattie et al., 1978 (misidentification)

IDENTIFICATION

Keys and descriptions of all life stages can be found in Contreras-Lichtenberg (1999). The adult male can be identified using Langton & Pinder (2007), the female using Contreras-Lichtenberg (1996), the exuviae using Langton (1991) and the larvae in second, third and fourth instar using Vallenduuk (1999) and Moller Pillot et al. (2000).

DISTRIBUTION IN EUROPE AND THE NETHERLANDS

G. paripes has been reported from the whole of Europe. The same species also lives in the Nearctic (Saether & Spies, 2004) and in all parts of the Netherlands (Nijboer & Verdonschot, unpublished data; Krebs, 1981, 1984; own data).

LIFE CYCLE

According to Bakhtina (1980) *G. paripes* has three generations in the Moscow area and two generations in ponds near Leningrad where the water temperature was 3.2–4.0 ° C. lower. The adults near Leningrad emerged in June and in July/August. In Germany the species has two or locally three generations (Reiss, 1968; Schleuter, 1985; Orendt, 1993). In some cases the second generation is much more numerous (Schleuter, 1985; Orendt, 1993; cf. Otto, 1991). In western Europe adults emerge from May to September (German data) or from the end of April to the end of October (Buskens, 1989a; Janecek, 1995). In Canada Danks (1971a) found larvae in fourth instar cocoons in winter, but in Florida lakes the larvae continue to feed in winter (Provost & Branch, 1959).

OVIPOSITION

McLachlan & McLachlan (1975) found the eggs of *G. paripes* equally spaced around the periphery of a bog lake, although the larvae lived only in a very restricted area of the bottom. Balushkina (1987) mentions 1400 eggs per egg mass for *G. paripes.*

FEEDING

The larvae spin a filter net within their tube (Provost & Branch, 1959). They generate a flow of water through the net by undulating their bodies day and night. The net and its contents are then consumed and a new net is spun. Their main food are blue-green and green algae, but many Crustacea were eaten during winter. The authors give no information about the possible consumption of dead organic particles (compare with *G. pallens*). An increase in plankton seems to increase the production of *G. paripes.* Wundsch (1943) found only planktonic algae in the gut. Experiments by Rasmussen (1985) showed that the addition of organic detritus and/or the presence and density of *Chironomus riparius* had no influence on the population of *G. paripes*. This shows that these two species utilise totally different food, *G. paripes* only filtering fine particles from the water column.

Beedham (1966) found a *Glyptotendipes* larva, possibly *G. paripes*, within the body of the swan mussel *Anodonta cygnea*, which was probably a case of facultative parasitism.

MICROHABITAT

The larvae live in tubes on the lake bottom (Provost & Branch, 1959). In lakes in Florida the authors found the larvae most abundantly on sand and absent from muck. In a bog lake in England the larvae occurred only on coarse sediments and were absent from the sheltered shore of the lake (McLachlan & McLachlan, 1975). The most important reason for this was that they preferred to build their tubes from organic particles between 1100 and 1700 µm in diameter (McLachlan, 1976). However, on bottoms directly affected by wave action they could mine in compact peat bottoms and were then independent of the presence of such particles. Wundsch (1943) found the tubes in clumps on empty scales of molluscs and on roots of *Typha*. Buskens (1983, 1987) found *Glyptotendipes* larvae only if enough organic matter was present on the bottom, but the genus was absent in some bog pools. Particle size or quality of food may be the reason for this (see also under Feeding).

The larvae sometimes occur on stones or on wood (e.g. Becker, 1994), but we did not find them on plants. However, Reiss (1968) observed incidental cases of larvae mining in plants. Janecek (1995) found the larvae mainly in algal coverts on rocks.

DENSITIES

McLachlan (1976) stated that the occurrence of the larvae was strongly influenced by particle size and wave action in lakes and so densities varied strikingly between different places within the same lake. The maximum density he found was 22,000 larvae/m^2. Charles et al. (1974) stated varying densities between two years in Loch Leven in Scotland: about 10,000 larvae/m^2 in spring 1970 and only 500/m^2 in 1971. Bakhtina (1980) found a maximum of about 3000 larvae/m^2 in a fish pond near Leningrad in July. Wundsch (1943) reported densities of 400 to 2112 larvae/m^2.

WATER TYPE

Current

Becker (1994) found *G. paripes* regularly in the river Rhine, but the species appears to be very rare in the river Fulda (Lehmann, 1971) and in the Meuse (Klink, 1991). In brooks and small streams in the Netherlands the larvae and exuviae are often collected in very slow-flowing water, but very rarely at velocities above 10 cm/sec (Limnodata.nl; own data).

Dimensions

In eutrophic lakes in Berlin *G. paripes* was collected significantly more in the larger lakes (Frank, 1987). In the Netherlands the larvae are sometimes numerous in small acid pools (less than 5 x 5 m^2 , e.g. Waajen, 1982). Lindeberg (1958) stated that the species is typical for the community of small rock pools in southern Finland, often in abundance.

In a eutrophic lake in Berlin *G. paripes* was numerous in the littoral zone and did not migrate to deeper layers when the deeper parts became more oxygenated (Neubert & Frank, 1980). Elsewhere the species is also scarce or absent from the deeper parts of storage reservoirs and sand pits (Mundie, 1957; Mol et al., 1982; Kuijpers, pers. comm.). In very large lakes the distance to the shore can be important; see under Dispersal.

Permanence

The larvae can be present in temporary pools, mainly conditional on the possibilities for egg deposition, which does not take place during drought (Werkgroep Hydrobiologie, 1993). The presence of permanent populations in the neighbourhood most probably also plays a role (see under Dispersal).

pH

McLachlan & McLachlan (1975) found the larvae abundantly in a bog lake with a pH in winter of 3.6–3.9. High densities in acid pools are also often found in the Netherlands (e.g. Vallenduuk, 1990; Werkgroep Hydrobiologie, 1993; Moller Pillot, 2003). However, the species is not at all confined to acid water and in the Netherlands is fairly common in lakes with a pH of 8 or higher (Limnodata.nl). Buskens (1983, 1987) found *Glyptotendipes* (in medium-sized water bodies) less common in acid pools and most common in moderately eutrophic pools. An important factor appeared to be the presence of organic matter on the bottom (see Microhabitat). Up to now it is not clear why a filter feeder can be so common in very acid water.

TROPHIC CONDITIONS AND SAPROBITY

The larvae occur mainly in mesotrophic and eutrophic lakes (Saether, 1979; Orendt, 1993). In Florida Provost & Branch (1959) found *G. paripes* only in eutrophicated or polluted lakes, where the oxygen content at the bottom could drop to zero. In the Netherlands, however, the larvae are especially abundant in acid conditions, where the orthophosphate content is sometimes very low (e.g. Werkgroep Hydrobiologie, 1993;

van Kleef, unpublished). Curiously, Saether (1979) does not mention the occurrence of the species in mesohumic and polyhumic lakes. Important factors seem to be the high nitrogen content and/or the rather high rate of decomposition of organic material in Dutch acid moorland pools and peat cuttings. In some cases the investigations were conducted shortly after restoration measures, which led to increased decomposition. Data from Limnodata.nl show that the species occurs in water with a low to very high chlorophyll-A content (> 100 µg/l).
Bazerque et al. (1989) found the larvae abundantly in rather seriously polluted water in the river Somme in France. The authors attributed the species to the group of moderately tolerant species. In the Netherlands there are no records from severely polluted water. The differences between our results and those of Provost & Branch (1959, see above) may be because we have no data from polluted lakes where the water column always contains enough oxygen. The difference from *G. pallens* may be because *G. paripes* is a typical bottom dweller. Ali & Baggs (1982) found larvae at higher densities in a water cooling reservoir (300–8000 larvae/m^2) than in a neighbouring lake (<100–6000 larvae/m^2), probably because of the higher temperature in the reservoir. For interpretation of the presence or absence of this species it can be important to realize that *G. paripes* is a filter feeder.

SALINITY

In the delta region of the Netherlands *G. paripes* is rather scarce and rarely found in water with chloride concentrations higher than 1000 mg/l (Krebs, 1981, 1984). According to Limnodata.nl it is a typical freshwater species. Tourenq (1975) did not find the species in the Camargue. Ruse (2002) collected *G. paripes* in low numbers in a brackish lake.

DISPERSAL OF LARVAE

The young larvae apparently disperse easily from the place of egg deposition: they were found by MacLachlan (1976) only in a very restricted area of the lake even though the eggs had been laid equally spaced along the periphery of the bog lake. In very large lakes the distance to the shore may be more important. For instance, Maenen (1983) collected many more *Glyptotendipes* larvae in the first 100 m from the shore in the IJssel and Marker lakes and hardly any larvae further than 500 m from the shore (the species were not identified in this investigation). There are no records indicating whether the dispersion of *G. paripes* larvae is confined to a narrow zone, as in *G. pallens*. The young larvae of the latter species settle in the vegetation zone and move a little further from the vegetation only in autumn.

DISPERSAL OF ADULTS

Ali & Fowler (1983) caught 95–100% of the adults very near to the lake from which they emerged. Less than 5% flew more than 200 m from the source.

Glyptotendipes salinus Michailova, 1987

SYSTEMATICS AND IDENTIFICATION

G. salinus is very similar to *G. barbipes*. The species has been reported from brackish water in Bulgaria. In the Netherlands the exuviae of one specimen that appears to match the description of the species was collected by A. Klink (Klink & Moller Pillot, 1996) in a freshwater sand pit in the Noordoostpolder. Because it is not clear if this specimen really is the same species, we do not describe here the biology and ecology of this species. *G. salinus* has not been reported from other West European countries

(Saether & Spies, 2004), with the exception of the British Isles (Langton & Pinder, 2007). The male, female, pupa and larva have been described by Michailova (1987) and Contreras-Lichtenberg (1999). The adult male can be identified using Langton & Pinder (2007 vol. 2: 164) and the exuviae using Langton & Visser (2003).

Glyptotendipes scirpi (Kieffer, 1915)

Glyptotendipes mancunianus Brundin, 1949: 749; Kalugina, 1975: 1830–1837; Moller Pillot et al., 2000: 27, 47.
? *Glyptotendipes foliicola* Walshe, 1951: 64 et seq.

IDENTIFICATION AND NOMENCLATURE

Identification of adult males, pupae and fourth instar larvae is possible using Contreras-Lichtenberg, 2001. Vallenduuk (1999) and Moller Pillot et al. (2000) keyed the fourth instar larva. In Moller Pillot (1984) the whole subgenus *Caulochironomus* is only referred to as *Glyptotendipes* cf. *caulicola*. As a result, in Dutch publications this name has often been used for this subgenus (e.g. Steenbergen, 1993). The female has been described and keyed by Rodova (1978: 72, 80, fig. 23).

DISTRIBUTION IN EUROPE AND THE NETHERLANDS

The species has been recorded in large parts of Europe except for the Mediterranean area (Saether & Spies, 2004). In the Netherlands an important proportion of all records of the subgenus *Caulochironomus* concern *G. scirpi* (own data). This also applies to many records registrated as *Glyptotendipes* cf. *caulicola* (see above). The species seems to be most abundant in the seepage regions in the fenland area in Holland-Utrecht, northwest Overijssel and in a comparable narrow zone along the rivers. There are few records from the Pleistocene area (Limnodata.nl).

LIFE CYCLE

Kruseman (1933) collected adult males from April to August. There are probably at least two generations.

FEEDING

Walshe (1951) found larvae of *G. foliicola* (actually *G. scirpi*?) mining in *Stratiotes*, where its diet consisted of phytoplankton. The larvae spin an asymmetrical net within their tube and the bulk of their food comes from plankton derived from the outside water.

MICROHABITAT

The larvae live in mines in firm plants like *Sagittaria*, *Typha* (Kalugina, 1975), *Alisma* and *Sparganium* (own data). In the Netherlands Henk Vallenduuk (pers. comm.) never found the species in *Typha*. In rearing experiments larvae sometimes left their mines in *Alisma* stems, probably because of oxygen shortage. One larva lived temporarily in a tube on the outside of the stem. As a rule, larvae and pupae remained within the stems until emergence, now and then making ventilation movements.

WATER TYPE

The species lives mainly in stagnant water and has been found scarcely also in very slowly flowing streams. They are most common in fen-peat lakes, but also live in different types of pools and ditches (Limnodata.nl; own data).

SOIL

Steenbergen (1993) collected the larvae of the subgenus *Caulochironomus* almost only on peat. Elsewhere there are also some records from sand and clay.

pH

The species has not been recorded in acid water. However, some records of the subgenus from water with a pH of around 6 probably apply to this species.

TROPHIC CONDITIONS AND SAPROBITY

Brundin (1949) collected the species mainly in the very eutrophic lakes Växjösjön and Bergundasjön, where in winter the oxygen content under the ice was very low. The records from the Netherlands are from mesotrophic to very eutrophic water (Limnodata.nl; own data). The figures in the matrix tables are tentative because of a lack of information from more polluted water.

SALINITY

There are no records from brackish water. Steenbergen (1993) collected the subgenus very rarely in slightly brackish water.

Glyptotendipes signatus (Kieffer, 1909)

Glyptotendipes varipes Shilova, 1976: 117; Rodova, 1978: 76–77, fig. 22 A-E
Glyptotendipes gr. *signatus* Moller Pillot, 1984: 203 et seq.; Moller Pillot & Buskens, 1990

SYSTEMATICS

G. signatus is the only European representative of the subgenus *Heynotendipes* (earlier *Trichotendipes*). Contreras-Lichtenberg (2001) stated the synonymy of the names *G. signatus* and *G. varipes* and redescribed the adult male, pupa and larva. The female has been described by Rodova (1978) as *G. varipes*.

DISTRIBUTION IN EUROPE AND THE NETHERLANDS

G. signatus has been reported from few European countries (Saether & Spies, 2004); these countries however are scattered over most of Europe and the species possibly lives in the whole European mainland. In the Netherlands the species has not been collected very often (about 80 localities); there are no records from South Limburg and Zeeland (Limnodata.nl; Nijboer & Verdonschot, unpublished data; own data).

LIFE CYCLE

Schleuter (1985: 157, fig. 29) recorded one generation in a pond in Germany, emerging in June and July. Shilova (1976, as *G. varipes*) collected pupae and adults from early June until the end of August in Russia (Rybinsk); she supposed there were two generations. Becker (1994) collected adult males along the Rhine from May to October. As far as is known, in winter the larvae are in second (?) and third instar (Kurazhskovskaya, 1971; Shilova, 1976; own data). These larvae can survive desiccation and freezing (Shilova, 1976).

FEEDING

The larvae feed on organic particles, which are collected between the zooids of Bryozoa (Kurazhskovskaya, 1971; Shilova, 1976).

MICROHABITAT

The larvae and pupae live in colonies of Bryozoa (Kurazhskovskaya, 1971; Shilova, 1976) on stones and plants.

WATER TYPE

The larvae are collected mainly in lakes, pools, broads and canals, in stagnant or slowly flowing water, and locally in the littoral zone of small lowland rivers, rarely in large rivers (Meuse). The species has not been collected in the river Fulda (Lehmann, 1971) or in the Rhine (Wilson & Wilson, 1984; Caspers, 1991). However, Becker (1994) collected adults rather often along the Rhine in light traps.

Permanence

Although Bryozoa are hardly found in temporary water, there is one record of *G. signatus* from a temporary ditch.

pH

We have one record in acid water (pH approx. 5); in all other cases the pH was about 7 or higher. The record from acid water is remarkable because Bryozoa cannot live there.

TROPHIC CONDITIONS AND SAPROBITY

The larvae can be found in mesotrophic and eutrophic water and are often collected in watercourses with fluctuating oxygen content. Probably the most important precondition is the presence of Bryozoa, which cannot live in polysaprobic water (Moog, 1995). The figures in the table in Chapter 4 are tentative.

SALINITY

There are no records from brackish water.

Glyptotendipes viridis (Macquart, 1834)

SYSTEMATICS AND IDENTIFICATION

Although *G. imbecilis* is possibly a synonym of *G. viridis* (see under *G. imbecilis*), both species are treated here separately. Identification of the adult males using Pinder (1978) or Langton & Pinder (2007) is provisional. The exuviae of these two species cannot be distinguished from each other. Larvae in fourth instar can be identified (fide Kalugina, 1975) using Vallenduuk (1999) and Moller Pillot et al. (2000).

DISTRIBUTION IN EUROPE AND THE NETHERLANDS

Because of the problems with identification, all records are doubtful (Saether & Spies, 2004). According to the larval keys mentioned above the species investigated by Dvořák (1996) belongs to *G. viridis*.

LIFE CYCLE

Kondo & Hakashima (1992) found the larvae in low numbers in spring and early summer and numerously only in late summer and autumn.

OVIPOSITION

The egg masses are laid underwater and attached to the edge of plants (Kondo & Hamashima, 1992).

FEEDING

Walshe (1951) observed many larvae of this species and never saw them mining or filter feeding. They fed from algae at the opening of the tubes or at holes temporarily cut through the walls of the tube. In the guts of the larvae Dvořák (1996) found many filamentous cyanophytes, some diatoms and much particulate organic matter (originating from phytoplankton or epipelic material). He considered the species to be a filter feeder living mainly on seston. Kondo & Hamashima (1992) supposed that the larvae are phytophagous and plankton feeders.

MICROHABITAT

Dvořák (1996) found the larvae mining in plants, especially *Typha angustifolia*, more rarely in *Phragmites australis* and *Nuphar lutea*. Kondo & Hamashima (1992) found the larvae mainly mining in stems and petioles of water plants, rarely in the leaves.

WATER TYPE

Gripekoven (1913: 180–182, as *Tendipes leucoceras*) collected this species only in flowing water. Other authors found the larvae in stagnant water (e.g. Kondo & Hamashima, 1992). The only Dutch record is from Lake Loosdrecht.

Graceus ambiguus Goetghebuer, 1928

SYSTEMATICS AND IDENTIFICATION

G. ambiguus is the only representative of the genus in Europe. The adult male can be identified using Cranston et al. (1989) or Langton & Pinder (2007), the pupa using Langton (1991). A description of the larva is in preparation.

DISTRIBUTION IN EUROPE AND THE NETHERLANDS

So far the species has only been known from the British Isles, Ireland and the Netherlands (Saether & Spies, 2004). In the Netherlands there are only two verified recent records, both in Twente, in the eastern part of the country (leg. H. Cuppen).

LIFE CYCLE

H. Cuppen (unpublished) collected pupae, exuviae and larvae in third and fourth instar in April.

WATER TYPE

The Dutch material was collected in temporary mesotrophic depressions in grassland.

Harnischia Kieffer, 1921

Cryptochironomus gr. *fuscimanus* Konstantinov, 1961; Chernovskij, 1949: 57–58, fig. 17
Tendipes (Harnischia) pseudosimplex Kruseman, 1933: 198–199

SYSTEMATICS AND IDENTIFICATION

The phylogeny of the genera related to *Harnischia* has been studied by Saether (1971). These genera are taken together as the *Harnischia* complex. In Kruseman (1933) and other older literature the genus was used in the wider sense. Only two species occur in the Netherlands and adjacent lowlands: *H. curtilamellata* (syn. *H. pseudosimplex*) and *H. fuscimana*. According to Biró (2000) the larvae of these two species can be distinguished by the length of the antennal blade, which does not reach the tip of the last antennal segment in *H. fuscimana*. Before 2000 all records of both species in Limnodata.nl were registered as *H. curtilamellata*. Rodova (1978) described the female of *H. curtilamellata*.

DISTRIBUTION IN EUROPE AND THE NETHERLANDS

Both *H. curtilamellata* and *H. fuscimana* live in nearly the whole European mainland (Saether & Spies, 2004). In the Netherlands the genus has been collected mainly in the Pleistocene areas of the country and in the large rivers (see Water type).

LIFE CYCLE

Pupae, exuviae and adults of both species have been collected in the Netherlands from the end of April to September (Kruseman, 1933; Kouwets & Davids, 1984; own data) and in Germany until November (Becker, 1994). In Sweden emergence may start only from early June (Brundin, 1949); Lehmann (1971) mentioned a male of *H. fuscimana* collected in Montenegro in March. There are probably two generations a year, at least in the Netherlands, in colder water possibly only one. In winter we collected only larvae in third instar. We found the third instar until early May.

FEEDING

According to Konstantinov (1961) the larvae are obligate predators. They eat no prey larger than their own body, mainly Chironomidae and Oligochaeta, but also Cladocera. It is not clear whether algae or detritus are also consumed, as stated for the genus *Cryptochironomus*.

MICROHABITAT

Both species are mainly collected on sandy bottoms, sometimes with gravel or detritus (Mol et al., 1982; Peeters, 1988; Palomäki, 1989; Buskens, unpublished). However Srokosz (1980) found *H. fuscimana* more in mud. The larvae are rare on pebbles and cobbles (e.g. Peeters, 1988), but Ertlová (1970) collected them often on Novodur plates (artificial substrate) in the river Danube. Brundin (1949) found the larvae mainly on Isoetid carpets. We were almost unable to find the larvae in winter, when they probably live deeper in the sandy bottom (cf. *Robackia*).

DENSITIES

Brundin found a mean density of 55 larvae/m² and a maximum of 380 larvae/m² in Lake Innaren in Sweden. Heinis (1993) reported maximum larval densities in December and a minimum density in July and August, corresponding with the summer emergence.

WATER TYPE

Current and dimensions

H. fuscimana occurs nearly exclusively in the lower parts of lowland brooks and streams, but there are some records from large lakes (Lehmann, 1971: 498; Prat, 1978a). The species seems to be less common in large rivers, but it has been collected in all stretches of the Rhine (Becker, 1994) and in the lower reaches of the Oder (Lenz, 1954–62: 222). The species is rare in faster running streams (Moog, 1995; Orendt, 2002a; Waterschap Roer and Overmaas, unpublished).

H. curtilamellata lives in lakes and streams. The exuviae have been collected sometimes in pools not more than 100 m wide (van Kleef, unpublished), but most records are from large lakes. The larvae live mainly in the littoral zone (Brundin, 1949; Mol et al., 1982), but when the oxygen content is high enough they can also be found at a depth of more than 20 m (e.g. Ketelaars et al., 1992; cf. Brundin, 1949). The species seems to prefer larger streams and rivers than *H. fuscimana* and is not rare in the rivers IJssel (Klink, 1985a) and Rhine (Wilson & Wilson, 1984; Caspers, 1991; Becker, 1994; cf. Klink & Moller Pillot, 1982). Orendt (2002, 2002a) recorded the species from some streams with a faster current in Bavaria.

Shade
There are no records from woodland brooks, but *H. curtilamellata* has been collected in two pools in woodland (van Kleef, unpublished).

Permanence
There are no records from temporary water.

pH

Koskenniemi & Paasivirta (1987) and Palomäki (1989) found the larvae in rather low numbers at pH 6.0 in lakes in Finland. Most other workers have not collected the larvae in acid lakes and ponds (e.g. Raddum & Saether, 1981; Barnes, 1983; Leuven, 1987; Buskens, unpublished). *H. curtilamellata* has rarely been found in acid moorland pools in the Netherlands, and then only single specimens at a pH of around 6 (Buskens, 1983; Leuven et al., 1987; van Kleef, unpublished).

TROPHIC CONDITIONS AND SAPROBITY

The larvae of *H. curtilamellata* occur mainly in mesotrophic to slightly eutrophic lakes (Brundin, 1949; Saether, 1979). This also applies to the records from the Netherlands, but both species are also collected in more heavily eutrophicated lakes and streams. In stagnant water, especially in smaller water bodies such as moorland pools, the larvae seem to be confined to water of better quality and with a rather high oxygen content (cf. Brundin, 1949). According to Zinchenko (1992) the numbers of larvae in the Kuibyshev water reservoir (regulated Volga) increased after a rise in the concentrations of all nutrients, but the most eutrophic areas contained lower numbers.

SALINITY

In the Netherlands the larvae have never been collected in brackish water. However, Tõlp (1971) found them in Estonia in water with up to 3500 mg Cl/l.

Kiefferulus tendipediformis (Goetghebuer, 1921)

Chironomus larva *biappendiculata* Kruglova, 1940: 220–222, fig. 1–4
Tendipes biappendiculatus Chernovskij, 1949: 70–71

NOMENCLATURE AND IDENTIFICATION

From the description and figures of Kruglova (1940) it appears that the larva *biappendiculata* (collected in Western Siberia) without doubt belongs to *Kiefferulus* and most probably it is *K. tendipediformis*. *K. tendipediformis* is the only species of the genus in Europe. Identification presents no problems, but the female has been described only briefly and only at the generic level (Saether, 1977). Rodova (1978) has keyed the female, but without a description.

DISTRIBUTION IN EUROPE AND THE NETHERLANDS

The species has been recorded in most European countries (Saether & Spies, 2004). The species is rather common in the Netherlands, but it seems to be scarce in the province of Zeeland and in the northern provinces (Krebs, 1981, 1984, 1985; Limnodata.nl; Nijboer & Verdonschot, unpublished; cf. Moller Pillot & Buskens, 1990: map 54, p. 45).

LIFE CYCLE

The species emerges from May to September (Lehmann, 1969; own data). The larvae winter in second (few?), third and fourth instar (own data).

FEEDING

The larvae are most probably detritus feeders, gathering fine particulate material and/or gnawing at decaying plant material. However, there are no exact data.

MICROHABITAT

Kruglova (1940) called the species phytopelophilous, living on bottoms with decaying plants. She found the larvae on silty bottoms among water plants, together with *Chironomus* species. At least some of the Dutch records are from similar situations, in stagnant as well in flowing water. The larvae seem to be absent from bottoms with silt layers of long-decayed organic material.

WATER TYPE

Current

The larvae are mainly found in stagnant water: ponds, pools and ditches. However, they are not rare in lowland brooks (Limnodata.nl; own data). Exuviae and adult males (possibly also larvae) have been collected in and near fast-flowing streams and even in the Upper Rhine (Orendt, 2002a; Scheibe, 2002; Caspers, 1991). Most probably in these cases the larvae live in sheltered localities or in pools along the stream.

Dimensions

We have no records from water bodies more than 100 m wide and no larvae have been collected in large rivers. The larvae are probably most common in water bodies 5–20 m wide (Limnodata.nl), but they also occur in narrower ditches (own data).

Permanence

There are only a few records from temporary pools (Fritz, 1981; own data).

D

pH

There are no Dutch records of *Kiefferulus* in water with a pH below 7. Barnes (1983) did not collect the species in acid ball-clay ponds.

TROPHIC CONDITIONS AND SAPROBITY

Lehmann (1969) supposed that the larvae live in clear, more or less eutrophic water without much pollution. Nevertheless he reported the presence of the larvae in the river Kossau in a location where the oxygen content is sometimes lowered. In the Netherlands the larvae have often been collected in brooks and ditches without anthropogenic pollution, but with locally much organic material. However, the species has also been found in water with a low oxygen content, a rather high BOD and more than 1 mg ammonium-N/l (Limnodata.nl; Grontmij | Aqua Sense, unpublished). The quality of food and absence of long periods of anoxia are possibly the most important preconditions.

SALINITY

Langton (1991) reported the species to be an inhabitant of fresh and brackish water.

Tourenq (1975) collected *K. tendipediformis* in the Camargue in water with a chloride content of about 2000 mg/l. In the Netherlands Steenbergen (1993) also collected the species in water with more than 1000 mg Cl/l. The scarcity of the species in the province of Zeeland is probably not caused by the brackish character of the greater part of this area (see under Microhabitat and Trophic conditions); the bottom qualities of most brackish waters may be unsuitable for this species.

Kloosia pusilla (Linné, 1767)

Cryptochironomus vytshegdae Pankratova, 1983: 166–168; Kownacki, 1989: 229

SYSTEMATICS

The genus *Kloosia* belongs to the *Harnischia* complex and is quite unlike other members of the complex, especially as a larva.

IDENTIFICATION

The adult male is absent from the keys by Pinder (1978) and Langton & Pinder (2007), but has been described and keyed by Kruseman (1933), Reiss (1988) and Cranston et al. (1989). The exuviae can be identified using Pinder & Reiss (1986, as Chironomini Genus E) and Langton (1991). The larva has not been treated by Moller Pillot (1984) and Pinder & Reiss (1983). It can be identified using Pankratova (1983: 166–168) and Klink & Moller Pillot (2003).

7 Kloosia pusilla: mentum (after Pankratova, 1983)

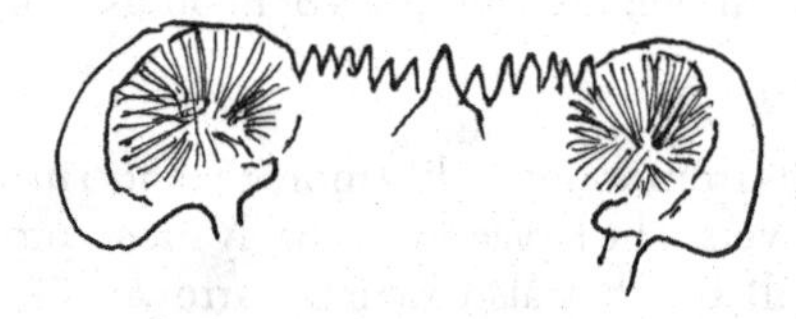

DISTRIBUTION IN EUROPE AND THE NETHERLANDS

The species has been found practically throughout Europe, from Russia to Italy and the Netherlands (Reiss, 1988). In the Netherlands the species was not collected between 1932 (Kruseman, 1933) and 1985 (Klink, 1985a). After 1985 the larvae were locally abundant in littoral sand habitats of the Lower Rhine, the Waal, the IJssel and the Rhine-Meuse estuaries. Some larvae have been collected in smaller streams: the Dinkel and the Swalm (B. Knol, unpublished; Waterschap Roer and Overmaas, unpublished). It is highly likely that the species survived in small rivers during the period when the large rivers were severe polluted in the 1960s and 1970s. The larvae may be overlooked if the sand is not carefully washed and if the mesh size is too large (Chernovskij, 1949; Smit et al., 1994).

LIFE CYCLE

Klink (1985a) collected adult males in May (few) and in July–August (many). There are probably two generations a year.

FEEDING

There are no data on larval feeding. Given the structure of the mouthparts and the shape of the body, they are without doubt carnivores.

DENSITY

The larvae are not rare if the appropriate habitat is present. A maximum of 938 larvae/m^2 was found in the lower course of the river Nida in Poland (a medium-sized lowland river) (Srokosz, 1980). Smit et al. (1994) recorded a local density in estuaries of 1613 larvae/m^2.

MICROHABITAT

The larvae are found in sandy sediments with little organic material (Pankratova, 1964; Reiss, 1988; Srokosz, 1980). The larvae belong to the interstitial invertebrate assemblage (Smit et al., 1994), living between the sand grains in rivers (psammorheobionts), with a very thin and highly flexible body (Chernovskij, 1949).

WATER TYPE

The larvae live exclusively in the sandy bottom of large rivers, rarely in smaller streams (cf. Distribution).

TROPHIC CONDITIONS AND SAPROBITY

Kownacki (1989) found the larvae in the Polish river Nida only after purification and the oxygen content of the river recovered to a high level. Pankratova (1964: 192) reported occurrence of the larvae also in the polluted part of the river Oka in Russia, together with *Robackia demeijerei*. In the Netherlands the species was not collected during the period when the large rivers were severely polluted (see Distribution), but they have been recorded there recently, although the rivers are still moderately polluted. The most important requirement for their presence is probably the existence of a well oxygenated sandy bottom (cf. *Robackia demeijerei*).

Lauterborniella agrayloides (Kieffer, 1911)

DISTRIBUTION IN EUROPE AND THE NETHERLANDS

The species has been recorded in many countries scattered over Europe, but is possibly absent from some parts of the Mediterranean area (Saether & Spies, 2004). In the Netherlands the species is scarce (collected at about 50 localities), and has been recorded mainly in the seepage zones between the Pleistocene and Holocene regions (Steenbergen, 1993; Limnodata.nl).

LIFE CYCLE

The species probably has two generations a year (see Brundin, 1949: 751).

MICROHABITAT

The larvae live in marsh vegetation (Zavrel, 1926; Thienemann, 1954). However, Whiteside & Lindegaard (1982) caught the larvae equally numerously on mud as in the isoëtides zone and the moss and *Nitella* zone in a Danish lake. Higler (1977) found the species in low numbers on *Stratiotes*.

FEEDING

According to Zavrel (1926) the larvae feed on organic particles on the soil. Monakov (2003) noted detritus and algae as food.

WATER TYPE

The larvae live exclusively in stagnant water, in lakes, marshes, alder carr, lakes and ditches (Zavrel, 1926; Brundin, 1949; Higler, 1977; Whiteside & Lindegaard, 1982). Tourenq (1975) found the larvae in the Camargue mainly in a rice field.

pH

Raddum & Saether (1981) collected the larvae at a pH of 6.25 and not below. In the Netherlands the larvae have been collected only in more or less alkaline water (Steenbergen, 1993; Limnodata.nl). Klink (1986) stated that in moorland pools in the Netherlands the species disappeared after acidification.

TROPHIC CONDITIONS AND SAPROBITY

Brundin (1949) collected the species sometimes numerously in oligotrophic lakes in Sweden. Orendt (1993) collected the species in two oligotrophic and one eutrophic lake in Bavaria. In the Netherlands many records are from fen-peat lakes with low orthophosphate and ammonium contents (Steenbergen, 1993), but in some cases the larvae were found in more or less eutrophic conditions (Limnodata.nl). In a large Dutch moorland pool (Beuven) the species disappeared after eutrophication (Klink, 1986). The larvae are rarely found in water with less than 80% oxygen saturation (Steenbergen, 1993; Rossaro et al., 2007; Limnodata.nl). Rossaro et al. stated that the ring organ of the pupa is relatively small, suggesting low tolerance to oxygen shortage. Besides the oxygen demand, the influence of saprobity is a little uncertain.

SALINITY

The larvae are not found in brackish water.

Lipiniella Shilova, 1961

The larva of *Lipiniella* is described and illustrated in Pinder & Reiss (1983) and Moller Pillot (1984: 117, 256). Larvae with tubuli (*L. moderata*) cannot be identified using the latter publication and should be included on p. 106 sub 7. Below we give a key for the two species of this genus because such a key is absent from most European identification books. A third species, *L. kanevi*, appears to be synonymous with *L. araenicola* (see Kiknadze et al., 1989: 116).

- Larvae without tubules. Frontal apotome without depression *L. araenicola*
- Larvae with one pair of short to moderately long ventral tubules at abdominal segment VIII. Frontal apotome with a rugose depression (see figure) *L. moderata*

The larvae of these two species do not differ regarding other characters (shape of the head, mentum and ventromental plates, etc.).

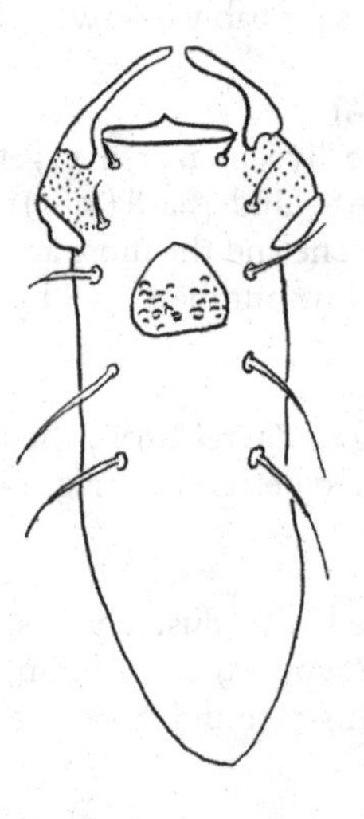

8 Lipiniella moderata: frontal apotome (from Kiknadze et al., 1989)

Lipiniella araenicola Shilova, 1961

Lipiniella arenicola Shilova, 1963: 71 et seq.; Moller Pillot, 1984: 256

NOMENCLATURE

Although Shilova, in her second article about the species (Shilova, 1963), spelled the species name as *arenicola*, the original spelling, *araenicola*, is valid.

IDENTIFICATION

The male and female are described in Shilova (1961). The exuviae are described and illustrated in Langton (1991) and Pinder & Reiss (1986). Shilova (1963) contains a description of the eggs and the larva in all four instars.

DISTRIBUTION IN EUROPE AND THE NETHERLANDS

In recent times *L. araenicola* has been found in Russia and in some West European and Scandinavian countries (Saether & Spies, 2004). In the Netherlands the species has been collected mainly in lakes and estuaries in the western part of the country (Klaren, 1987; Moller Pillot & Buskens, 1990; Limnodata.nl). Before 1940 the larvae were also present in Dutch rivers (Klink, 1986a) and without doubt in other European countries too. It is not known where the species survived between 1940 and 1979.

LIFE CYCLE

Shilova (1961, 1976) caught adults in Russia from June to August and supposed there were two generations. Klaren (1987) found two generations in the Haringvliet, emerging in May and July–August; in winter the larvae were in diapause in fourth instar.

SWARMING AND OVIPOSITION

Shilova (1976) observed the cone-shaped swarms in the evening at a height of 7–10 m, sometimes descending to 2 m. Small larvae of *L. araenicola* appeared in late summer at the highest sites on the mud flats in the Haringvliet and in time appeared at lower sites. This suggests that oviposition took place on one of the exposed sites on the flats and that the larvae then migrated to other areas (Smit et al., 1996).

MICROHABITAT

Smit et al. (1994) called the larvae characteristic of fine sand (grain size mostly 63–210 µm). They build remarkably strong tubes. The larvae are restricted to sands with low silt contents (Shilova, 1976) and are almost entirely absent from sandy habitats with 5–10% silt (Klaren, 1987; Smit et al., 1992). Klink et al. (1995) called the larvae characteristic of sandy soils where conditions prevent the sedimentation of much silt. Klaren (1987) stated about the same density of larvae within the sediment at depths of 0–5 cm and 5–10 cm, but second and third instar larvae were collected mainly in the upper 5 cm of the sediment, and the same was true for fourth instar larvae in the higher parts, which dried up every day during low tide. The larvae were almost absent at depths below 10 cm.

DENSITIES

In fine sand Smit et al. (1994) reported densities of approx. 1450 larvae/m^2. In shallow sandy parts of the Haringvliet Klaren (1987) found densities of less than 100 to more than 1000 larvae/m^2. At deeper, more silted localities the densities were much lower. The highest densities recorded by Shilova (1976) were about 6500 larvae/m^2.

FEEDING

According to Smit et al. (1992) *L. araenicola* is a selective grazer on benthic algae; planktonic algae are hardly consumed. Shilova (1976) noted that more blue-green algae or diatoms and green algae were eaten, depending on the season; she sometimes also found fine organic particles and sand in the gut.

WATER TYPE

Shilova (1976) reported the occurrence of the species in a number of lakes and rivers in Russia. The larvae were only present in the littoral zone. Becker (1994) collected some adult males along the river Rhine in Germany. Janzen (2003) mentioned the species as an inhabitant of several small streams in Germany, at slow and rather fast currents. In the Netherlands the larvae are often present and abundant in the littoral zone of large lakes and estuaries with sandy bottoms. Limnodata.nl also contains some records at a depth of more than 10 m. They are presently rare or are absent from the large rivers, but have been collected in low numbers in some smaller streams. Klaren (1987) also collected the larvae in the tidal part of the estuary, which dried up for a long period every day . Here growth was notably retarded and only one generation could be observed.

pH

The larvae are especially common in alkaline lakes (Smit et al. 1992).

TROPHIC CONDITIONS AND SAPROBITY

The larvae of *L. araenicola* are less tolerant of mild hypoxia than several *Chironomus* species, but more than *Stictochironomus* larvae (Smit et al., 1992). They can be found in slightly polluted lakes, but in these cases they live in the sandy bottoms.

SALINITY

The larvae are only collected in fresh water or in water with a slightly raised salt content.

Lipiniella moderata Kalugina, 1970

IDENTIFICATION

The adult male has been described by Kalugina (1970). The larva and cytology have been described and illustrated by Kiknadze et al. (1989, 1991). Klink & Moller Pillot (2003) have keyed the larva. For the differences between the two species of the genus see above.

DISTRIBUTION IN EUROPE AND THE NETHERLANDS

L. moderata has been recorded from Eastern Europe and from only few places in Central and Western Europe (Norway, Hungary, Serbia and the Netherlands) (Biró, 2000; Saether & Spies, 2004). In the Netherlands the larvae have been collected in several branches of the rivers Rhine and Waal (Klink et al., 1995; T. van Haaren, pers. comm.).

LIFE CYCLE

Kiknadze et al. (1989) collected prepupae and adults in West Siberia in May.

MICROHABITAT

Kiknadze et al. (1989) collected the larvae in slightly silted sandy sediments in the littoral zone of a river and a reservoir. Biró (2000) found the larvae in Hungary and Serbia in sandy mud and mud. A Dutch population lived in a river branch with a sandy bottom and only limited deposition of silt (Klink et al., 1995).

WATER TYPE

Adults and larvae have been collected in Russia, Belarus and Siberia in large rivers and reservoirs (Kiknadze et al.. 1989). In the Netherlands populations have been found only in branches of the Lower Rhine (Klink et al., 1995) and the Waal (Klink, 2002; T. van Haaren, pers. comm.).

Microchironomus deribae (Freeman, 1957)

Leptochironomus deribae Ringe, 1970: 315–317, fig. 3; Pinder, 1978: 126, fig. 61B, 157B

IDENTIFICATION

Adult males can be identified using Pinder (1978) and Langton & Pinder (2007), and exuviae using Langton (1991). Ringe (1970) published information about synonyms and illustrated the male hypopygium.

DISTRIBUTION IN EUROPE AND THE NETHERLANDS

M. deribae has been recorded from many European countries along the sea coasts and also from some countries further inland, e.g. Hungary (Saether & Spies, 2004). In the Netherlands there are many reliable records from the delta region and along the North Sea coast. Most of the records from other parts of the country (Nijboer & Verdonschot, 2001; Limnodata.nl) are probably based on misidentifications.

LIFE CYCLE

In the Camargue the adults emerged from March until September, with maxima in May and July (Tourenq, 1975). In one year Krebs (1978) found two generations, emerging in May and August/September; in another year he recorded only a spring generation (see also under Trophic relations). In winter most larvae were in third instar and only a few in fourth instar.

FEEDING

The guts of the larvae usually contain many diatoms and other small particles, but also much detritus (own observations). The larvae are probably fairly selective grazers, but see under *M. tener*.

MICROHABITAT

The larvae live on clayish, sandy, muddy or silty bottoms in tubes (Dejoux, 1971; Krebs, 1979, 1981).

DENSITY

Krebs (1978) found densities from 300 to 1600 larvae/m^2 in winter and in summer a maximum of 900 larvae/m^2 .

WATER TYPE

The larvae have been collected in shallow lakes, pools and ditches; more rarely in estuaries (Ringe, 1970; Dejoux, 1971; Tourenq, 1975; Krebs, 1978, 1979, 1981).

TROPHIC CONDITIONS AND SAPROBITY

Krebs (1978), Parma & Borghouts-Biersteker (1978) and Merks & Rijstenbil (1981)

studied a ditch which in summer had a high oxygen demand at the bottom, long nightly periods without oxygen and high sulphide contents near the bottom. As a rule *M. deribae* lived in this ditch the whole year round, but sometimes not in summer. Although the ditch fauna was not investigated in the year in which the chemical properties of the ditch were determined (cf. Krebs, 1985), low oxygen contents were also reported in a year that the larvae were also present in summer (Krebs, 1985). The species is known to be quite tolerant of low oxygen levels, but the conditions mentioned here may be just beyond what the larvae can survive.

SALINITY

This species is confined to brackish water regions. The larvae were found incidentally in Lake Volkerak-Zoom in the Netherlands for four years after the freshening of the lake (van der Velden et al., 1995). Ringe (1970) recorded the species from two slightly brackish lakes on the island of Fehmarn (Northern Germany). The larvae survive large fluctuations in salinity and maximum chloride concentrations of 15,000 mg/l (Tourenq, 1975). In regions with brackish water Tourenq also reported the occurrence of larvae in fresh water. This has also been found in the Netherlands (Moller Pillot & Buskens, 1990).

Microchironomus tener (Kieffer, 1918)

Tendipes (Parachironomus) tener Kruseman, 1933: 190–191
Leptochironomus tener Ringe, 1970: 317–318, fig. 4; Pinder, 1978: 126, fig. 61A, 157A

IDENTIFICATION

Males of both species of the genus *Microchironomus* have been keyed by Pinder (1978) and Langton & Pinder (2007). Figures of the hypopygium can be found in Ringe (1970) and Prat (1978a). The female has been described and keyed by Rodova (1978). The female of the genus has also been keyed by Saether (1977).

DISTRIBUTION IN EUROPE AND THE NETHERLANDS

M. tener has been recorded throughout Europe (Saether & Spies, 2004). The species can be found everywhere in the Netherlands, but it appears to be much more common in the Holocene part of the country (Nijboer & Verdonschot, unpublished data; Limnodata.nl).

LIFE CYCLE

Kruseman (1933) collected the adults in the Netherlands from May to September. Otto (1991) found two generations a year in northern Germany, emerging mainly in June and August. In Russia (Rybinsk) Shilova (1976) found only one generation a year.

FEEDING

Prat (1980) mentioned that *Microchironomus* larvae are predators of tubificids, but it is not clear whether this was based on observations or only suppositions. We found mainly diatoms and detritus in the guts of the larvae.

MICROHABITAT

The larvae inhabit the lake bottom in open water and are rare in vegetation (Bijlmakers, 1983; Otto, 1991). Pagast (1931) found them on open gyttja deposits in Lake Usma. Bijlmakers collected them more on sand than on silt, but they also live on silt and clay and even on thick detritus layers (Shilova, 1976; Verstegen, 1983; Rieradevall & Prat,

1989; Otto, 1991), on artificial substrate (Prat, 1980) and in low numbers on stones (Palomäki, 1989; own data). The larvae appear to be totally absent from sand pits in the Netherlands, which are usually very poor in silt (Buskens, unpublished data).

DENSITIES

Shilova (1976) recorded the highest densities (160 larvae/m^2) at a depth of 15 m. Rieradevall & Prat (1989) found approx. 10–100 larvae/m^2 during the year in profundal samples from a Spanish lake. Bijlmakers (1983) collected only 14–75 larvae /m^2 in Belversven, a eutrophic shallow moorland pool in the Netherlands. Prat (1980) found more than 500 larvae/m^2 on an artificial substrate in winter. He reported that the larvae only gradually colonised an artificial substrate, probably because their food (mainly tubificids) was scarce in the first phase.

WATER TYPE

Current and dimensions

In flowing water the larvae are found scarcely in large and small rivers, but never in narrow brooks (Klink & Moller Pillot, 1982; Klink, 1985a; Klink, 1986a; Caspers, 1991; Smit, 1982; cf. Lehmann, 1971; Limnodata.nl). The larvae are collected mainly in lakes and large pools and very rarely in small pools and ditches less than 5 m wide (e.g. Steenbergen, 1993; H. Cuppen, pers. comm.). They appear to live mainly in deeper water: Tõlp (1971) collected them at depths from 2 to 16 m and Shilova (1976) from 3 to 18 m, with the highest densities at a depth of 15 m. Kuijpers (pers. comm.) found them down to a depth of 24 m. Their presence in Dutch lakes has probably been underestimated because investigations of Dutch lakes have been restricted mainly to the littoral zone.

Permanence

The species has never been recorded in temporary water.

pH

There are no records of *M. tener* at pH values lower than 6.

TROPHIC CONDITIONS AND SAPROBITY

The larvae are collected in rather oligotrophic to eutrophic water and rarely in hypertrophic water (Brundin, 1949; Prat, 1978a; Verstegen, 1985). However, there are a number of records from slow-flowing hypertrophic water (e.g. Brabantse Delta, unpublished). In the Dutch province of Groningen the larvae are collected more in the eutrophic lakes (D. de Vries, in litt.). Steenbergen (1993) found the species significantly more in water with a low orthophosphate content (< 0.05 mg P/l) and more than 80% saturation with oxygen (cf. Limnodata.nl). Although the larvae are red they appear to need a rather high oxygen content in the water column: they are rare on silty bottoms with vegetation (see Microhabitat) and were collected at a great depth only in oligotrophic Spanish reservoirs (Prat, 1978a).

SALINITY

M. tener is not a rare species in slightly brackish water (Limnodata.nl). The species was much more common in lakes and canals near Amsterdam in Kruseman's time, when most were brackish, than they are today (compare Kruseman, 1933, 1934 with Steenbergen, 1993). However, Steenbergen collected the larvae significantly more often in fresh water and never in water with more than 1000 mg Cl/l.

Microtendipes Kieffer, 1915

SYSTEMATICS AND NOMENCLATURE

Pinder & Reiss (1983, 1986) divided the genus into two groups based on the characters of the larvae and pupae. Cranston et al. (1989) stated that a final division requires a generic revision. Here we use the preliminary division into two groups: the *rydalensis* group and the *pedellus* group. The *rydalensis* group (= *tarsalis* group Moller Pillot, 1984) contains the species *M. rydalensis* and *M. tarsalis*. The *pedellus* group contains *M. britteni*, *M. chloris*, *M. confinis*, *M. diffinis* and *M. pedellus* (see Pinder & Reiss, 1983: 324; Moller Pillot, 1984: 135). The larvae of *M. confinis* and *M. nitidus* are unknown.
Many authors suppose that *M. chloris* and *M. pedellus* are seasonal forms of the same species, the dark coloured form (*M. chloris*) flying in spring and the pale form (*M. pedellus*) in summer (see Shilova, 1973; Goddeeris, 1983). These two authors call the species *M. pedellus* De Geer, 1776. However, some authors collected one or both forms both in spring and in summer (e.g. Lehmann, 1971; Becker, 1994). In the Netherlands there are two larval types within the *pedellus* group: one type with pale median mental teeth and the labral lamella with 12–14 teeth, the other type with usually dark median mental teeth and a labral lamella with 19–28 teeth. The former has been named *M. pedellus* agg. because a reared male (in spring!) was pale. Gouin (1936) identified a separately reared larva as *M. pedellus* (labral lamella with 12 teeth). The other larval type has been named *M. chloris* agg. (Moller Pillot, 1984: 135).

The species *chloris* and *pedellus* are so far possibly indistinguishable as adult males. Although Langton & Pinder (2007) give an additional character to distinguish between them, these two species are possibly not the same as the aggregates of Moller Pillot (1984). A further problem is that most authors use the name *M. pedellus* for both species because they suppose the name *M. chloris* is only a synonym (see above). In this book we maintain the names of Moller Pillot (1984) because the differences in morphology and ecology in the larvae are clear (see however under *M. diffinis*) and we have no alternative for these names.

IDENTIFICATION

As mentioned above, not all keys distinguish between the species *M. chloris* and *M. pedellus*. The male adults can be identified best using Langton & Pinder (2007). The female of *M. pedellus* s.l. has been described and illustrated by Rodova (1978). Larvae can be identified to aggregate level using Moller Pillot (1984).

FIRST AND SECOND INSTAR LARVA

The larva of *M. chloris* has two narrow central mental teeth already in first instar, in contrast to most other Chironomini (Baz', 1959; Kalugina, 1959: 94). The second lateral teeth are a little longer and wider than the first. The antenna is five-segmented in first instar. In second instar the larval characters are the same as in later instars and the body is already red (Baz', 1959).

The genus will be treated below under the following names:
Microtendipes chloris agg.
Microtendipes confinis
Microtendipes diffinis
Microtendipes pedellus agg.
Microtendipes rydalensis
Microtendipes tarsalis agg.

Microtendipes chloris agg. sensu Moller Pillot, 1984

DISTRIBUTION IN EUROPE AND THE NETHERLANDS

According to Saether & Spies (2004) both species *M. chloris* (Meigen, 1818) and *M. pedellus* (De Geer, 1776) are widely distributed over the whole European mainland. *M. chloris* agg. has been collected nearly everywhere in the Netherlands, but is very scarce in the province of Zeeland (Nijboer & Verdonschot, unpublished).

LIFE CYCLE

The first generation of adults ('*M. chloris*') emerges in Western Europe in spring, from the end of March until the end of May (Macan, 1949; Mundie, 1957; Reiss, 1968; Goddeeris, 1983; Jónsson, 1987). The second generation emerges in summer ('*M. pedellus*'). Janecek (1995) collected adults of *M. chloris* from April to June and *M. pedellus* from May to October. In urban waters we found the first emergence as early as the middle of March. Goddeeris (1983: 35 et seq.) recorded a partial third generation from fish ponds in the Ardennes, emerging in late summer. Most probably this was also the case in Lake Belau in northern Germany (Otto, 1991: fig. 46). The larvae have an overwintering diapause in fourth instar (Goddeeris, 1983; 1986); in autumn larvae in third instar can also be collected. Collé (1983) noted that the larvae in the Netherlands exhibited substantial growth already in March.

SWARMING AND OVIPOSITION

The adults swarm above the water surface and near the banks. Baz' (1959) observed that the females landed on the water surface in the evening, deposited their eggs and flew away. The egg masses become attached to algae near the banks. According to Sokolova (1971) an egg mass contains 750–800 eggs; Nolte (1993) mentioned records of 328 and 350 eggs.

FEEDING

Larvae in third and fourth instar seek food from their tubes, without leaving the tube entirely. They feed mainly on the substrate near the tube, without much selectivity (Walshe, 1951; Baz', 1959). Walshe found mainly detritus and Characeae in the guts; Baz' found, besides sand, all types of green and blue-green algae and diatoms. According to Gouin (1936) the larvae of *M. pedellus* agg. feed on dead leaves.

MICROHABITAT

The first and second larval instar are positively phototactic and live near the surface in tubes, for example between algae (Baz', 1959). Older larvae are tube-building bottom dwellers, living on organic and mineral bottoms, but more usually on a combination of these. They can also be found on or in wood, under the epidermis of plants, on stones and on molluscs. (e.g. Mol et al., 1982; Brodersen et al., 1988; Palomäki, 1989). The larvae construct their tubes horizontally in the top 1 cm of the sediment and do not colonise deeper layers, like most other Chironomini (Olafsson, 1992). Drake (1982) observed that plants were colonised by *M. pedellus* larvae when *Cladophora* became entangled around the plants.

WATER TYPE

Current

Lehmann (1971) and Becker (1994) stated that *M. chloris* is relatively rare in rivers (much scarcer than *M. pedellus*, both identified from adult males). Our experience in the Netherlands is that larvae of *M. chloris* agg. can also live in brooks and streams (even sometimes in moderately fast currents), but are relatively scarcer (cf. Orendt, 2002a).

These larvae are more typical inhabitants of stagnant water.

Dimensions
Although the larvae are sometimes collected in rather narrow ditches and upper courses, they are much more common in wider canals and lakes (Steenbergen, 1993; Limnodata.nl). In lakes they do not live in the profundal zone and occur most commonly at a depth of 1–3 m (Mundie, 1957; Mol et al., 1982). However, Koskenniemi (1992) noted that *Microtendipes chloris*-type larvae can sometimes live in deeper bottom areas. Only very few larvae and exuviae have been collected in large rivers (Wilson & Wilson, 1984; own data).

Permanence
Wotton et al. (1992) found the larvae in low numbers in sand filter beds temporarily filled with water, where only very few specimens could complete their life cycle. Ten Cate & Schmidt (1986) collected one larva in a temporary upper course. The species is without doubt particularly rare in narrow temporary ditches and upper courses (see also under Dimensions).

SOIL
Steenbergen (1993) collected the larvae significantly less on clay than on sand and peat.

pH
Roback (1974) reported the occurrence of the larvae of *M. pedellus* in water with a pH of 5.6–8.4. Leuven et al. collected the larvae only at pH > 5. Buskens (1989) found very few larvae in a moorland pool (Beuven: Lobeliabaai) with a low pH (about 5), but they can be numerous at this pH according to unpublished data from van Kleef. In some cases larvae and exuviae have been collected at pH values between 4 and 4.5 (Bryce, 1965; van Kleef, unpublished). Palomäki (1989) collected the larvae in goodish numbers in Lake Alajärvi in Finland at pH 6.0. However Klink (1986) mentioned *Microtendipes* gr. *chloris* as a taxon that disappears after acidification. Steenbergen (1993) and Limnodata.nl found no clear preference between pH 7 and 9.

TROPHIC CONDITIONS AND SAPROBITY
Steenbergen (1993) collected the larvae significantly more in water with low phosphate (< 0.15 and especially < 0.05 mg P/l) and low ammonium content; the larvae were scarce in water with less than 40% oxygen saturation. Buskens (1989; 1989a) observed a sharp decline of *Microtendipes* larvae after eutrophication, increasing again after restoration, when the orthophosphate content again decreased from 1.94 to 0.03 μmol/l. Brodersen et al. (1998: 583, 590) stated that the genus is an indicator of oligomesotrophic lakes and Langdon et al. (2006) found the larvae scarcely in eutrophic conditions. However, the larvae live mainly in places with some decomposition of organic material and seem to be scarce in oligosaprobic water bodies (own data). Brundin (1949) reported their occurrence in extremely polyhumic lakes.
Rossaro et al. (2007) and Grontmij | Aqua Sense (unpublished) sometimes found larvae of *M. chloris* group in water with an oxygen content of only 0.5 mg/l. However, Rossaro et al. noted that the ring organ of the thoracic horn of the pupae of three *Microtendipes* species is very small, suggesting a low ability to respond to oxygen shortage.

IRON
Rasmussen & Lindegaard (1988) stated that *Microtendipes* larvae are susceptible to high concentrations of Fe^{2+}-ions. Such high concentrations are possible in seepage-water (see Chapter 2).

SALINITY

Steenbergen (1993) collected the larvae significantly more in water with a chloride content < 150 mg/l and very rarely when it was higher than 1000 mg/l. Tõlp (1971), however, found them in Estonia in water with up to 3400 mg Cl/l.

PARASITISM

Kouwets & Davids (1984) stated that adults of *Microtendipes* were parasitised by larvae of water mites unusually often.

Microtendipes confinis (Meigen, 1830)

IDENTIFICATION

Adult males can be identified using Langton & Pinder (2007) and exuviae using Langton (1991). The larva is unknown, but will resemble *M. chloris* or *M. pedellus.*

DISTRIBUTION IN EUROPE

M. confinis has been collected in many countries in Central and Western Europe, but there are no records from the Netherlands (Saether & Spies, 2004).

LIFE CYCLE

The life cycle is most probably no different from that of *M. chloris* (two generations a year and wintering in fourth instar). Becker (1994) collected adults from May to September. In the Pyrenees the exuviae were collected very numerously at the end of October (Gendron & Laville, 1992).

MICROHABITAT

In the river Rhine this species was only reared from stones (Becker, 1994). It is very likely that in lakes the larvae will also live on other substrates.

WATER TYPE

The species is known from lakes and rivers (Langton, 1991). In the Upper and Middle Rhine the larvae were locally more numerous than other species of the genus; in the Lower Rhine the species was scarce (Becker, 1994). Gendron & Laville (1992) collected large numbers of exuviae in the river Aude in the Pyrenees with current velocities between 1 and 2 m/sec. The species has also been collected in fast-flowing streams in Germany (Orendt, 2002a; Michiels, 2004).

Microtendipes diffinis (Edwards, 1929)

IDENTIFICATION

The adult male can be identified using Langton & Pinder (2007) and the exuviae using Langton (1991). Bryce (1960) described the larva, which can be identified by the labral lamella with about 16 teeth. However, this character is most probably not sufficient to determine the species name with certainty. All Dutch records are based on identification of larvae.

DISTRIBUTION IN EUROPE AND THE NETHERLANDS

The species has been recorded in a number of countries in Western Europe, but in few other countries (Saether & Spies, 2004). Occurrence in the Netherlands has not been definitely stated. There are about 20 records of larvae scattered over almost the whole country.

WATER TYPE

According to Fittkau & Reiss (1978) the species lives in lakes. However, Langton (1991) mentions only rivers. Scheibe (2002) collected the species along a summercold brook in the Taunus mountains. The Dutch larvae were collected in lowland brooks, canals and small lakes.

Microtendipes pedellus agg. sensu Moller Pillot, 1984

For systematics and identification see under the genus; for biology see *M. chloris* agg.

Distribution in the Netherlands

There are no data about the occurrence of this species in other countries than the Netherlands. The larvae have been collected mainly in the Pleistocene areas of the Netherlands, in the eastern and southern provinces. Nijboer & Verdonschot (2001) mention 193 localities, several of which are in the Holocene areas, but most of these have yet to be verified (see under Water type).

LIFE CYCLE

Fourth instar larvae have been collected mainly from November to April, but also sometimes in summer. The life cycle is most probably the same as in *M. chloris* agg.

MICROHABITAT

The larvae live in lowland brooks on sand and organic bottoms. Becker (1994) reared adults of *M. pedellus* in the river Rhine mainly from stones.

WATER TYPE

Current

The larvae live nearly exclusively in flowing water. It is not clear why they are rare in regulated streams, because they are collected mainly in slow-flowing lowland brooks and in places where they are protected from strong currents, for example above a weir. Nevertheless, they also live in streams with faster currents (Orendt, 2002a; Janzen, 2003). Becker (1994) collected and reared *M. pedellus* much more than *M. chloris* from the river Rhine in Germany. The larvae are also found in stagnant water, but as far as is known only in water bodies with groundwater seepage.

Dimensions

Most records are from brooks 2–10 m wide. In stagnant water they are to be expected in somewhat larger water bodies.

Permanence

There are no records from temporary water and the larvae probably occur there only rarely (cf. *M. chloris* agg.).

TROPHIC CONDITIONS AND SAPROBITY

The larvae have been collected mainly in brooks with better water quality. They seem to prefer water with higher oxygen content.

Microtendipes rydalensis (Edwards, 1929)

IDENTIFICATION OF THE LARVA

The larva can be distinguished from other species of the *rydalensis* group by the pecten epipharyngis with 7–9 teeth of different length, the labral lamella with 13–14 teeth and the three equal-sized median mental teeth (figs. see Pinder, 1976; Pinder & Reiss, 1983). An aberrant larva (probably another species) was collected in a Dutch moorland pool; this larva has a pecten epipharyngis with 9 teeth, but the central mental tooth is smaller, and a labral lamella with about 19 teeth (coll. Buskens).

DISTRIBUTION IN EUROPE AND THE NETHERLANDS

The species has been recorded in a number of countries scattered over Europe (Saether & Spies, 2004). However, only few localities are known outside Germany. In the Netherlands both the larva and exuviae have been collected at one place only (leg. B. Knol, J. Meeuse), in the Twenthe region and the province of Drenthe respectively.

MICROHABITAT

Pinder (1976) collected the larvae in the rooting zone of *Ranunculus* in the river Stour in England. A Dutch larva has been collected on a sandy bottom, probably under an overhanging bank (B. Knol, pers. comm.).

WATER TYPE

The species has been collected only in running water (see however under Identification). The records from England (Pinder, 1976), Ireland (Douglas & Murray, 1980) and two Dutch localities are from more or less natural lowland brooks. However, Janzen (2003) found the larvae regularly in less natural brooks and streams in Germany, where there are also several records from faster running streams (Orendt, 2002a; Scheibe, 2002; Janzen, 2003). Casas & Vilchez-Quero (1989) collected the exuviae in a fast-flowing stream with carbonate incrustations (A. Blancas) in the Sierra Nevada in Spain.

Microtendipes tarsalis agg. sensu Moller Pillot, 1984

Chironomus formosus Goetghebuer, 1912: 19–20, Pl. V fig. 9–11

IDENTIFICATION OF THE LARVA

The larva of *M. tarsalis* agg. has a pecten epipharyngis with three teeth and a labral lamella with about 30 teeth (Moller Pillot, 1984; see description and figures in Goetghebuer, 1912). In *M. tarsalis* the central tooth is smaller than the adjacent teeth. It cannot be excluded that more than one species belong to this aggregate, also because the Dutch records are from a different type of water.

DISTRIBUTION IN EUROPE AND THE NETHERLANDS

Saether & Spies (2004) state at least ten European countries where *M. tarsalis* has been collected. In the Netherlands there are three unpublished records of the larva, from lakes in Noord-Holland and the northwestern part of Overijssel. Only one of them has been verified and seems to belong to the species *M. tarsalis*.

WATER TYPE

Goetghebuer (1912, 1928) collected *M. tarsalis* near running water in the Belgian Ardennes. Douglas & Murray (1980) collected the species in an Irish lake near the outflow of a brook. The Dutch records apply to larvae collected in fen-peat lakes (see above).

Nilothauma brayi (Goetghebuer, 1921)

Kribioxenus brayi Moller Pillot, 1984: 138–139, figs. IV.6

SYSTEMATICS

N. brayi is the only species of the genus in Europe, except for *N. hibaratertium* in Portugal (see Adam & Saether, 1999).

IDENTIFICATION

The adult male and female have been described and keyed by Adam & Saether (1999). The exuviae have been keyed by Langton (1991). The larva has been described by Brundin (1949) and keyed by Pinder & Reiss (1983) and Moller Pillot (1984).

DISTRIBUTION IN EUROPE

N. brayi has been collected in most countries in Western and Northern Europe (Saether & Spies, 2004). In the Netherlands there is only one record: Klink collected a larva in the 'Grensmaas' stretch of the river Meuse (on the Netherlands–Belgium border in the southern part of the province of Limburg) near Linne after the high water levels in 1995 (Klink & Moller Pillot, 1996). This specimen may have been brought down from rivers in the Ardennes.

LIFE CYCLE

The species flies in Northern Europe in July and August (Brundin, 1949; Paasivirta, 1976). There seem to be two generations a year.

MICROHABITAT

Brundin (1949) found the larvae on organic silt at a depth of 4–19 m (see also Water type).

WATER TYPE

Nearly all records of the species are from lakes (Brundin, 1949; Paasivirta, 1976), but Goetghebuer (1936) reported the species living in rivers in the Ardennes and Janzen (2003) collected the larvae in fast-flowing streams in Germany. The Dutch specimen was collected in the Grensmaas stretch of the river Meuse (see above).

The larvae live at different depths in the lakes; Paasivirta (1976) collected them in the littoral zone at a depth of 1 and 2 m, and Brundin (1949) at a depth of 4–19 m. According to Adam & Saether (1999) the larvae are most common in the open gyttja bottom of the lower littoral zone.

TROPHIC CONDITIONS

Brundin (1949) and Paasivirta (1976) collected the species in oligotrophic lakes.

Omisus caledonicus (Edwards, 1932)

Microtendipes rezvoi Chernovskij, 1949: 92, fig. 70; Pankratova, 1983: 266, fig. 211
Microtendipes caledonicus Pinder, 1978: 128, fig. 158D
Paratendipes spec. Kreuzer, 1940: 470 (fide Berezina, 1998)

SYSTEMATICS AND IDENTIFICATION

The genus *Omisus* probably has only one European species. The adult male can be identified using Langton & Pinder (2007), the pupa using Langton (1991). The larva has been described and illustrated by Berezina (1998) and keyed by Pinder & Reiss (1983).

DISTRIBUTION IN EUROPE AND THE NETHERLANDS

The species has been found in many countries in Western and Northern Europe (Saether & Spies, 2004). In the Netherlands only subfossil remains (until 1970) are known from the province of Gelderland at Gerritsfles and from the province of Noord-Brabant at Groot Huisven and Achterste Goorven (Klink, 1986). In the 1930s the species was still present in a small peaty pool in northern Germany (Kreuzer, 1940: 470; as *Paratendipes*).

pH

Berezina (1998) found the larva in a Russian dystrophic forest pool with a pH of 5.8–6.8 in summer and 4.0–4.5 in winter. According to Klink (1986) and van Dam & Buskens (1993), however, the species most probably disappeared from the Dutch moorland pools due to acidification.

MICROHABITAT

The larvae found by Berezina (1998) formed mucous cases attached to the middle third section of moss plants, 15–20 cm from the bottom.

WATER TYPE

According to Berezina (1998) the larva inhabits mainly dystrophic lakes and pools, but less frequently also eutrophic lakes.

TROPHIC CONDITIONS AND SAPROBITY

According to Klink (1986) *O. caledonicus* is an inhabitant of oligotrophic water. Klink noted that it is a widespread phenomenon that oligotraphent species have disappeared from Dutch moorland pools due to acidification. (See however under *Pagastiella orophila*).

Pagastiella orophila (Edwards, 1929)

Genus ? larva *minuta* Kruglova, 1940: 224–226, fig. 8–11

DISTRIBUTION IN EUROPE AND THE NETHERLANDS

Pagastiella orophila is a common species in Scandinavian lakes (e.g. Brundin, 1949) and less common in Central and Western Europe (e.g. Goddeeris, 1983; Serra-Tosio & Laville, 1991; Orendt, 1993). There are no records from a number of other European countries (Saether & Spies, 2004). Until the second half of the 20th century *P. orophila* was a normal inhabitant of many Dutch moorland pools in the eastern and southern part of the country (Klink, 1986). Kruseman (1933) collected 14 adults in 1931 in Oud-Loosdrecht in the central peaty district. In more recent times larvae have been found only in the northeastern part of the country in at least five moorland pools (Moller Pillot & Buskens, 1990; Duursema, 1996; J. Kuper, pers. comm.).

LIFE CYCLE

Goddeeris (1983) reported one generation a year in the Belgian Ardennes, emerging from the end of May until July, with possibly a small partial second generation in August. Orendt (1993) concluded the same for oligotrophic lakes in Bavaria. Goddeeris also discussed data from other European studies.

Most larvae hibernate in fourth instar and only a few in third instar (Goddeeris, 1983). Armitage (1968) observed that the larvae in a Finnish lake built cocoons or sealed off their tubes in winter. Little or no food was taken in the first three-quarters of the winter, during ice cover.

FEEDING

In a Finnish woodland lake diatoms formed the most important food source, but utilisation of some degenerating blue-green algae could not be excluded (Armitage, 1968).

MICROHABITAT

The larvae are bottom dwellers, living mainly on open bottoms in the littoral zone of lakes (Kruglova, 1940; Brundin, 1949). Paasivirta (1976) collected the larvae in winter at somewhat greater depths than in summer. Kruglova (1940) collected them from a sediment of silt and undecayed plant material and reported their occurrence on other such water bottoms elsewhere in Russia.

DENSITIES

Brundin (1949) found mean densities of 1000–2000 larvae/m² and a maximum of 4500 larvae/m² in the lower littoral zone of the oligotrophic Lake Innaren in Sweden. Mossberg & Nyberg (1980) recorded densities from 20 to 2320 larvae/m² in another Swedish lake.

WATER TYPE

The larvae are inhabitants of the littoral and sublittoral zones of lakes (Kruglova, 1940; Paasivirta, 1976; Pinder & Reiss, 1983). Johnson (1989) found them also in the profundal zone of eight Swedish lakes; Brundin (1949) noted that the larvae were absent from the hypolimnion of stable stratified oligohumic lakes. All Dutch records are from shallow moorland pools. In view of its life cycle, the species cannot live in water bodies that dry out completely in late summer.

pH

The larvae have been recorded very often at pH 4–5 (Mossberg & Nyberg, 1980; Raddum & Saether, 1981; Raddum et al., 1984; Duursema, 1996). *P. orophila* increased in numbers during acidification of a Scottish lake, possibly because of oligotrophication of the lake (Brodin & Gransberg, 1993). Macan (1949) found the species in water with a pH of 6.7. Orendt found them in Bavaria at pH 8.1–8.3. At Mirwart in the Belgian Ardennes the species was common in a pond with a pH of 6.5–7.5 and rare in a pond with a pH of 6.9–9.5.

In most of the Dutch moorland pools he investigated, Klink (1986) found the remains of this species only during palaeolimnological analysis. He inferred that *P. orophila* disappeared due to acidification of the pools. However, under the influence of agricultural acidification the concentrations of sulphate, nitrate and ammonium have also increased strongly (van Dam & Buskens, 1993). The pH is probably not very important and the fact that in the Netherlands the species seems to be restricted to acid water may be due to the low oxygen demand of these water bodies.

TROPHIC CONDITIONS AND SAPROBITY

According to Saether (1979) the larvae are (at least in the littoral zone) characteristic inhabitants of oligohumic (not extremely) oligotrophic lakes and mesohumic to polyhumic lakes. Orendt (1993) found the species in Bavaria only in oligotrophic lakes. In the Netherlands all recent records are from undisturbed or only slightly disturbed moorland pools. Ruse (2002) stated that *P. orophila* is a characteristic species of water with low conductivity.

However, Johnson (1989) collected the species in the profundal zone of oligotrophic lakes in Sweden and stated that the species lived only in lakes which did not have too little phytoplankton and were not very poor in nutrients. Janecek (1995) found the species in Austria even in a fertilised carp pond. Brundin (1949) supposed that the larvae are not very sensitive to changes in oxygen content. It is possible that the seeming contradictions in preference for more or less oligotrophic water are attributable to differences in water temperature between boreo-alpine and lowland lakes.

Parachironomus Lenz, 1921

SYSTEMATICS

The genus *Parachironomus* should be placed in the *Harnischia* complex (Saether, 1977, 1977a). However, as Saether (1971) suggested, the genus may be the sister group of the other genera combined. The larva, in particular, is quite different from the other genera in the complex; for example the typical pecten epipharyngis carries several teeth of various lengths. Three provisional groups have been identified within the genus (Moller Pillot, 1984; Pinder & Reiss, 1983). These groups have no systematic value and the same is probably true for the groups suggested by Pinder & Reiss (1986), which are based on rather arbitrary characters.

In the Netherlands the *P. frequens* group and the *P. biannulatus* group have only one representative each; the other species belong to the *P. arcuatus* group. (For a discussion of the confusion between *P. biannulatus* and *P. vitiosus* see the entry for the former species.) In second instar the mentum and ventromental plates of *P. arcuatus* and *P. frequens* are hardly different (Shilova, 1968). Lehmann (1970) described 17 European species. Twelve of them will be treated here.

IDENTIFICATION

The adult males can be identified using Lehmann (1970) and most species also using Langton & Pinder (2007). Some species have been redescribed by Spies (2000). The most common females have been described and illustrated by Rodova (1978). Not all species are known as pupa and Langton (1991) had to use provisional names. The situation is even worse for the larvae: within gr. *arcuatus* identification is at present impossible. The species names used in the key by Pankratova (1983) are completely useless.

BIOLOGY AND ECOLOGY

In contrast to most other members of the *Harnischia* complex, the larvae of *Parachironomus* are only partly predaceous and some of them are parasites. Contrary to what is stated by Pinder & Reiss (1986) and Cranston et al. (1989) most species do not live in soft sediments, but can be found either in the water column on plants or other firm substrates or living as parasites on snails or in colonies of Bryozoa. No species in Western Europe is known as a typical leaf or stem miner, but (in contrast to many other species of the *Harnischia* complex) most species build tubes stuck to the substrate (Lenz 1954–62; Shilova, 1968). Reiss (1968a) suggested that the genus is probably more or less thermophilous and does not occur in alpine lakes above 1000 m. However, some species also live in subarctic lakes in Sweden (Brundin, 1949).

DISPERSAL

The larvae of most species often swim, at least in second instar (Shilova, 1968) and often settle on plants or other substrates near the water surface (e.g. Lenz, 1954–62; Reiss, 1968). In drift samples in the Polish Warta River, Grzybkowska (1992) found *Parachironomus* larvae more abundantly than any other Chironomini species. The adults of *P. arcuatus* often colonise new water bodies and Dettinger-Klemm (2003: 74) considered *P. parilis* to be a ubiquitous coloniser. The genus as a whole contains a range of species with a high capacity to find a suitable habitat, either in the adult or larval stage.

Parachironomus gr. arcuatus

Cryptochironomus gr. *pararostratus* Chernovskij, 1949: 66–67, fig. 32

Most species of the genus belong to this larval group (see above). Lenz (1954–62: 199 et seq.) distinguished three other groups within this group, mainly differing in antennal and labral characters. The species will be treated separately, as far as possible.

Parachironomus arcuatus Goetghebuer, 1921

DISTRIBUTION IN EUROPE AND THE NETHERLANDS

The species occurs in most of Europe, except for some parts of the Mediterranean area (Fittkau & Reiss, 1978; Saether & Spies, 2004). The species has been found throughout the Netherlands, but is less common in regions where fresh, non-acid stagnant water is scarce (Moller Pillot & Buskens, 1990; Limnodata.nl).

LIFE CYCLE

Shilova (1968) observed two generations a year in Russia, emerging in June and August. Otto (1991: fig. 47) observed at least three emergence maxima in northern Germany. Fritz (1981) reported emergence in a German pool in April, June/July and September (only one specimen). Lenz (1954–62: 205) supposed that *Parachironomus* does not have a well-defined emergence period. He found pupae and adults from early May until early October, with only slightly higher numbers of emergences in some periods. This is consistent with the data from the Netherlands.

Larval development is completed in not much more than one month in summer, as indicated by investigations by Wotton et al. (1992). In sand filter beds, temporarily filled with water for 16–77 days, these authors recorded the first emergence peak after 21–38 days.

In winter most larvae are in diapause in second and third instar, but some fourth instar larvae have also been found (own data). Third instar larvae can be present until the middle of April.

FEEDING

The larvae build a tube of algae and detritus and filter the water by undulating movements. They often leave the tube and gather food in the area around the tube (Shilova, 1968). The food consists of algae and detritus, but they also consume eggs and pupae of other chironomids (own observations; Lenz, 1954–62: 204). However, according to Konstantinov (1961) the larvae of gr. *arcuatus* (as *Cryptochironomus* gr. *pararostratus*) are facultative predators, feeding mainly on small chironomid larvae and Oligochaeta. Konstantinov supposed that *P. arcuatus* larvae are mainly predatory when living on plants, but when living between algae or in the mud they switch to vegetable food because they are less able to find their prey.

MICROHABITAT

The larvae live mainly on plants and firm substrates and (in summer) rarely on or in sandy or silty bottoms (Brundin, 1949; Lenz, 1954–62; Reiss, 1968; Lehmann, 1971: 530; Mackey, 1976). They can be found in high densities on nymphaeids, *Stratiotes*, *Ceratophyllum* and *Glyceria maxima*, but less numerously on *Phragmites*. They are rarely found in mines in water plants (Urban, 1975: 421; Michailova, 1989) and such mines are without doubt made by other species. Higler (unpublished data) collected the larvae on artificial (plastic) 'Stratiotes' plants at least as numerously as on true plants. Mol et al. (1982) collected them in more or less the same numbers on plants and artificial plants. In winter, larvae have been found on dead plant material and some of them may then live on or in the bottom.

In the littoral zone of lakes and in rivers the larvae need some shelter to avoid being washed away. Mol et al. (1982) rarely collected a larva on the open bottom in Lake Maarsseveen. Meuche (1939) rarely found *Parachironomus* larvae between filamentous algae in the littoral zone of lakes. They are often present on stones in lakes or rivers (Klink & Moller Pillot, 1982; Peeters, 1988; Brodersen et al., 1998). Smit (1982) found them in the river Meuse mainly on the undersides of stones in the littoral zone. Brundin (1949: 160) supposed that older larvae often swim and thus reach firm substrates in the water layer (see also under *P. vitiosus*). Lenz (1954–62) argued that the larvae are not planktonic, but are probably forced to change location relatively often (see Dispersal).

DENSITY

Koehn & Frank (1980, as *P. cryptotomus*) found 2000 larvae/m^2 locally in a polluted slow-flowing channel in Berlin.

EGGS

The egg mass has been described by Shilova (1968). Balushkina (1987) reported 560 eggs per egg mass for *P. pararostratus*.

WATER TYPE

Current

The larvae live in stagnant as well as flowing water. Lehmann (1971) called *P. arcuatus* more or less rheophilous, but he collected the species only in the lower reaches (potamal zone) of the river Fulda. Orendt (2002a) recorded the species in several fast-flowing streams in Bavaria, but Bazerque et al. (1989) hardly found any *Parachironomus* larvae at stations with a fast current and Braukmann (1984) collected them only in some lowland brooks. The larvae appear to be rare or absent from the Dutch fast-flowing streams Geul and Gulp (Cuijpers & Damoiseaux, 1981; Waterschap Roer en Overmaas, unpublished). The species is rather common in larger rivers, at least in the rivers Meuse (Smit, 1982; Peeters, 1988) and Rhine (Becker, 1994), where the larvae live where the current is weaker (e.g. the underside of stones, see under Microhabitat).

Shade

The larvae seem to be rare or absent in brooks and ditches shaded by trees (e.g. Tolkamp, 1980). The reason for this may be the scarcity of plants (as habitat) and/or algae (as food).

Dimensions

The larvae are often absent from small pools and narrow ditches (Kreuzer, 1940; Schleuter, 1985; Krebs & Moller Pillot, in prep.). However, they have been encountered regularly in water bodies less than 4 m wide and less than 30 cm deep (Steenbergen, 1993). We found hardly any larvae in polluted ditches when only a water layer of less than 10 cm was left above a thick layer of anaerobic silt. In very large lakes the densities are often low, either because of a lack of substrate or the influence of water movement. Tõlp (1971) found the larvae down to 5 m deep. In Lake Vechten (Netherlands) the larvae living on *Ceratophyllum demersum* were only numerous in a zone with a depth of 2.5–4 m (Dvořák & Best, 1982). The larvae are not found at a greater depth in storage reservoirs (Mundie, 1957; Kuijpers, pers. comm.). The species is common in large rivers (see, however, under Microhabitat and Current).

Permanence
P. arcuatus occurs often in temporary pools, especially larger pools (Wotton et al., 1992; own data). The larvae are absent from such pools in spring if they are only filled with water after the previous October. Moreover, the presence of enough adults in the neighbourhood is a significant factor. The figures given for temporary water in the matrix table in Chapter 4 are influenced by the fact that many temporary waters are small, shaded or acid.

pH

The larvae are nearly absent from very acid waters (Raddum & Saether, 1981; Barnes, 1983; Buskens, 1983, 1987; Leuven et al., 1987). The lowest pH mentioned in the literature is 4.6 (Werkgroep Hydrobiologie, 1993). Steenbergen (1993) found significantly more larvae at pH >7.5.

TROPHIC CONDITIONS AND SAPROBITY

Larvae of *Parachironomus* gr. *arcuatus* can live in both very eutrophic and in oligotrophic/mesotrophic conditions (Brodersen et al., 1998; Langdon et al., 2006). Information in Limnodata.nl and Steenbergen (1993) does not indicate a significant preference for more or less eutrophicated or more or less oxygenated water. However, the larvae are often absent and never numerous in oligotrophic/mesotrophic conditions (Brundin, 1949; Buskens, 1983; van Kleef, unpublished; Orendt, 1993; Schmale, 1999). In the river Meuse Smit (1982) found the larvae most abundantly in polluted stretches. From our own investigations it appears that the larvae are often absent in polysaprobic conditions, but that they are sometimes most numerous in stretches with rather severe pollution, even if the oxygen content drops down every night to nearly zero. As far as has been verified, this was always the case for the species *P. arcuatus*. Bazerque et al. (1989) found *Parachironomus* larvae in heavily polluted stretches of the rivers Selle and Somme in France. Wilson & Ruse (2005) also considered *P. arcuatus* to be pollution tolerant (tolerance group D), more than other species of this genus. They noted that the presence of exuviae is often associated with organic enrichment in canals. Because the occurrence of the larvae is in some degree dependent on the presence of a suitable substrate (e.g. plants or stones), pollution may be only an indirect cause of the absence of the species. The larvae take advantage of the richness in food in hypertrophic water and colonise the places where oxygen deficiencies are infrequent or have little effect. Their presence indicates that there are spots where enough food is present and the larvae are protected against the greatest negative effects of pollution and fast currents.

SALINITY

Parachironomus arcuatus larvae live in fresh and slightly brackish water (Krebs, 1982), but are less common in brackish water. In the Camargue Tourenq (1975) found the species only in fresh water and Steenbergen (1993) collected significantly fewer larvae in water with more than 1000 mg Cl/l. In the Netherlands there are several records of larvae in water with chloride contents up to 2000 mg/l and few records from water with chloride contents up to 4810 mg/l (Schreijer, 1983; Krebs & Moller Pillot, in prep.). Tõlp (1971) recorded broadly the same maximum in Estonia: a total salt content of 7200 mg/l.

DISPERSAL

As mentioned under the genus, the larvae of some *Parachironomus* species seem to swim more than other genera. This enables them to reach suitable places near the water surface more often. The fact that many temporary water bodies are occupied indicates rather frequent migration of the females by air. Although dispersal of adults and larvae appears to be the most conspicuous aspect of the strategy of the species, *P. arcuatus* does not fit exactly into one of the life-history tactics described by Verberk (2008). The species displays a clear choice for a strategy with low food specialisation, low concurrence power and a high dispersal capacity to find a place where there is much food in summer but no extreme and harmful conditions.

Parachironomus biannulatus (Staeger, 1893)

Parachironomus vitiosus Shilova, 1965: 102 et seq. (only larva)
Parachironomus gr. *vitiosus* Moller Pillot, 1984: 112, 231; Pinder & Reiss, 1983: 329, fig. 10.51 B, D, F–H; Steenbergen, 1993: 541
nec *Parachironomus biannulatus* Shilova, 1968: 114 et seq.

SYSTEMATICS AND NOMENCLATURE; IDENTIFICATION

Shilova (1968) identified *P. biannulatus* incorrectly. The hypopygium (fig. 11b) refers to another species. Also the pupa (fig. 8a, b and 10b) does not correspond with *P. biannulatus* in Langton (1991: 270, fig. 110e–h) because the pupa of Shilova has a transverse spine band on sternite II and the hook row on tergite II interrupted. The larva and pupa are described in Shilova (1965) as *P. vitiosus* (see under this species). Larvae identified using Moller Pillot (1984) or Pinder & Reiss (1983) as *P. vitiosus* belong to *P. biannulatus* because their descriptions were based on Shilova (1965). As far as is known, no larvae have been incorrectly identified as *P. biannulatus* based on the description given by Shilova (1968).
Identification of females is possible using Rodova (1978: 95 et seq., fig. 28 pro parte).

DISTRIBUTION IN EUROPE AND THE NETHERLANDS

P. biannulatus has been recorded throughout Europe, except parts of the Mediterranean area and parts of Scandinavia (Fittkau & Reiss, 1978; Saether & Spies, 2004). The species is widely distributed in the Netherlands, but seems to be absent from the province of Zeeland (Nijboer & Verdonschot, unpublished; Limnodata.nl).

OVIPOSITION

The eggs may be deposited on the water surface (Munsterhjelm, 1920: 126). This author counted approx. 250 eggs in an egg mass.

MICROHABITAT

The larvae stick their tubes to a firm substrate, either on the bottom or on plants (Shilova, 1965, as *P. vitiosus*). Smit (1982) found the larvae in low numbers on stones in the river Meuse. Peeters (1988) collected the species in this river not on stones in the littoral zone, but on the bottom in places with gravel or pebbles.

FEEDING

According to Shilova (1965, as *P. vitiosus*) the larvae build tubes and nets using excretions from their salivary glands. They filter the water through the net by moving their bodies and catch algae.

SOIL

Steenbergen (1993) collected the species (as *P. vitiosus*) significantly more often on peat than on clay. In the river Meuse the larvae live on bottoms with gravel or pebbles (Peeters, 1988). Becker (1994) reared a male adult from stones in the Rhine.

WATER TYPE

Current

The species is a common inhabitant of lakes, watercourses and canals (Kruseman, 1933; Limnodata.nl) and scarcely found in brooks and rivers (Smit, 1982; Moller Pillot & Buskens, 1990; Caspers, 1991; Limnodata.nl). Lehmann (1971) did not report finding the species in the river Fulda. According to Shilova (1965) the larvae avoid strong currents. Steenbergen (1993) did not find the larvae in flowing water.

Dimensions

Steenbergen (1993) collected the species significantly more often in larger water bodies, but the larvae were sometimes also present in ditches less than 4 m wide. In the Rybinsk reservoir the larvae live mainly at a depth of 1.5 – 2 m, more scarcely up to 15 m deep (Shilova, 1976). They are scarce in large rivers, but this may be due to currents (Smit, 1982; Peeters, 1988; own data).

Permanence

The larvae have not been found in temporary water.

pH

Although Steenbergen (1993) mainly investigated basic water, he found *P. biannulatus* (as *P.* gr. *vitiosus*) slightly more frequently in water with pH <7.5 and significantly more frequently in water with low calcium and magnesium contents (see also under Trophic conditions). In investigations in more or less acid water (Buskens, 1983; Verstegen, 1985; Leuven, 1987; van Kleef, unpublished) the species was never found at pH values lower than 6.95.

TROPHIC CONDITIONS AND SAPROBITY

Steenbergen (1993) found the larvae significantly more often in water with a low phosphate content and with a lower chlorophyll-a content (< 50 µg/l). Limnodata.nl also gives a low mean orthophosphate content (0.038 mg P/l). Schmale (1999) reported the species from mesotrophic dune lakes in the Netherlands (Berkheide), but in very low numbers. According to Ruse (2002) the species is characteristic of water with relatively high conductivity, but this does not mean that such water is severely eutrophicated.

SALINITY

Steenbergen (1983) found *P. biannulatus* significantly more often in fresh water and never in water with a chloride content above 1000 mg/l. There are no other records from brackish water.

Parachironomus cinctellus Goetghebuer, 1921

IDENTIFICATION

The adult male is absent from Langton & Pinder (2007) and can only be identified using Lehmann (1970). The characteristic hypopygium has been illustrated also by Goetghebuer (1937–54: fig. X:142).

DISTRIBUTION IN EUROPE

The species has been recorded in very few European countries, including Belgium and Germany (Seather & Spies, 2004). In the Netherlands some male adults were collected by Kouwets & Davids (1984) near Lake Maarsseveen from May to September.

Parachironomus danicus Lehmann 1970

IDENTIFICATION

The male is not keyed in Pinder (1978), but can be identified using Langton & Pinder (2007).
The pupa is described and keyed in Langton (1991). The larva is unknown.

DISTRIBUTION IN EUROPE

P. danicus has been recorded only in the British Isles, Denmark, Germany and Poland (Saether & Spies, 2004).

WATER TYPE

According to Langton (1991) the larvae live in ponds. Exuviae have been collected in a fast-flowing brook in Bavaria (Orendt, 2002a).

Parachironomus digitalis (Edwards, 1929)

DISTRIBUTION IN EUROPE AND THE NETHERLANDS

P. digitalis has been recorded in large parts of Europe, except the Mediterranean area (Fittkau & Reiss, 1978; Saether & Spies, 2004). There are two records from the Netherlands: one male collected by Kouwets & Davids (1984) near Lake Maarsseveen in August 1979 and one male collected by Klink (1985a) near Kampen (along the river IJssel).

WATER TYPE

According to Fittkau & Reiss (1978) the species lives in small to large stagnant and flowing waters. Brundin (1949) reported the occurrence of this species in lakes in Sweden and in the river Kossau in Germany. Becker (1994) reared three males from stones in the river Rhine near Koblenz in Germany, and adult males have been collected along fast-flowing streams in Central Europe (Lehmann, 1971; Orendt, 2002a). Paasivirta & Koskenniemi (1980) found the larvae of *P. digitalis* in the moss-grown littoral zone of a polyhumic lake in Finland.
Schnabel & Dettinger-Klemm (2000) reared the species from two temporary pools in the floodplain of the river Lahn in Germany.

Parachironomus frequens (Johannsen, 1905)

D

Parachironomus longiforceps Ertlová, 1974: 869 et seq.; Shilova, 1968: 104, 117–123
Parachironomus gr. *longiforceps* Moller Pillot, 1984: 227–229
Parachironomus spec. Kampen Moller Pillot, 1984: 109, 230, fig. IV 2.h.4, 25c, 25d

SYSTEMATICS AND IDENTIFICATION

Janecek (pers. comm.) suggested that *P.* spec. Kampen is not the same species as *P. frequens*. However, at least some larvae have no notch in the central mental tooth by abrasion. In second instar the larva of *P. frequens* can hardly be distinguished from *P. arcuatus* (Shilova, 1968). Older larvae are red-green in colour (Ertlová, 1974).

DISTRIBUTION IN EUROPE AND THE NETHERLANDS

P. frequens has been recorded throughout Europe (Saether & Spies, 2004). In the Netherlands the species lives in suitable water bodies throughout the country, but may be absent from the province of Zeeland (Limnodata.nl; Nijboer & Verdonschot, unpublished data; own data).

LIFE CYCLE

Ertlová (1974) found the species in the Danube emerging from May until late September, most numerously in July and August. Lehmann (1971: 512) suggested that *P. frequens* is a univoltine summer species.

EGGS

The egg mass and eggs have been described by Shilova (1968). The eggs are most probably deposited near Bryozoa colonies because the second instar larvae do not swim to seek a place to live, as opposed to other *Parachironomus* species.

MICROHABITAT

The larvae live between Bryozoa (*Plumatella repens*) on stones (Ertlová, 1974). This author found them mainly in the lower part of the littoral zone of the river Danube. Peeters (1988) collected them on the bottom of the river Meuse where this consisted of pebbles and cobbles. Klink (1989) pointed out that this species could settle in the Rhine and became numerous in the last few centuries because the banks were protected by stones to prevent erosion.

DENSITY

Ertlová (1974) found a maximum of 6–8 larvae or pupae in 1 cm^3 of bryozoan colonies.

WATER TYPE

Current

Parachironomus frequens is a typical inhabitant of the lower courses ('Barbenregion') of rivers (Lehmann, 1971; Ertlová, 1974). The larvae are common and often numerous in the rivers Rhine, IJssel and Meuse (Caspers, 1980, 1991; Klink & Moller Pillot, 1982; Smit, 1982; Klink, 1985a; Peeters, 1988). The larvae have been collected scarcely in the slow-flowing lower courses of brooks and small rivers in the Netherlands, but very rarely in the narrower and faster flowing upper courses. Orendt (2002a) reported occurrence in three rather small rivers in Bavaria.
The larvae are usually absent from stagnant water bodies such as lakes and ponds, but they can be present there, sometimes numerously (Reiss, 1968; Tourenq, 1975). In the Netherlands there are only a few records from stagnant water (usually canals).

Dimensions

Although the majority of records in Europe are from large rivers, Tourenq (1975) found *P. frequens* numerously in a canal less than 4 m wide. In the Netherlands the larvae have been found in some small streams (e.g. Linge) and even (rarely) in ditches. The presence of Bryozoa is the most determining factor.

Permanence

Tourenq (1975) also reported emergences in temporary marshes, especially when the marsh dried up only after the end of August. Temporary water bodies where this species can live are unlikely to exist in many places in the Netherlands.

pH

The larvae are rarely found in acid water, but they were rather common in a Finnish reservoir at pH 5.5–6 (Koskenniemi & Paasivirta, 1987).

TROPHIC CONDITIONS AND SAPROBITY

Without doubt *P. frequens* is a highly pollution tolerant species. The larvae were already present in the polluted stretches of the Rhine and Meuse in the 1970s, when the pollution was rather severe (Caspers, 1980; Smit, 1982; own observations). Ruse (2002) found *P. frequens* only in water with high a conductivity (mean 751 μS/cm). Their scarcity in stagnant water is probably due to their rather high oxygen requirements. However, our data do not permit us to estimate any preference for or tolerance of pollution with any accuracy. Tourenq (1975) demonstrated that the oxygen content often fluctuated sharply, but he did not measure the oxygen content at night. The figures in table 2 (Chapter 4) are provisional.

SALINITY

The larvae have been found mainly in fresh water, but they can survive oligohaline and probably also mesohaline conditions (Tourenq, 1975). Ruse (2002) collected the species mainly in water with a relative high conductivity, but not higher than 1000 μS/cm.

Parachironomus mauricii Kruseman, 1933

Chironomus varus limnaei Guibé, 1942: 1–15
Parachironomus sp. Pe Langton, 1991: 268, fig. 110 a–d

IDENTIFICATION

The male is not keyed in Pinder (1978), but can be identified using Langton & Pinder (2007).
The pupa is described and keyed in Langton (1991) as ? *Parachironomus* sp. Pe. Guibé (1942) gave a short description of the larva, from which it can be concluded that the pecten epipharyngis has few teeth and the labral setae are shorter than in other species of the genus.

DISTRIBUTION IN EUROPE AND THE NETHERLANDS

The species probably lives in the whole European mainland except for the Mediterranean area, but is rather rarely found (Krebs, 1988; Saether & Spies, 2004). In the Netherlands there are records from the provinces of Noord-Holland, Zuid-Holland, Friesland, Overijssel and Limburg (Kruseman, 1933; Krebs, 1988; Limnodata.nl; Vallenduuk, pers. comm.; van Kleef, unpublished).

LIFE CYCLE

Adults are collected from May until September; the species is probably at least bivoltine (Krebs, 1988). Young (1973) suggested only one generation, but most probably incorrectly (see *P. varus*).

FEEDING

Guibé (1942) observed that the larvae fed on the body fluid of Lymnaeidae. In no cases did he observe any visual harm to the host.

MICROHABITAT

The larvae are parasites of snails of the family Lymnaeidae. H. Vallenduuk found them on *Radix*; the larvae lived within the shell without visual harm to the snail (compare Feeding). Young larvae were found on young snails. Guibé (1942) stated that the young larvae infected the snail and in some cases came near the head of the snail and were devoured. Older larvae appeared to be unable to infect a host. The prepupae leave their host before pupation.
Young (1973) also collected *Parachironomus* larvae in low numbers on other snails, and even on the bivalve *Sphaerium corneum*. This may also apply to *P. mauricii* or *P. varus*. He found up to 70% infection on *Lymnaea peregra*.

WATER TYPE

P. mauricii has been collected in dune lakes, ponds, pools, ditches and lowland brooks (Shilova, 1976; Krebs, 1988; Limnodata.nl; Vallenduuk, pers. comm.). Ruse (2002a) stated that the species settled in gravel lakes within three years. Because they depend on the presence of Lymnaeidae, the larvae do not occur in acid water.

Parachironomus monochromus (van der Wulp, 1874)

Parachironomus monochromus Lehmann, 1970: 146–147
Tendipes monochromus Kruseman, 1933: 192, fig. 53; Goetghebuer, 1937–1954: 46, fig. 149
nec *Tendipes monochromus* Edwards, 1929
nec *Chironomus monochromus* Iovino & Miner, 1970

IDENTIFICATION

In the older literature (before 1933) this species is confused with *P. tenuicaudatus* (Malloch), but this had only influence on identifications after 1933 in North America. Male adult, pupa and larva have been redescribed by Spies (2000). The larva belongs to gr. *arcuatus* sensu Moller Pillot (1984), see Spies (2000). Identification of females is possible using Rodova (1978: 95, 99).

DISTRIBUTION IN EUROPE AND THE NETHERLANDS

P. monochromus has been recorded in large parts of Europe, except for parts of the Mediterranean area and parts of Scandinavia (Fittkau & Reiss, 1978; Saether & Spies, 2004). Data from North America are based on misidentifications (Spies, 2000).
In the Netherlands Kruseman (1933) collected adult males at five localities in different parts of the country; there are very few recent records of adult males and exuviae.

LIFE CYCLE

Kruseman (1933) collected adults in the Netherlands in May and September. Schleuter (1985) reported emergence near Bonn, Germany, from April to September, with the largest numbers in summer.

WATER TYPE

Kruseman (1933) caught the adults along lakes in the fenland districts and in the dunes. Brundin (1949) found the species in low numbers in three oligotrophic and one eutrophic lake in Sweden. Lehmann (1970) and Langton (1991) recorded *P. monochromus* mainly in small pools and ditches. Schleuter (1985) and van Kleef (unpublished) also collected the species in pools. There are few records from flowing water (Rossaro, 1987; Spies, 2000).

pH

P. monochromus emerged in large numbers from a pool with a pH varying between 5 and 6 (Schleuter, 1985). Most other records are from water bodies with a higher pH.

Parachironomus parilis (Walker, 1856)

IDENTIFICATION

Adult males have been described and/or keyed by Kruseman (1933), Lehmann (1970), Pinder (1978) and Langton & Pinder (2007). In Pinder (1978) the species can be confused with *P. swammerdami* or *P. mauricii* (see Langton & Pinder, 2007: 180). Females differ from *P. arcuatus* mainly by the pale apodemes (Rodova, 1978: 95, 99, fig. 30). The exuviae have been keyed by Langton (1991: 272). Larvae belong to gr. *arcuatus* sensu Moller Pillot, 1984.

DISTRIBUTION IN EUROPE AND THE NETHERLANDS

P. parilis most probably lives in the whole European mainland and the British Isles (Fittkau & Reiss, 1978; Saether & Spies, 2004). In the Netherlands adult males, pupae and exuviae have been collected in or near more than 30 water bodies throughout the country (e.g. Kruseman, 1933; van Kleef, unpublished).

LIFE CYCLE

The life cycle is most probably no different from that of *P. arcuatus*. Adults and exuviae have been collected from May to early October (Kruseman, 1933; Brundin, 1949; Smith & Young, 1973; Schleuter, 1985: 136, 158; own results).

MICROHABITAT

The larva probably behaves no differently from *P. arcuatus*. Michailova (1989) mentioned that *P. parilis* also mines in leaves and stems of submerged plants (but see under *P. arcuatus*). Smith & Young (1973) collected the larvae from their emergence traps.

OVIPOSITION

One egg mass contains about 500 eggs (Munsterhjelm, 1920; Nolte, 1993).

WATER TYPE

Current and dimensions

P. parilis is much less common in flowing water than *P. arcuatus*. Caspers (1980, 1991) did not find the species in the Rhine and Klink & Moller Pillot (1982) did not find it in the Rhine or the Meuse. Lehmann (1971) collected few adults along the lower reaches of the river Fulda. Occurrence in slow-flowing streams has been recorded in Langton (1991) and Limnodata.nl, but the majority of records are from pools, ditches and lakes. The species appears to be more common in moorland pools than *P. arcuatus* (van Kleef, unpublished). Possibly *P. parilis* is more common in smaller water bodies.

Permanence

The species seems to be rare in temporary water. We found only one record by Dettinger-Klemm (2003: 71).

D

pH

P. parilis was reared and collected from a few pools with a pH of about 5 by Schleuter (1985) and by H. van Kleef (unpublished). Most other records are from less acid or even basic water.

TROPHIC CONDITIONS AND SAPROBITY

In contrast to *P. arcuatus*, Brundin (1949) collected *P. parilis* only in oligotrophic lakes in Sweden. In the Netherlands the species has not been found in very polluted water. Except for large rivers, the numbers of collected pupae and adults of *Parachironomus*, identified to species level, are low. *P. parilis* appears to be the more common species in mesotrophic environments, for instance in the moorland pools investigated by H. van Kleef (unpublished results). There is one record from very eutrophic water (Krebs, 1984: 92). *P. parilis* most probably prefers a lower trophic level than *P. arcuatus*, which is reflected by the lower mean pH values because in the Netherlands mesotrophic pools are often more or less acid.

SALINITY

P. parilis is probably rare in brackish water. It has been found once in water with 1300 mg chloride/l in the Dutch province Zeeland (Krebs, 1984: 92), but has not been collected in other brackish water bodies in the province (Krebs, 1981, 1984). Tourenq (1975) also found the species only in fresh water.

Parachironomus swammerdami (Kruseman, 1933)

IDENTIFICATION

The male is not keyed in Pinder (1978), but can be identified using Langton & Pinder (2007).

The pupa and larva are unknown.

DISTRIBUTION IN EUROPE AND THE NETHERLANDS

The species has been recorded only in the British Isles, the Netherlands, Germany and Poland (Saether & Spies, 2004). The only Dutch record concerns the type specimen, collected by Kruseman (1933) near Amsterdam.

Parachironomus tenuicaudatus (Malloch, 1915)

Parachironomus baciliger Kruseman, 1933: 193, fig. 53a
Parachironomus bacilliger Mundie, 1957: 193; Lenz, 1954–62: 199 et seq., figs 224, 237–240

SYSTEMATICS AND IDENTIFICATION

Adult male, pupa and larva have been redescribed by Spies (2000). The larva belongs to gr. *arcuatus*; Lenz (1954–62) attributes the species to his gr. *varus*. At present the larva cannot be identified to species level. The female is unknown.

DISTRIBUTION IN EUROPE AND THE NETHERLANDS

P. tenuicaudatus has been found in many countries, scattered over nearly the whole European mainland (Saether & Spies, 2004). In the Netherlands 13 records are known from all parts of the country.

LIFE CYCLE

Adults have been collected from the end of May or early June to the end of September (Mundie, 1957; Reiss, 1968; Janecek, 1995; Kouwets & Davids, 1984). The last authors found the species most numerously in July and August. Mundie (1957) noted that the species probably has two generations a year, but locally only one.

FEEDING

Lenz (1951) observed four larvae eating a dead pupa of *Chironomus* and noted that they also attack and eat pupae of their own species. Lenz supposed that this behaviour is not specific to *P. tenuicaudatus*, but thinks that most or all *Parachironomus* larvae can adapt themselves to the available food.

MICROHABITAT

The larvae most probably live mainly on submerged macrophytes (Spies, 2000). In Lake Constance the larvae lived on glass traps near the water surface (Reiss, 1968).

Verneaux & Aleya (1998) reared the species in low numbers from artificial substrate at the bottom of the littoral zone of Lake Abbaye in the Jura. Janecek (1995) collected the larvae between epiphytic algae and Bryozoa.

pH

P. tenuicaudatus has not been found in acid water.

WATER TYPE

The larvae live in flowing and stagnant water (Spies, 2000). The species has been collected rarely in fast-flowing streams (Orendt, 2002a). Several pupae have been collected in lowland brooks (partly det. P. Langton, partly Limnodata.nl). Other Dutch records are from wider ditches and lakes and one from a smaller pool (H. van Kleef, unpublished).

Parachironomus varus (Goetghebuer, 1921)

IDENTIFICATION

The male has been keyed by Pinder (1978) and Langton & Pinder (2007). The female can be identified using Rodova (1978). The pupa was not known with certainty (see Langton, 1991: 275), but Langton & Visser (2003) no longer make this reservation. The larva has been described by Guibé (1942); in particular, the pecten epipharyngis is different from that of other *Parachironomus* species. Lenz (1954–62) also gives a short description.

DISTRIBUTION IN EUROPE AND THE NETHERLANDS

P. varus has been found on nearly the whole European mainland, but records are rather scarce (Saether & Spies, 2004). In the Netherlands the species has been collected in the Twenthe district, the northwestern part of the province of Overijssel and near Amsterdam (Kruseman, 1933; van Benthem Jutting, 1938; Tj.H. van den Hoek, unpublished).

LIFE CYCLE

Young (1973) suggested there is only one generation of *Parachironomus*, living on *Physa*, and only emerging in spring (May–June). However, Kruseman (1933) collected adult males from June to September. Young's results (continuous presence of infected snails in summer) can probably be explained by the poor synchronisation of the second generation.

FEEDING

The larva lives on the snail *Physa fontinalis* and eats from the tissues of the living host (van Benthem Jutting, 1938; Guibé, 1942).

MICROHABITAT

The larva lives on living snails of the genus *Physa* and rarely on other snails (van Benthem Jutting, 1938; Guibé, 1942; Vallenduuk, pers. comm.). They spin a mucous tube on the outer side of the shell, usually along the pulmonary cavity. In contrast to *P. mauricii*, the prepupa spins a tube for pupation on the snail, which dies as a result (van Benthem Jutting, 1938).

WATER TYPE

As far as is known, the species is found in stagnant water (van Benthem Jutting, 1938; Guibé, 1942). Verneuax & Aleya (1998) reared the species from an artificial substrate on the bottom of Lake Abbaye in the Jura.

Parachironomus vitiosus (Goetghebuer, 1921)

Parachironomus vitiosus (adult male and female) Shilova, 1965: 102 et seq.
nec *Parachironomus vitiosus* (larva and pupa) Shilova, 1965: 102 et seq.
nec *Parachironomus* gr. *vitiosus* Moller Pillot, 1984: 112, 231; Pinder & Reiss, 1983: 329, figs 10.51 B, D, F–H

IDENTIFICATION AND NOMENCLATURE

The male can be identified using Pinder (1978) and Langton & Pinder (2007), the female using Rodova (1978). Shilova (1965) described adults (from Rybinsk reservoir), eggs and larvae, but the larvae of the second to fourth instar and the pupae were sampled separately in the river Volga. Without doubt these pupae belong to *P. biannulatus*. A. Klink found a pupa of *P. biannulatus* with a larval skin as described by Shilova as *P. vitiosus*. (See also under *P. biannulatus*). Because many authors have identified the larvae incorrectly on the basis of Shilova's description, all identifications of *P. vitiosus* larvae made using keys published before 2004 must in fact be *P. biannulatus*. Pupae and exuviae can be correctly identified using Langton (1991). The larva is not known and possibly belongs to gr. *arcuatus*.

DISTRIBUTION IN EUROPE AND THE NETHERLANDS

The species lives throughout Europe, except Spain and Portugal (Saether & Spies, 2004).
In the Netherlands *P. vitiosus* has been collected by Kruseman (1933) at three places, in the provinces of Noord-Holland, Zuid-Holland and Utrecht, and by H. van Kleef (unpublished) in two pools in the southern and eastern part of the country.

LIFE CYCLE

In English storage reservoirs there appeared to be two generations a year (Mundie, 1957), emerging from early June to September. Shilova (1965) also found two generations in the Rybinsk reservoir in Russia. In the Boden See Reiss (1968) observed emergence from May until the middle of October. However, Lehmann (1971) supposed that in the Fulda region *P. vitiosus* had to be attributed to the univoltine summer species.

EGGS

The females have been reported as depositing egg masses with about 300 eggs (Shilova, 1965; most probably this refers to *P. vitiosus*).

MICROHABITAT

In Lake Constance the larvae lived on glass traps near the water surface and probably also on plants (Reiss, 1968). In Lake Belau in Germany the species only emerged from the reed zone, not from the open water (Otto, 1991: 43). Verneaux & Aleya (1998) reared the species in low numbers from an artificial substrate at the bottom of Lake Abbaye in the Jura.

WATER TYPE

P. vitiosus lives in lakes, medium-sized pools and streams (Fittkau & Reiss, 1978). In the storage reservoirs investigated by Mundie (1957) this species was the most abundant of the genus. H. van Kleef (unpublished) found the exuviae in two pools in the Pleistocene part of the Netherlands; the records by Kruseman (1933) are from the Holocene fen-peat area. The species is scarce in streams and rivers (Lehmann, 1971; Caspers, 1991); there are no records from streams in the Netherlands.

pH

The species is most probably scarce in acid water, but in two cases *P. vitiosus* has been recorded in water with a pH between 5.5 and 6 (Koskenniemi & Paasivirta, 1987; van Kleef, unpublished).

TROPHIC CONDITIONS AND SAPROBITY

Brundin (1949) collected *P. vitiosus* in oligotrophic/dystrophic lakes. Most other records are from more or less mesotrophic environments, for instance in the fen-peat area in the Netherlands in the 1930s (Kruseman, 1933) and two restored pools investigated by H. van Kleef (unpublished). However, the volume of data is insufficient to allow definite conclusions to be drawn.

Paracladopelma Harnisch, 1923

SYSTEMATICS AND NOMENCLATURE

According to Cranston et al. (1989) the European species belong to three groups: *P. mikianum* group, *P. nigritulum* group and *P. camptolabis* group. The last group contains *P. camptolabis* and *P. laminatum*. Owing to a mistake, the name *P. camptolabis* agg. was used for *P. nigritulum* in Moller Pillot (1984). This was corrected in Moller Pillot & Buskens (1990: 24, 86–87). In Dutch keys the name *P. laminatum* agg. is used for *P. camptolabis* gr.

IDENTIFICATION

A key to the species groups for adult males can be found in Cranston et al. (1989) and the hypopygia are illustrated in fig. 10.46. Very good figures of the hypopygia of *P. camptolabis* and *P. laminatum* have also been published in Lehmann (1971: 542–543). Langton & Pinder (2007) made a key for the three British (and Dutch) species. Exuviae of these three species can be identified using Langton (1991).
The larval groups are easiest to identify using Moller Pillot (1984); however, see above for the mistake in nomenclature. The larva of *P. camptolabis* can probably be distinguished in fourth instar from *P. laminatum* because the former has a brown frontal apotome, genae and gula, and the latter does not.

FEEDING

Konstantinov (1961) considered the larvae of the genus *Paracladopelma* to be obligate predators.

IRON

Rasmussen & Lindegaard (1988) collected larvae of the genus *Paracladopelma* only in water with less than 0.2 mg Fe^{2+} per l, less than for any other species of chironomids. However, the genus was scarce in the investigated river system. Moller Pillot (2003) also collected the larvae of *P. camptolabis* on soils of rusty silt (details unpublished). Ferrous ions can be expected only if there is upward seepage of ground water or in very oxygen poor circumstances (see Chapter 2).

Paracladopelma camptolabis (Kieffer, 1913)

NOMENCLATURE AND IDENTIFICATION

For mistakes in Dutch nomenclature, and for identification, see the introduction under the genus.

DISTRIBUTION IN EUROPE AND THE NETHERLANDS

The species has been recorded in many countries over nearly the whole European mainland. In the Netherlands larvae and exuviae have been collected at many localities in the eastern and southern provinces.

LIFE CYCLE

Pinder (1983) found three or four generations a year, emerging from April until early autumn. Lindegaard & Mortensen (1988) reported only two generations a year in a lowland brook in Denmark with about the same temperature regime. The larvae of *P. camptolabis* have an overwintering diapause in second and third instar from early autumn (Goddeeris, 1983; 1986; Lindegaard & Mortensen, 1988). Pinder also collected some fourth instar larvae in winter.

MICROHABITAT

The larvae appear to have no preference for mineral or organic sediments. For instance, Pinder (1980) often collected the larvae in sandy deposits, but elsewhere or at other times more in organic sediments. We found this species sometimes (even numerously) on fine detritus, but Verdonschot & Lengkeek (2006) collected *P. camptolabis/laminatum* more often on sand. Tolkamp (1980) found *P. camptolabis/laminatum* about equally abundantly on sand and sand with fine detritus.

DENSITIES

In most cases the larvae occur in low densities. Lindegaard & Mortensen (1988) reported mean densities of about 75 larvae/m^2 and in winter maximum densities of about 400 larvae/m^2. The maximum density in the Roodloop (a small upper course in the Netherlands) was 100 larvae/m^2 and as a rule less than 10 larvae/m^2 were collected (Moller Pillot, 2003).

WATER TYPE

Current

P. camptolabis is a normal inhabitant of slow-flowing lowland streams in Western Europe (Lindegaard-Petersen, 1972; Pinder, 1980; Lindegaard & Mortensen, 1988). In the Netherlands larvae of *P. camptolabis/laminatum* are common in lowland brooks; the majority of these (based on identification of exuviae) belong to *P. camptolabis*. The species is rare in fast-flowing streams, but is reported there more often than other species of the genus (Lehmann, 1971; Moog, 1995; Scheibe, 2002; Michiels, 2004). The larval type is very scarce in the Geul and the Gulp in South Limburg (Cuijpers & Damoiseaux, 1981; Waterschap Roer en Overmaas, unpublished) and also in the large rivers in the Netherlands (Klink, 2002). Klink (1985a) collected two males along the river IJssel and Caspers (1991) reported *P. camptolabis* from the Middle and Lower Rhine. The species is often found in lakes (see below), but has not yet been recorded in Dutch lakes or other stagnant waters.

Dimensions

According to Lenz (1959, 1954–62) the larvae are rather common inhabitants of the littoral zone of lakes, but they are absent from the profundal zone. Brundin (1949: 762) mentioned at least one record from the profundal zone. The larvae do not live in small ponds, but Macan (1949) collected them in a small lake about 200 m in length. According to Verdonschot et al. (1992), the larvae occur more often in the middle courses of lowland streams, but there are also many records from upper courses (own data).

Permanence
We have only one record from a narrow temporary upper course (Moller Pillot, 2003: 119).

pH

Nearly all known records are from water with a pH between 7 and 8, but as for *P. nigritulum* at least one population has been found at a pH of around 6 and hardly any specimens have been found in water at a pH of approx. 5 (Moller Pillot, 2003). However, it is not certain that pH is a limiting factor (see under *P. nigritulum*).

TROPHIC CONDITIONS AND SAPROBITY

Lenz (1959: 447) stressed that *P. camptolabis*, in contrast to *P. nigritulum*, lives mainly in the littoral zone (generally not in the profundal zone) of lakes, but that both species need a rather high oxygen content. However, Rossaro et al. (2007) found both species sometimes in water with only 1 mg O_2/l (see under *P. nigritulum*). The Dutch data do not suggest a difference between both species regarding saprobity or oxygen content.

Paracladopelma laminatum (Kieffer, 1921)

DISTRIBUTION IN EUROPE AND THE NETHERLANDS

P. laminatum is known from many countries over nearly the whole European mainland (Saether & Spies, 2004). In the Netherlands the species has been reported from more than 60 localities in the eastern and southern part of the country and from the coastal dune region. Some of these records are probably of larvae identified as *P. laminatum* agg. The species is not rare in the Twenthe district, the province of Drenthe, in the Veluwe district and in the dune region.

LIFE CYCLE

The life cycle is most probably no different from that of *P. camptolabis* (see under this species). Orendt (1993) reported three generations a year in a lake in Bavaria. Schmale (1999) collected many exuviae in spring and only relatively few in summer and early autumn.

MICROHABITAT

According to the literature summarised by Becker (1994) the larvae prefer sandy bottoms. Buskens (unpublished) collected the larvae of *P. laminatum* (mainly *P. camptolabis/laminatum*) in Dutch sandpits, nearly exclusively on sand. See, however, under *P. camptolabis*.

WATER TYPE

Current
There are few Dutch records from lowland streams (H. Cuppen, pers. comm.), but the species is possibly not rare in slow-flowing water (larvae have not been identified to species level). In the literature the adults and exuviae are rarely reported from fast-flowing water (Lehmann, 1971; Orendt, 2002a; Michiels, 2004). Becker (1994) collected the adults regularly along the river Rhine; Klink (1985a) caught an adult male along the river IJssel. The species is probably more common in lakes (Lenz, 1959; Orendt, 1993; Schmale, 1999; Buskens, unpublished). In the Netherlands most (or all) larvae of *Paracladopelma* occurring in dune lakes and sand pits belong to this species.

Dimensions
The scarce data from streams vary from a narrow upper course (leg. H. Cuppen) to the lower course of the river Fulda in Germany (Lehmann, 1971). In stagnant water the species seems to be confined to lakes.

pH
Buskens (unpublished) collected larvae and exuviae in sand pits with a pH of 6.3–7.5. In dune lakes the exuviae have been collected at pH 7.5–9 (Schmale, 1999). Only a few records from lowland brooks have been identified to species level. The larvae probably do not live in very acid water, but see under *P. nigritulum*.

TROPHIC CONDITIONS AND SAPROBITY
Orendt (1993) collected the species mainly in a mesotrophic lake in Bavaria and in small numbers in a slightly eutrophic lake (not in other oligotrophic and eutrophic lakes). The species is not rare in Dutch eutrophic dune lakes, but the exuviae are rarely collected in lakes with a somewhat lower oxygen content (Schmale, 1999). Lenz (1959) supposed that the larvae cannot live in the profundal zone of lakes when the oxygen content is low. The Dutch data that give any indication about saprobity suggest that *P. laminatum* tends to be more confined to water with hardly any pollution than *P. nigritulum*.

Paracladopelma mikianum (Goetghebuer, 1937)

IDENTIFICATION
The hypopygium of the adult male is illustrated in Cranston et al. (1989: fig. 10.46 D, as *P. schlitzensis*). The exuviae can be identified using Langton (1991).

DISTRIBUTION IN EUROPE
The species has been collected in several European countries, including Germany and France (Michiels, 1999; Saether & Spies, 2004). There are no records from the lowlands in these countries (cf. Ashe & Cranston, 1990; Serra-Tosio & Laville, 1991).

LIFE CYCLE
Michiels (1999) collected relatively large numbers of exuviae of this species in the lower course of the river Salzach in Bavaria. The exuviae were collected mainly from August to early October (one specimen in early May).

WATER TYPE
The species has been collected nearly exclusively in fast-flowing streams and rivers such as the Salzach in Bavaria (Michiels, 1999) and the Rhine between Lake Constance and Bingen (Caspers, 1991).

Paracladopelma nigritulum (Goetghebuer, 1942)

Paracladopelma obscura Brundin, 1949: 763
Paracladopelma camptolabis agg. Moller Pillot, 1984 (see introduction under the genus)

DISTRIBUTION IN EUROPE AND THE NETHERLANDS
P. nigritulum has been recorded in many European countries, but is possibly absent

from parts of Eastern Europe (Saether & Spies, 2004). In the Netherlands the species has been collected in the whole Pleistocene part of the country, at more than 150 localities (Moller Pillot & Buskens, 1990; Nijboer & Verdonschot, unpublished).

LIFE CYCLE

The life cycle is probably not very different from that of *P. camptolabis* (see there). We collected some juvenile larvae in winter, but by March the majority of larvae appeared to be in fourth instar. In Lake Constance the adults emerge from early April until the middle of November (Reiss, 1968).

FEEDING

See under the genus.

MICROHABITAT

Tolkamp (1980) collected the larvae (as *P. camptolabis*) mainly in bare sand and in sand with fine detritus, and rarely in gravel or coarse detritus. Verdonschot & Lengkeek (2006) found this species more on sand. See also under *P. camptolabis*.

WATER TYPE

Current

P. nigritulum lives in lowland brooks and small rivers in the Netherlands (Moller Pillot & Buskens, 1990; Verdonschot et al., 1992) and also in lakes in the Alps and in Scandinavia (Brundin, 1949; Lenz, 1959; Reiss, 1968; Paasivirta, 1974). The species is rarely collected in more or less fast-flowing streams (Braukmann, 1984, as *P. camptolabis*; Casas & Vilchez-Quero, 1989: nearly 1 m/sec; Orendt, 2002; Waterschap Roer en Overmaas, unpublished; own data). The figures in the matrix table in Chapter 4 only apply to brooks and small rivers.

Dimensions

The species has not been collected in large rivers. According to Verdonschot et al. (1992) *P. nigritulum* is more common than other species of the genus in upper courses (possibly because the larvae are more or less cold stenothermous, see below). There are many records from narrow upper courses 1 to 1.5 m wide (Limnodata.nl; own data). In lakes the larvae appear to be most abundant in the profundal zone, but they also live in the littoral zone (Brundin, 1949; Lenz, 1959; Paasivirta, 1974). They are not collected in small pools.

Permanence

The larvae are rare in temporary upper reaches of lowland brooks (Verdonschot et al., 1992; Moller Pillot, 2003).

pH

Most Dutch records are from brooks with a pH between 6.5 and 8 (Limnodata.nl). Moller Pillot (2003) also collected the larvae regularly at a pH around 6 and single specimens have been found at a pH near 5. The absence from many acid upper reaches may be due to their temporary character. However, Raddum & Saether (1981) did not collect any *Paracladopelma* in very acid lakes (with pH < 6). The acidity may become limiting at such a low pH.

TROPHIC CONDITIONS AND SAPROBITY

The larvae are characteristic of oligotrophic conditions in the littoral as well as the profundal zone of lakes (Saether, 1979). According to Lenz (1959) the larvae need a rather

high oxygen content. Rossaro et al. (2007) sometimes found *P. nigritulum* in water with only 1 mg O_2/l. However, the ring organ of the pupae is relatively small, suggesting a low ability to respond to oxygen shortage. The many Dutch records are from well oxygenated, although sometimes very eutrophic and rarely slightly organically polluted lowland brooks (Limnodata.nl; unpublished data).
Brundin (1949) considered the species to be cold stenothermous.

DISPERSAL
In the Roodloop (a small upper course) relatively large numbers of larvae of this species were collected in drift samples (Moller Pillot, 2003).

Paralauterborniella nigrohalteralis (Malloch, 1915)

Lauterborniella brachylabis Chernovskij, 1949: 88, fig. 62; Izvekova, 1980: 73, 78–79, figs 25–9

IDENTIFICATION
The female has been described by Rodova (1978); the female genitalia are also described and illustrated in Saether (1977).

DISTRIBUTION IN EUROPE AND THE NETHERLANDS
The species is widely distributed over Europe, but it is probably more common in the east; there are no records from Belgium, the northern part of France and the Iberian peninsula (Saether & Spies, 2004; Serra-Tosio & Laville, 1991). The species is scarce in the Netherlands (about 30 localities) and has been collected mainly in the central part of the country and in the southern and eastern provinces (e.g. Limnodata.nl; D. Tempelman, unpublished). In view of the above-mentioned distribution in Europe the rarity of the species in the Netherlands may be due to climatological reasons. The first record in the Netherlands is from 1997; previously the species seemed to be absent.

LIFE CYCLE
Tõlp (1958) supposed that the species has at least two generations a year in Estonia. Shilova (1976) reported only one generation. Adults have been collected in Sweden and in Russia from early June until the middle of August (Brundin, 1949; Shilova, 1976).

FEEDING
Izvekova (1980) observed that the larvae build tubes in places where they can find enough food. They gather small particles from both sides of their tubes, selecting and consuming mainly detritus, but also dead planktonic algae.

MICROHABITAT
Tõlp (1956, 1958) attributed the species to the psammopelorheophilous fauna. The larvae live in the mud, especially between the vegetation in the riparian zone of the river. In the autumn the larvae move to the more central parts (Tõlp, 1958). Brundin (1949) and Shilova (1976) found the larvae mainly on sapropel, and locally also between algae on stony bottoms.

WATER TYPE
The larva inhabits lakes, old river branches and shallow parts of slow-flowing rivers (Shilova, 1976). The species is usually absent from more or less fast-flowing streams and rivers, for instance in the pre-Alpine rivers treated by Orendt (2002a) and in the

river Fulda (Lehmann, 1971). Braukmann (1984) collected the species in none of the investigated German brooks. However, Becker (1994) reared an adult male from the Upper Rhine and also caught some males with a light trap. In the past the larvae also lived in the Dutch part of the Rhine (Klink, 1989). In the Emajõgi, a slow flowing river in Estonia, it is one of the most numerous Chironomidae (Tõlp, 1958). According to Pankratova (1964) the larvae can also live at sites with moderate or rather fast currents. In the Netherlands the species has been collected in lowland brooks and old branches of rivers in which the current is usually very slow (Klink, 2008; Limnodata.nl), in a lower stretch of the large rivers (Wantij, near Dordrecht; pers. comm. D. Tempelman) and only once in an isolated stagnant pool.

TROPHIC CONDITIONS AND SAPROBITY

Brundin (1949) collected the species in many oligotrophic lakes in Sweden, although in low numbers. The larvae lived only in the deeper parts of these lakes when the oxygen content remained rather high. Rossaro et al. (2007) also collected the larvae only in water with a high oxygen content; the ring organ of the pupa is relatively small and therefore the species is likely to be intolerant of oxygen shortage. Zaćwilichowska (1970) did not find the larvae in the polluted zone of the river San in Poland. In the Netherlands the larvae have been collected in very eutrophic brooks and streams. However, the only record from stagnant water is a pool with exceptionally good water quality (H. Cuppen, pers. comm.).

Paratendipes Kieffer, 1911

EUROPEAN SPECIES

Ashe & Cranston (1990) listed four species in Western Europe: *P. albimanus*, *P. nubilus*, *P. nudisquama* and *P. plebeius*. The last species differs from *P. albimanus* only in the colour of the first tarsomere of the male (Langton & Pinder, 2007). Brundin (1949) considered *P. plebeius* to be a synonym for *P. albimanus*. Goetghebuer (1937–1954) and Rossaro & Mietto (1998) also suggested that these are the same species; they seem to be indistinguishable as larvae (cf. Pankratova, 1983). Langton (1991) gives only one pupal type: *Paratendipes* Pe 1. Both species will be treated here under *P. albimanus*. *Paratendipes albitibia* may also be a separate species, but more probably it is a synonym for *P. albimanus*. The name is treated below. In the Netherlands *P. nubilus* is only known as the adult and *P. intermedius* only as the larva. *P. intermedius* is a nomen dubium and appeared to be a synonym for *P. nubilus*.
The three species mentioned above are not the only species present in Europe. Ashe & Cranston (1990) and Saether & Spies (2004) reported two further species from Romania. Chernovskij (1949) reported a species (easy to recognise as a larva) called 'connectens No.3', living in river sand. It does not seem to be one of the above-mentioned 'adult species'.

D

Paratendipes albimanus (Meigen, 1818)

? *Paratendipes plebeius* Langton & Pinder, 2007: 182 (see above)

SYSTEMATICS AND IDENTIFICATION

As stated under the genus and under *Paratendipes albitibia*, there is some doubt about the value of some characters of the adult male. Identification of males is possible using Langton & Pinder (2007), but these authors do not mention all the European species

of the genus; see also Goetghebuer (1937–1954) and Cranston et al. (1989). Lehmann (1971: 544) published a more detailed figure of the male hypopygium. The exuviae can be identified using Langton (1991) as *Paratendipes* Pe 1, including probably *P. plebeius* and *P. albitibia*.
Using Moller Pillot (1984) the larvae are identified as *Paratendipes* gr. *albimanus*. The insertion 'group' can be omitted if one does not consider *P. plebeius* and *P. albitibia* to be separate species. The number of median teeth in the larva (3, 4 or 5) has no systematic value. Information about the possible confusion of third instar larvae with *P. nudisquama* can be found in the entry for that species.

DISTRIBUTION IN EUROPE AND THE NETHERLANDS

P. albimanus lives nearly everywhere on the European mainland (Saether & Spies, 2004). In the Netherlands the species is common and widespread in the southern and eastern provinces, but very rare in nearly the whole Holocene area, except the province of Flevoland and along the inland fringe of the coastal dunes (Moller Pillot & Buskens, 1990; Steenbergen, 1993; Nijboer & Verdonschot, unpublished).

LIFE CYCLE

Ward & Cummins (1978) found only larvae in first and second instar in a brook in Michigan in winter. At this site the species was univoltine, emerging in June/July. Low temperature and poor food quality probably prevented a second generation. The authors mentioned the possibility of a second generation in other cases, especially in standing waters (Ward & Cummins 1978, 1979). In southern England Pinder (1983) also found one generation a year, fourth instar larvae being present only in July and August. In the Netherlands adults emerge from the end of April until the middle of September; a second generation may develop locally (Moller Pillot & Buskens, 1990; Moller Pillot, 2003). Third instar larvae are already numerous in winter; in April larvae are in third and fourth instar.

FEEDING

The larvae are collector-gatherers of fine particulate detritus (Ward & Cummins, 1978). Growth velocity increases with the microbial density of the substrate (Ward & Cummins, 1979). In a small Dutch upper course, the Roodloop, the population seemed to disappear when no organic silt was present (Moller Pillot, 2003: 43, 120).

OVIPOSITION

Nolte (1993: 29, fig. 27) found egg masses attached to submersed stones in a stream which were most probably of *P. albimanus*, considering the habitat and the identification of a larva. Each mass contained 408–445 elliptical to reniform eggs.

MICROHABITAT

Ward & Cummins (1978) contains a summary of data from the literature on the microhabitat of this species. *P. albimanus* is a typical bottom dweller, most abundant on muddy bottoms, but also living on sand and gravel. The species is rarely found on plants (Drake, 1982; Soszka, 1975; Pankratova, 1964; Srokosz, 1980). However, Verdonschot & Lengkeek (2006) reported finding the species relatively often on or between vegetation. Brodersen et al. (1998) found the larvae rarely on stones.

DENSITIES

Ward & Cummins (1978) recorded maximum densities of 10,000 larvae/m^2 in riffles and 30,000 larvae/m^2 in pools. In Dutch lowland brooks the densities are usually less than 100 larvae/m^2, rarely up to 4000/m^2 (Ellenbroek & Hendriks, 1972). In stagnant water bodies in the Netherlands the number of larvae is always very low. In lakes in Germany higher densities are possible (e.g. Reiss, 1968).

WATER TYPE

Current

P. albimanus has been found most abundantly in slack flow regions (Ward & Cummins, 1978), but also sometimes lives in brooks and small streams with a moderate current (Orendt, 2002; Michiels, 2004). In the Netherlands the larvae are very common in lowland brooks and relatively scarce in the hilly parts of South Limburg (Waterschap Roer & Overmaas, unpublished). In a small upper course the larvae seemed to disappear when the current became very slow, but also when organic silt disappeared owing to fast currents (Moller Pillot, 2003; see under Feeding).
Although Caspers (1991) and Becker (1994) reported the species for the whole German part of the river Rhine, the larvae appeared to be scarce (but probably increasing) in the large rivers in the Netherlands, with the exception of the estuarine stretches (e.g. Smit, 1982; Smit et al., 1994; Klink, 2002). In the Netherlands there are just a few records from stagnant water; the bottoms of the eutrophic canals, ditches and pools are possibly too poor in oxygen.

Dimensions

As stated above, the larvae are rare in large streams, but can be scarcely found in the estuarine stretches. The important reasons for this seem to be the presence of a suitable microhabitat and current velocity. The larvae were sometimes numerous in the Roodloop, a narrow upper course (width 1 m) in the Dutch province of Noord-Brabant (Moller Pillot, 2003).
In stagnant water the species is more or less confined to lakes and larger ponds, but Schleuter (1985) collected the species in a wheel rut 25 cm wide and 6 m long. The oxygen content in small pools is probably the limiting factor. In lakes the larvae live in the littoral, sublittoral and profundal zones (Reiss, 1968; Ward & Cummins, 1978).

Permanence

The larvae are rarely found in temporary pools or upper courses (Schleuter, 1985; ten Cate & Schmidt, 1986; Moller Pillot, 2003). In many cases the absence of larvae will be caused by the univoltine life cycle and flight period in summer.

D

pH

According to Schleuter (1985), ten Cate & Schmidt (1986) and Moller Pillot (2003) the larvae are rare at pH 5 in winter and can often be found at pH > 5.7. Raddum & Saether (1981) collected *P. albimanus* only at pH > 6. In the Netherlands most records are from water bodies with a pH > 6.85 and many at more than 7.60 (Limnodata.nl).

TROPHIC CONDITIONS AND SAPROBITY

P. albimanus has been recorded in oligotrophic as well as eutrophic lakes (Brundin, 1949; Brodersen et al., 1998).
Sládeček (1973) considered *Paratendipes* larvae to be specific to alpha- and beta-mesosaprobic water. However, Wilson & Ruse (2005) considered the genus to be intolerant of organic pollution. In Dutch brooks and streams the larvae are found in water with a rather high BOD and ammonium content, but not in oxygen poor conditions (Limnodata.nl). Ellenbroek & Hendriks (1972) found the larvae numerously in a slightly polluted part of the Teelebeek and also in a community of *Chironomus* and Tubificidae just after a heavy discharge of organic material (where the oxygen content was still high). As pointed out by Ward & Cummins (1979), the larvae benefit from

high microbial densities in the substrate. Despite their high haemoglobin content, all data indicate that they need a rather good oxygen supply (see also under Water type).

IRON

Rasmussen & Lindegaard (1988) found the larvae only in water with less than 2 mg Fe^{2+}/l. The species seems to be less tolerant of iron than *Macropelopia nebulosa* and *Prodiamesa olivacea.*

SALINITY

There is one record from brackish water: Tõlp (1971) collected the larvae in the Gulf of Finland in Estonia at a chloride content of about 3000 mg Cl/l.

Paratendipes albitibia Kieffer, 1922

According to Ashe & Cranston (1990) *Paratendipes albitibia* is a nomen nudum. Goetghebuer (1937–1954: 66) and Prat (1978a) give several differences between this species and *P. albimanus*: the colour of the legs, the leg ratio, the antennal ratio and the shape of appendage 1a (digitus). Because the last difference is infinitesimally small and the other characters are relative, it seems probable that *albitibia* is synonymous with *albimanus*. The species is known from lakes in Germany, Austria, France and Spain (Fittkau & Reiss, 1978; Prat, 1978a).

Paratendipes connectens No.3 Lipina Chernovskij, 1949

IDENTIFICATION

The larva has been described by Chernovskij (1949) and Pankratova (1983).

DISTRIBUTION IN EUROPE

The larvae have been collected in Russia and Bulgaria (Pankratova, 1983) and palaeo-ecologically in the Rhine in the Netherlands (Klink, 1989).

ECOLOGY

The larvae live in the sandy bottom of rivers.

Paratendipes nubilus (Meigen, 1830)

Paratendipes intermedius Chernovskij, 1949: 90, fig. 214

IDENTIFICATION

The adult male is not described in Langton & Pinder (2007), but is easily recognisable by its marked wings; see Kruseman (1933), Goetghebuer (1937–1954) and Cranston et al. (1989: fig. 10.50). The pupa has been described by Biro & Klink (2005). The larva, described by Chernovskij (1949) as *Paratendipes intermedius,* appeared to belong to *P. nubilus* based on associated larvae and pupae among which a pupa with marked wings and an attached larval skin (Klink, unpublished).

DISTRIBUTION IN EUROPE AND THE NETHERLANDS

Adult males have been collected in Germany, the Netherlands, Italia, Romania and Macedonia (Saether & Spies, 2004). In the Netherlands the adult male was collected near Rotterdam by Snellen in the 19th century and near Bodegraven by de Meijere early in the 20th century (Kruseman, 1933).
The larvae have been collected in Russia, Estonia, Romania, Poland and the Netherlands (Tõlp, 1956; Srokosz, 1980; Pankratova, 1983; Klink, 2002; H. Cuppen, T. van Haaren, pers. comm.). In the Netherlands the records are almost entirely confined to the middle and lower reaches of the large rivers, with one record from the eastern part of the country (see Water type). It is possible that the species died out in the 20th century owing to water pollution and returned after water quality improved a few decades ago.

MICROHABITAT

Pankratova (1964) found the larvae in sand. In de river Nida in Poland the larvae were common in muddy sand in the middle section of the river course (Srokosz, 1980). According to Klink (2008) the species is a characteristic inhabitant of shifting sand. In the river Pripyat in Belarus and in Dutch rivers most larvae were collected in sand in the deeper parts of the river (RIZA, unpublished).

WATER TYPE

The larvae are collected in the middle and lower reaches of large rivers (Pankratova, 1964; Srokosz, 1980; H. Cuppen, T. van Haaren, A. Klink, H. Vallenduuk, pers. comm.). Cuppen (pers. comm.) found a larva in the Dinkel, a small river in the Netherlands.

SAPROBITY

The larvae live in sandy bottoms and are most probably susceptible to the presence of anaerobic silt. The first record following the improvement in the water quality of the Rhine was in 1988 (H. Cuppen, pers. comm.).

Paratendipes nudisquama (Edwards, 1929)

Paratendipes transcaucasicus Chernovskij, 1949: 90, fig. 66; Moller Pillot, 1984: 130, 132

SYSTEMATICS

P. nudisquama and some species not known from Western Europe are attributed to a separate group within the genus, characterised by the bare squama and hypopygium characters (Cranston et al., 1989).

IDENTIFICATION

Fourth instar larvae can be identified using Moller Pillot (1984: 129). However, third instar larvae of *P. albimanus* can be mistaken for *P. nudisquama* because the first lateral teeth in third instar are lower (Ward & Cummins, 1978: fig. 7). A more reliable character is that in *P. nudisquama* the first and second lateral teeth are never grown together.

DISTRIBUTION IN EUROPE AND THE NETHERLANDS

Paratendipes nudisquama has been collected in nearly the whole of Europe (Saether & Spies, 2004). In the Netherlands the larvae have been collected from locations dispersed over the fen-peat area, the whole Pleistocene area and Southern Limburg, but they appear to be scarce everywhere (Nijboer & Verdonschot, unpublished; own unpublished data).

LIFE CYCLE

Kruseman (1933) caught a male in the Netherlands in June. Brundin (1949) collected adults in Sweden in June and July. Most larvae collected in the Netherlands were in third instar and most of them were collected from February to May. A prepupa was collected in July. At the end of September larvae were found in third and fourth instar. Most probably the species has (as a rule?) one generation a year, as in *P. albimanus*.

WATER TYPE

Although Brundin 1949 collected the species once from a location at a depth of 6 m in a Swedish lake, he found larvae of *P. nudisquama* mainly in the lagg-zone of peat moors. There are more than 20 records from the Netherlands, at first glance from very different habitats. The larvae probably live mainly in peaty moorland pools and peat cuttings (e.g. Duursema, 1996) and more or less peaty helocrene springs or peaty places with groundwater seepage. However, they have also been collected in the upper reaches of natural streams (Verdonschot, 1990) and in canals in the fenland region in the province of Overijssel (Janse & Monnikendam, 1982). Langton (1991) collected the exuviae in streams.

pH

Many records, especially records with high numbers, are from peaty environments with a pH lower than 5 (e.g. Verstegen, 1985; Duursema, 1996). However, the larvae can also live in springs and marshes with a rather high pH and thus appear to be more or less pH independent.

TROPHIC CONDITIONS AND SAPROBITY

According to Limnodata.nl the larvae have been collected sometimes in water with low oxygen content and/or high ammonium content. This is not surprising for a species living in moorland. There are too few data for definite conclusions.

SALINITY

There are no records from brackish water.

Phaenopsectra Kieffer, 1921

SYSTEMATICS

According to Ashe & Cranston (1990) and Saether et al. (2000) there are two species in Europe: *P. flavipes* and *P. punctipes*. However, Pinder (1978) and Langton & Pinder (2007) noted that the hypopygia of *P. flavipes* and *P. punctipes* are indistinguishable, so that the species only can be discerned by colour. The same applies to the females of these two species (Rodova, 1978). Langton (1991) also distinguished two species (*P. flavipes* and P. f. Bala), differing only in the extent of the armament on the pupal tergite VI. It cannot be excluded that these differences do not merit a division into two species. At least one other species is present in Europe. Many authors (e.g. Koskenniemi & Paasivirta, 1987; Serra-Tosio & Laville, 1991) have mentioned provisionally the occurrence of *P. albescens* or *P. albiventris* (nomina nuda according to Ashe & Cranston, 1990). Rodova (1978) keyed four species as a female.
Larvae from the Netherlands with mentum and mandible resembling *Phaenopsectra* sp. as illustrated by Pinder & Reiss (1983: figure 10.58 B,E) belong to *Sergentia* (see under this genus).

Phaenopsectra flavipes (Meigen, 1818)
? Phaenopsectra punctipes (Wiedemann, 1817)

Pentapedilum exsectum Chernovskij, 1949: 81, fig. 51; Ertlová, 1970; Shilova, 1976; Holzer, 1980: 113–114 (nec aliis)
Lenzia Brundin, 1949: 115 et seq.; Lenz, 1954–62: 246–248, figs 380–383

IDENTIFICATION OF THE LARVAE

Living larvae of *Phaenopsectra flavipes* are deep red and in alcohol remain redder than most other genera. This can help in identification.

DISTRIBUTION IN EUROPE AND THE NETHERLANDS

Phaenopsectra flavipes and *P. punctipes* have been collected almost throughout Europe (Saether en Spies, 2004). In the Netherlands the genus is common nearly everywhere, but there are no records from the province of Zeeland and the Frisian Islands.

LIFE CYCLE

In Sweden *P. flavipes* flies from May to the middle of October and *P. punctipes* from May to September (Brundin, 1949). In the Netherlands pupae and exuviae can be found from the end of March to September. The data indicate two or three generations (Kouwets & Davids, 1984; Otto, 1991; own unpublished data). In winter few larvae are in fourth instar and the majority are in second and third instar; in April larvae are found in third and fourth instar.

FEEDING

The larvae appear to be mainly detritivores (Wilson & Ruse, 2005). Mackey (1979) found much very fine detritus and some diatoms in the gut of the larva. Brock (1984) found the larvae fairly numerously in litter bags with decomposing leaf blades of *Nymphoides peltata*. Crawford & Rosenberg (1984) reported *P. punctipes* larvae consuming decaying needles. Thorp & Bergey (1981) considered the American species of *Phaenopsectra* to be collector-gatherers and scrapers. Moog (1995) also mentioned filtering behaviour, but the origin of this is not clear.

MICROHABITAT

The larvae display a preference for stems and leaves of plants and are found only scarcely on stones and sandy bottoms (Brundin, 1949; Lehmann, 1971; Mol et al., 1982; Becker, 1994; own observations). However, they are less common in dense vegetation (Steenbergen, 1993). Lenz (1954–62) mentioned that the larvae of *P. punctipes* build tubes on leaves of *Potamogeton lucens*. Occurrence of the larvae on plants, especially in lowland streams, is likely to be irregular and patchy because of the dynamic nature of these systems (Tokeshi & Townsend, 1987).
Holzer (1980) collected larvae in sand and gravel, even to a depth of more than 30 cm in the substrate. In shaded Dutch lowland brooks Tolkamp (1980) collected the larvae much more numerously on detritus than on sandy bottoms.

WATER TYPE

Current
According to Hawtin (1998) *P. flavipes* is an indicator of slowly flowing or stagnant water. The larvae are widespread but never numerous in the lowland brooks of the Dutch province of Noord-Brabant (Limnodata.nl; own data). Steenbergen (1993) collected the larvae in many lakes and stagnant water courses in the Dutch province of Noord-Holland.

According to Lehmann (1971), Orendt (2002) and Michiels (2004) *P. flavipes* is a regular, although not abundant inhabitant of streams, and not only at low water velocities. Ertlová (1970) collected some larvae on an artificial substrate at a current of about 200 cm/sec. However, the larvae are rare in the faster flowing brooks and streams in the Dutch province of Limburg (e.g. Geul, Gulp and Geleenbeek) and Braukmann (1984) did not collect any larvae in the brooks and streams in Germany.
It seems to be probable that many records from fast-flowing streams are based on specimens originating from slow-flowing or stagnant sites in or near the river bed. It must be noted that Rossaro & Mietto (1998: 205) characterise *Phaenopsectra* as a genus living in moderately to fast-flowing water (> 10 cm/sec) in contrast to many bottom dwellers, for which the optimum current velocity is less than 5 cm/sec. These authors measured the velocity at the spot where the larvae live, while in the Netherlands the velocity is measured in the water column (see Chapter 4, table 4). *Phaenopsectra* larvae creeping on plants in the middle of a stream will experience stronger currents than bottom dwellers.

Shade
In contrast to *Parachironomus arcuatus*, for example, the presence of the larvae seems to be influenced little or not at all by shading from trees. The amount of available food (all year around) is probably a more important factor.

Dimensions of ditches and streams
In the Dutch province of Noord-Holland the larvae are very scarce in water bodies (usually ditches) less than 10 m wide (Steenbergen, 1993). However, they are regularly present in narrow (1 or 2 m wide) upper courses of brooks (e.g. Moller Pillot, 2003). The larvae are common in wider lowland brooks and small rivers (e.g. Verdonschot, 1990; Verdonschot et al., 1992). The larvae are very scarce in the river Meuse (Smit, 1982; Peeters, 1988) and rather scarce in the river Rhine (Wilson & Wilson, 1984; Becker, 1994). The scarcity in large rivers seems to be caused more by the current than by the width of the stream. Current and oxygen seem to be the key factors, as discussed in the relevant paragraphs.

Lake dimensions
In Swedish lakes the genus belongs to the most common and abundant chironomids (Brundin, 1949). The larvae live here only in the littoral zone. Mundie (1957) did not collect the genus at all in his reservoirs. Shilova (1976, as *Pentapedilum exsectum*) found them scarcely near the shores of the Rybinsk reservoir (Russia). Also, Mol et al. (1982) collected the larvae rarely in Lake Maarsseveen (Netherlands), although Steenbergen (1993) found them in this part of the Netherlands more frequently in lakes (water > 100 m wide and > 1.5 m deep) than in smaller water bodies. The differences in reported presence between different workers can probably be explained by differences in protection against waves in different lakes and the presence of suitable microhabitats. In the profundal zone of lakes the larvae are able to survive only in oligotrophic to mesotrophic conditions (Wiederholm & Eriksson, 1979).

Permanence
The larvae are not rare in temporary upper courses, but the numbers are low (ten Cate & Schmidt, 1986; Moller Pillot, 2003). Dettinger-Klemm (2003) found *P. punctipes* rarely in temporary pools and not at all in years with long dry periods. Schleuter (1985) also found the genus rarely in temporary water and Cuppen (1980) did not encounter the larvae at all in temporary ditches and pools in the Dutch province of Gelderland. The larvae of the American species *P. pilicellata* can aestivate in a cocoon in second and

third instar (Grodhaus, 1980), but there is no indication that the European species can aestivate in this manner.

pH

Henrikson et al. (1982) stated that *Phaenopsectra* had increased in numbers in acidified lakes (pH < 5) in Sweden. They supposed that this could be the result of cessation of fish predation because the larvae seem to be predated very strongly by fish. Raddum et al. (1984) found the larvae of the genus in Lake Boksjø in Norway at pH 4.5–4.8, but they almost disappeared after liming when the pH increased to 6.7–7.0. In the Netherlands, however, the genus is scarce in acid pools (e.g. Buskens, 1983; Leuven, 1987; Duursema, 1996; van Kleef, unpublished), although the lowest pH at which the genus was found in the Netherlands is 4.08 (van Kleef, unpublished). Steenbergen (1993) recorded a mean pH of 8.0 for 93 water bodies in the province of Noord-Holland (including many lakes and canals). Without doubt the mean pH in the water bodies in the Pleistocene areas is much lower.

TROPHIC CONDITIONS AND SAPROBITY

Brundin (1949) found *Phaenopsectra* to be common and sometimes abundant in oligotrophic lakes, but not in most eutrophic lakes. In the profundal zone of lakes the genus disappears during eutrophication (Wiederholm & Eriksson, 1979). Wilson & Ruse (2005) considered the genus to be pollution intolerant. Bazerque et al. (1989) did not find the exuviae in a stretch of the river Somme in France with heavy organic and metal pollution, but collected them in fairly large numbers above and below this stretch. However, Moller Pillot & Buskens (1990) supposed that the genus is about equally abundant in organical polluted and unpolluted water, based on several records from lowland brooks in the Netherlands which were heavily polluted (but not poor in oxygen). Reassessment of these data revealed that the numbers in such cases were very low and the larvae were absent when the water was badly oxygenated. Steenbergen (1993) mentioned scarce occurrence in water poor in oxygen or water with more than 1 mg orthophosphate/l. In the province of Noord-Holland the larvae were most common in water bodies with low concentrations of phosphate, nitrogen and chlorophyll-a.

SALINITY

Steenbergen (1993) mentioned few records in water with more than 300 mg Cl/l. Apart from those, there is only one record of *Phaenopsectra* in brackish water: Tõlp (1971) collected the larvae (as *Pentapedilum exsectum*) in the Bay of Kassari in Estonia at a chloride content of more than 4000 mg Cl/l.

D

Polypedilum Kieffer, 1912

SYSTEMATICS

Until far into the 20th century (e.g. Pinder, 1978) the present subgenus *Pentapedilum* was considered to be a separate genus because of the macrotrichia on the wing membrane. At present four subgenera are recognised in Europe: *Polypedilum* s.str., *Uresipedilum* Sasa & Kikuchi, *Pentapedilum* Kieffer and *Tripodura* Townes (Saether, 1977; Pinder & Reiss, 1983). Their systematic relations and characters are described in Saether & Sundal (1999). Värdal et al. (2002) divided the subgenus *Tripodura* into several groups, of which the *halterale* group, the *pullum* group, the *apfelbecki* group and the *aegyptium* group are represented in Europe. For groups of species distinguished by other authors, see under Identification.

IDENTIFICATION

Identification of adults

There were many problems in the past with the identification of adult males of *Polypedilum*. The key by Pinder (1978) has often been used since 1978, but the subgenus *Tripodura* could not be identified well using this key because figures 4 and 5 had been interchanged; see also under *P. bicrenatum, P. pullum* and *P. scalaenum*. This problem has been corrected in the new key by Langton & Pinder (2007). There are also problems within the subgenus *Pentapedilum*; see under *P. uncinatum*. A new key to all instars of *Pentapedilum* by Oyewo & Saether is in preparation.

Rodova (1978) described and keyed the females of nine species.

Identification of larvae

Partly because of difficulties with identifying adults, the larval keys also present many problems and it is still not possible to identify all the species. The classification into groups by Chernovskij (1949) and Lenz (1954–62) has lead to confusion, which is still partly present in Moller Pillot (1984): see under *P. bicrenatum*, *P. convictum*, *P. laetum* and *P. scalaenum*. The division into groups by all these authors has no systematic value and leads only to confusion. The same is true for gr. *nubeculosum* and *uncinatum* agg., as used by Moller Pillot (1984) and Moller Pillot & Buskens (1990); see under the species names. A new key has been made by Tempelman (in prep.).

In many species some specimens have been found with one or three central mental teeth instead of two. Moreover, many abnormal specimens have been found, two of which have been described in Moller Pillot (1984).

Living larvae are slender in comparison with most other genera. Nearly all species are deep red; an exception is the larva of *P. cultellatum*, which is hardly reddish (see under Haemoglobin).

MICROHABITAT

Chernovskij (1938) and van de Bund (1994) observed that the larvae of *Polypedilum* species lived in the upper layers of the mud, in contrast to *Chironomus plumosus*. In sand-gravel substrates Holzer (1980: 113) found the larvae of *Polypedilum*, especially *P. bicrenatum* (as *P.* gr. *scalaenum*), often deeper than many other species (up to more than 50 cm deep). This author could not compare his findings with typical burrowing genera such as *Chironomus* and *Stictochironomus*.

Some species (*P. cultellatum*, *P. sordens*) live on water plants and are rarely found on the bottom.

HAEMOGLOBIN

The larvae of most species are fiery red and contain much haemoglobin (see e.g. Heinis, 1993: 53). Nevertheless, many species are not found in water with low oxygen content. This may be caused by the fact that many species burrow in the bottom. However, the haemoglobin concentration of the larvae of *P. nubeculosum* appear not to be constant within the species: Int Panis et al. (1996) reported a concentration of 23 µg Hb/mg body weight, whereas Rossaro et al. (2007) found only 9 µg Hb/mg. The fact that *Polypedilum* is rarely present in the Alps above 1000 m (Reiss, 1968a) may be a result of lower oxygen pressure.

The larvae of the plant-inhabiting *P. cultellatum* are only slightly reddish.

Polypedilum acifer Townes, 1945

SYSTEMATICS

The species belongs to subgenus *Tripodura* Townes, 1945.

IDENTIFICATION

The larva is described and keyed in Rossaro (1985) and Tempelman (in prep.).

DISTRIBUTION IN EUROPE AND THE NETHERLANDS

P. acifer has been collected in many countries in Central and Southern Europe (Saether & Spies, 2004). In the Netherlands the species has been collected only in the province of Limburg (A. van Nieuwenhuijzen, unpublished; H. Cuppen, unpublished).

MICROHABITAT

Becker (1994) reared several males from stones and gravelly and sandy substrate in the river Rhine.

WATER TYPE

Langton (1991) stated that the characteristic habitat for this species is a stream with a moderate to rapid current. Rossaro (1987) collected the species in the river Po, mainly at a site with a relatively fast current. Becker (1994) reared the species from the Upper Rhine. The Dutch records are from two more or less natural lowland brooks in woodland and some fast-flowing small streams (Geul and Gulp).

Polypedilum acutum Kieffer, 1915

DISTRIBUTION IN EUROPE AND THE NETHERLANDS

The species has been collected in several countries, scattered across almost the whole of Europe (Saether & Spies, 2004). In the Netherlands Kruseman (1933) collected a male near Valkenswaard. A probable second record concerns a male caught near a bog pool in the Groote Peel (Moller Pillot, unpublished).

WATER TYPE

The species is known from brooks and lakes (Lenz, 1954–62; Fittkau & Reiss, 1978; Douglas & Murray, 1980). Casas & Vilchez-Quero (1989) collected the species near a stream in the Sierra Nevada with a current velocity of 0.6 m/sec. Scheibe (2002) caught adult males in the Taunus mountains along a summercold brook.

Polypedilum aegyptium Kieffer, 1925: 270

Polypedilum vetterense Brundin, 1949: 772, 837–839, figs 211, 212 ; Shilova, 1955: 313 et seq.

SYSTEMATICS

The species belongs to subgenus *Tripodura* Townes, 1945.

IDENTIFICATION

The adult male is not included in the key by Langton & Pinder (2007), but the male and female can be identified using Värdal et al. (2002). The exuviae are described and keyed in Langton (1991). Shilova (1955) also described the larva (as *P. vetterense*); the four-segmented antenna seems to be characteristic.

D

DISTRIBUTION IN EUROPE

P. aegyptium has been encountered scarcely, but scattered over nearly the whole of Europe (Saether & Spies, 2004). The species may be present in the Netherlands.

LIFE CYCLE

Shilova (1955) stated that adults were flying along the Amu-Darya during the whole summer.

MICROHABITAT

Shilova (1955) found the larvae in great numbers on floating plant material in the river Amu-Darya; only few larvae were found on the bottom.

WATER TYPE

Brundin (1949) collected the species in Lake Vättern in southern Sweden. Elsewhere the species has been found in streams and rivers (e.g. Shilova, 1955; Michiels, 1999; Orendt, 2002a). Some records are from moderately to fast-flowing streams.

Polypedilum albicorne (Meigen, 1838)

IDENTIFICATION

Lehmann (1971) treated the differences of the adults and pupae between *P. albicorne* and other species, with a figure of the wing of *P. albicorne*. The problems with identifying the larvae have not been fully resolved and some doubt still exists about identification of the exuviae (see under Trophic conditions).

DISTRIBUTION IN EUROPE AND THE NETHERLANDS

The species has been collected in the whole of Europe (Saether & Spies, 2004). Distribution in the Netherlands seems to be confined to the Veluwe and the eastern part of the country.

LIFE CYCLE

In an English chalk stream the adults emerged in May and July (Drake, 1982). Lehmann (1971) caught adults from April until October/November.

MICROHABITAT

The larvae live in helocrenes in the silt and between mosses on stones (Lehmann, 1971). Becker (1994) reared adult males from stones and stony substrate. The larvae (of this species?) collected by van Kleef (unpublished) lived on organic bottoms.

WATER TYPE

Fittkau & Reiss (1978) noted the occurrence of *P. albicorne* in brooks, rivers, lakes and pools. However, the occurrence of the species appears to be confined to special circumstances. In the Fulda river system the larvae live in springs and the upper rhithral zone in the silt of helocenes and between mosses on stones (Lehmann, 1971). There are several records from fast and moderately flowing streams (e.g. Orendt, 2002a; Michiels, 2004), but Caspers (1991) reported the species also for the Rhine upstream of Basle and for the Lower Rhine. In the Netherlands the larvae have been collected in small upper courses of lowland brooks, but also in some larger lowland streams in the same regions (A. Klink, unpublished; H. Cuppen, unpublished).
The species has been found only scarcely in stagnant water: for example in oligotrophic and mesotrophic lakes in Sweden (Brundin, 1949) and Ireland (Douglas & Murray,

1980). In the Netherlands and Belgium the exuviae and larvae have been collected in three moorland pools (van Kleef, unpublished). See, however, under Identification.

pH

The larvae of *P. albicorne* are moderately tolerant of acid: Orendt (1999) found the species in streams with a pH of 4.8 to 6.8. In the Netherlands the exuviae have been found in a moorland pool with a pH of 7.5 (van Kleef, unpublished) and larvae have been collected in several brooks at pH 7 and higher (H. Cuppen, A. Klink: unpublished records).

TROPHIC CONDITIONS AND SAPROBITY

Reiss & Fittkau (1971: 102) considered the larvae to be not only polyoxybiontic, but also crenophilous and cold stenothermous. The last attribute was not fully confirmed by Lehmann (1971: 526). Moog (1995) considered *P. albicorne* to be one of the best indicators for unpolluted water among the Chironomini. In the Netherlands most records are from clear upper courses of lowland brooks. The records from moorland pools (van Kleef, unpublished) indicate that the larvae can endure some organic loading and give cause to doubt their high oxygen need. However, these specimens may possibly belong to an (as exuviae) undescribed species. The figures in the matrix tables should be used with some care.

Polypedilum amoenum Goetghebuer, 1930

According to Saether & Spies (2004) *P. amoenum* has been found in Germany, Italy, Romania and Russia. Fittkau & Reiss (1978) mentioned its occurrence in lakes.

Polypedilum apfelbecki (Strobl, 1900)

SYSTEMATICS

Within the subgenus *Tripodura* the species is not related to other European species (gr. *apfelbecki*, see Värdal et al., 2002: 352 et seq.).

IDENTIFICATION

The adult male can be identified using Langton & Pinder (2007). The exuviae are absent from Langton (1991), but they are described in Langton & Visser (2003). Värdal et al. (2002) give a number of characters of the larva, which resembles the larva of *P. scalaenum*: antennal segments 3–5 reduced and antennal blade much longer than the flagellum. However, the Lauterborn organs are small or absent and posterior lobes of the ventromental plates are absent.

DISTRIBUTION IN EUROPE AND THE NETHERLANDS

P. apfelbecki has been collected in only a few countries in Western, Central and Southern Europe. Its presence in the Netherlands has not yet been ascertained.

WATER TYPE

According to Fittkau & Reiss (1978) the larvae live in brooks and small streams. Lehmann (1971) collected adults possibly belonging to this species in lentic parts of spring brooks. Scheibe (2002) caught adult males along a summercold brook in the Taunus mountains. However, Prat (1978a) caught a female along a Spanish reservoir.

TROPHIC CONDITIONS AND SAPROBITY

Moog (1995) considered *P. apfelbecki* to be one of the best indicators for unpolluted water among the Chironomini.

Polypedilum arundineti (Goetghebuer, 1921)

Polypedilum arundinetum Pinder, 1978: 138

IDENTIFICATION

The larva can be identified using Tempelman (in prep.), based on material reared by H. Vallenduuk.

DISTRIBUTION IN EUROPE AND THE NETHERLANDS

Saether & Spies (2004) reported *P. arundineti* from many countries in Western and Central Europe and Scandinavia, but not from the Mediterranean area. Very few records are from the Netherlands and these are from the eastern and southern part of the country (of which three in the Twenthe district). The species seems to be scarce everywhere, but according to Langton (1991) it is more common in northern and montane regions.

LIFE CYCLE

Adult males have been collected in June and July in Sweden (Brundin, 1949) and Ireland (Douglas en Murray, 1980).

MICROHABITAT

In the river Cole in England the larvae were found on a bottom with water-worn stones (4–8 cm in diameter) and fine gravel, heavily organically polluted (Davies & Hawkes, 1981). In a pool in Twenthe (the Netherlands) the larvae lived in a 15 cm thick layer of organic and loamy mud and were absent on the firm bottom and between vegetation.

DENSITY

Davies & Hawkes found a mean density of 211 larvae/m^2 at the optimal site.

WATER TYPE

The larvae have been found in lakes, pools and streams. According to Langton (1991) the species is an inhabitant of stagnant water. Fittkau & Reiss (1978) also mentioned marshes or bogs.

TROPHIC CONDITIONS AND SAPROBITY

In the river Cole in England the larvae of *P. arundineti* were numerous in an organically polluted stretch, where Tubificidae and *Chironomus riparius* dominated and where the oxygen concentration was about 5 mg/l (Davies & Hawkes, 1981). Brundin (1949) collected the species in three oligotrophic lakes in Sweden. In the Netherlands and Belgium the larvae have been collected in four mesotrophic to eutrophic pools (see, however, Microhabitat).

Polypedilum bicrenatum Kieffer, 1921

Polypedilum gr. *bicrenatum* Moller Pillot, 1984: 150, 156; Moller Pillot & Buskens, 1990: 12, 24, 49; Heinis, 1993: 43–57
Pentapedilum uncinatum Beattie, 1978a: 109–113 (misidentification)
Polypedilum gr. *scalaenum* Chernovskij, 1949: 79, fig. 48
Polypedilum scalaenum Beattie, 1981: 153–154; Beattie, 1982: 287 et seq. (misidentification)

SYSTEMATICS

The species belongs to subgenus *Tripodura* Townes, 1945.

IDENTIFICATION

As stated under the genus in the key for adult males by Pinder (1978: 136) figures 4 and 5 have been interchanged, so that *P. bicrenatum* was identified as *P. scalaenum*. The key by Langton & Pinder (2007) presents no problems. An additional description can be found in Hirvenoja (1962). The female has been described by Hirvenoja (1962) and Rodova (1978).
The key for larvae by Moller Pillot (1984) gives more or less the correct characters, but it is possible that other species have sometimes been identified as *P.* gr. *bicrenatum*. His figure of the antenna (fig. IV.11.p) does not quite correctly represent the illustrated specimen. Very good figures of the larval antenna and mouth parts can be found in Kiknadze et al. (1991: fig. 45). In third instar the gula is usually very slightly darkened, so that the key by Tempelman (in prep.) cannot be used. In third instar the antenna is already characteristic. The larvae are fiery red in second, third and fourth instar (even more intensely red than in *P. nubeculosum*).

DISTRIBUTION IN EUROPE AND THE NETHERLANDS

The species has been reported from the whole of Europe (Saether & Spies, 2004). In the Netherlands the species occurs throughout the country, most commonly in the Pleistocene regions (Limnodata.nl; Nijboer & Verdonschot, unpublished data). The map of group *bicrenatum* in Moller Pillot & Buskens (1990) gives a good indication.

LIFE CYCLE

According to Beattie (1978, as *Pentapedilum uncinatum*; 1981, as *P. scalaenum*) *P. bicrenatum* has two generations a year in Lake Tjeuke, the Netherlands, emerging in June and August. Only one generation emerged if too little organic mud was present (see under Microhabitat). In Hjarbæk Fjord, Denmark and in lake Belau in Germany *P. bicrenatum* produced a large generation in early summer and a smaller late summer generation in August and early September (Lindegaard & Jónsson, 1987). However, in another Danish lake the species seemed to be univoltine (Lindegaard & Brodersen, 2000: 320).
The larvae hibernate in second and third instar in diapause. Larvae in third instar can be found until mid May.

FEEDING

Izvekova (1980) observed that the larvae gather detritus and benthic algae around and within their tubes. They have no filtration system, but food is supplied by the water current.

MICROHABITAT

The larva is a typical soil inhabitant. According to Beattie (1978, as *Pentapedilum unci-*

natum; 1981, as *P. scalaenum*) the larva can live in mud and in sand. Development of the larvae was slower in sand without much organic material, so that only one generation emerged. Other authors (Srokosz, 1980; Klink, 1983; Otto, 1991) found *P. bicrenatum* mainly on sandy bottoms without much organic silt. Otto (1991: 43) noted that more adults of *P. bicrenatum* emerged from within the *Phragmites* zone than from open water. In Lake Maarsseveen the larvae were also present on stones and rarely on plants and artificial plants (Mol et al., 1982). However, the larvae found on plants may have been another species.
The larvae live in tubes of mud in the upper layers of the sediment (own observations, see also Izvekova, 1980). In the laboratory they made tubes standing out 2–3 mm above the surface. Heinis (1993) observed that the very small second instar larvae were present relatively deep in the sediment (up to 6 cm deep), contrary to the common rule that bigger larvae burrow deeper than younger ones. Holzer (1980: 113–114, as *P.* gr. *scalaenum*) found the larvae up to more than 50 cm deep if the sediment in a stream consisted of sand and gravel.

WATER TYPE

Current and dimensions

According to most of the literature (e.g. Brundin, 1949; Fittkau & Reiss, 1978; Langton, 1991) *P. bicrenatum* is a species of stagnant water. The species was not collected by Lehmann (1971) in the river Fulda or by Braukmann (1984) in German streams. Only some adults have been collected along the river Rhine (Caspers, 1991; Becker, 1994). However, Orendt (2002a) reported the species from several streams in Bavaria and Shilova (1976) reported its presence in small brooks. In the Netherlands there are many records from brooks and streams, even fast-flowing brooks, based on identification of larvae as gr. *bicrenatum* (e.g. Limnodata.nl; Waterschap Roer & Overmaas, unpublished). The presence in slow-flowing streams has been confirmed by rearing.
In stagnant water the species is rather common in large lakes (even in Lake IJssel in the Netherlands) and estuaries, and most common in ditches (Brundin, 1949; Maenen, 1983; Lindegaard & Jónsson, 1987; Steenbergen, 1993; own data). In the Netherlands Heinis (1993: 28) collected the larvae mainly in the littoral zone of Lake Maarsseveen, but not deeper than 10 m. In the storage reservoirs in the Biesbosch the larvae are scarcely collected at greater depths (Kuijpers, pers. comm.).

Permanence

There are some records from watercourses which rarely dry up in summer.

SOIL

Steenbergen (1993) found the larvae rarely on clay and more on peaty and sandy bottoms.

DENSITIES

In the mud zone near the shore of Lake Tjeuke in the Netherlands Beattie (1978) found about 4000 second instar larvae/m^2, with a maximum of 7349 larvae/m^2. In summer the fourth instar was present, with a maximum of 1679 larvae/m^2. In other situations he found much lower densities. In 1993 the mean density in Lake Volkerak-Zoommeer in the Netherlands was 1323 larvae/m^2 (van der Velden et al., 1995). In Hjarbæk Fjord, Denmark, densities up to more than 10,000 larvae/m^2 were recorded (Lindegaard & Jónsson, 1983).

pH

In Hjarbæk Fjord, Denmark, the larvae lived in high densities in water with a pH of

10.5–11 (Lindegaard & Jónsson, 1983). In the Netherlands most records are from water with a pH of 7 to 9. However, in one case the species was found in a moorland pool at pH 4.27 (van Kleef, unpublished, det. P. Langton). Raddum & Saether (1981) collected larvae of the '*Tripodura simulans*' type in different Norwegian lakes at pH 4.5 and 5.5.

TROPHIC CONDITIONS AND SAPROBITY

Brundin (1949) collected *P. bicrenatum* often in Swedish oligotrophic lakes, but the species was also present in eutrophic lakes. In stagnant water in the Dutch province of Noord-Holland the larvae are most common when the phosphate, nitrogen and chlorophyll-a content are low (Steenbergen, 1993). They were not found when the oxygen saturation percentage was lower than 40. Limnodata.nl give nearly the same results for different water types in the Netherlands (mainly lowland streams). Peters et al. (1988) stated that in slow-flowing regulated lowland streams the larvae were characteristic for water with very low organic loading; however, a small number of larvae were found in more polluted water with a moderate current. The larvae have a high haemoglobin content and will be better able to live at low oxygen concentrations than other chironomids (Heinis, 1993: 53). The main advantage of this is probably that it enables the larvae to live deeper in the substrate (see Microhabitat).

SALINITY

In Hjarbæk Fjord, Denmark, the larvae were absent in water with a salinity of 1‰ and appeared only after further freshening (Lindegaard & Jónsson, 1983). However, in Estonia the larvae occurred in water with up to 3000 mg Cl/l (Tõlp, 1971, as *P.* gr. *scalaenum*). Hirvenoja (1962) also investigated specimens from brackish water in Finland with approx. 3000 mg Cl/l. In the Dutch province of Noord-Holland Steenbergen (1993) collected the larvae only rarely in water with a chloride content > 300 mg/l.

Polypedilum convictum (Walker, 1856)

nec *Polypedilum* gr. *convictum* Chernovskij, 1949: fig. 45; Pankratova, 1983: 247–248, fig. 196; Lenz, 1954–62: 240–243
Polypedilum laetum agg. Moller Pillot, 1984: 160–161 (pro parte); Tolkamp, 1980: 151 et seq.

SYSTEMATICS AND IDENTIFICATION

The species belongs to subgenus *Uresipedilum*. As stated earlier (Moller Pillot, 1984), the use of a group *convictum* is doubtful because it has a different meaning in Chernovskij (1949) and Lenz (1954–62). It appeared later that both these authors had made a mistake. The larva has been described correctly by Soponis & Russell (1982) and Rossaro (1985). First authors also described the second and third instar. In Moller Pillot (1984) *P. convictum* is included in *P. laetum* agg. Many old data have been verified later, for example material collected by Tolkamp (1980).
There has never been a problem with identification of the adult male (e.g. using Pinder, 1978) or the exuviae (using Langton, 1991). The characteristic male hypopygium is also illustrated in Pinder (1974) and in Prat (1978a). In the latter publication, however, the figures of *P. cultellatum* and *P. convictum* have been interchanged.

DISTRIBUTION IN EUROPE AND THE NETHERLANDS

P. convictum has been found in the whole of Europe (Saether & Spies, 2004). In the Netherlands the species seems to be confined to the eastern and southeastern part of the country.

LIFE CYCLE

Pinder (1983) stated at least two generations a year. He caught fourth instar larvae only from April until early October, and in late autumn only larvae in second instar. In Lake Belau in northern Germany the second generation was rarely represented (Otto, 1991: fig. 49). Adults are collected from May until October (Lehmann, 1971; Pinder, 1983; Michiels, 1999), but in Lake Belau almost exclusively in June.

EGGS

Williams (1982) observed large numbers of egg masses drifting in a river just after dusk. A high percentage belonged to *P. convictum*. The egg masses were capable of floating for several hours, supported by a gas vacuole within the egg mass.

MICROHABITAT

Different authors mention totally different substrates. Lehmann (1971) found the larvae mainly on mosses and stones, Becker (1994) on stones and gravel, Pinder (1980, 1983) and Pinder & Farr (1987) on coarse flint, but also on organic detritus. The larvae can also live on the stems and leaves of *Potamogeton* and *Ranunculus penicillatus* (Tokeshi & Pinder, 1985, 1986); in this case the larvae show significant departures from random colonisation. Drake (1982) found the larvae rarely on plants. Cuppen (pers. comm.) collected the larvae in fast-flowing brooks, especially between detritus. In the river Meuse the larvae have been found on artificial substrate (unpublished data). Tolkamp (1980, as *P. laetum* agg.) reported a preference for detritus in spring, whereas in winter coarse gravel was preferred. Overall, the preferred microhabitat can be summarised best as 'coarse substrate' (see Hawtin, 1998).

DENSITY

Pinder (1983) found large fluctuations in the density of larvae, only partly related to emergence of adults, with a maximum of 140 larvae/m^2 .

WATER TYPE

Current and dimensions

P. convictum is a species of running water; records from lakes are restricted to Scandinavia (Brundin, 1949). The larvae are mainly collected in upper and middle courses of streams and rivers with fast or moderate currents (Lehmann, 1971; Caspers, 1991; Becker, 1994; Michiels, 1999, 2004; Orendt, 2002a). In lowland brooks the species appears to be most common in highly oxygenated small brooks (Hawtin, 1998; H. Cuppen, pers. comm.), although Langton (1991) also mentioned larger streams or rivers. In the Netherlands large populations have also been found in the middle courses of lowland brooks in woodland (e.g. Tolkamp, 1980).

Permanence

The species has not been collected in temporary water.

pH

Most investigated streams mentioned in the literature have a rather high pH; ten Cate & Schmitt (1986) found the larvae in upper courses in Twenthe (the Netherlands) with a pH of 6.5–8.1.

TROPHIC CONDITIONS AND SAPROBITY

P. convictum appears to be a species of highly oxygenated brooks and streams without much fine particulate organic matter. However, there are not enough data on moderately polluted brooks with high oxygen contents. Therefore some caution is necessary in using the matrix figures.

Polypedilum cultellatum Goetghebuer, 1931

SYSTEMATICS AND IDENTIFICATION

P. cultellatum belongs to the subgenus *Uresipedilum*, but the larva is very different from the larva of the other European species, *P. convictum*. Identification of the larva was only possible after the description by Rossaro (1985). In Moller Pillot (1984: 166) the species keys out as *P.* cf. *uncinatum*. Good figures of the larval mouthparts and antenna can be found in Kiknadze et al. (1991). The characteristic male hypopygium is illustrated in Langton & Pinder (2007) and also in Pinder (1974) and in Prat (1978a). In the last publication, however, the figures of *P. cultellatum* and *P. convictum* have been interchanged. The female has been described by Rodova (1978).

DISTRIBUTION IN EUROPE AND THE NETHERLANDS

P. cultellatum has been found in nearly the whole of Europe (Saether & Spies, 2004). In the Netherlands the species has been collected mainly in the Pleistocene part of the country (Limnodata.nl).

LIFE CYCLE

Pinder (1983) found two generations in spring and early summer and a third generation in late summer in a small chalk stream in southern England. Lehmann (1971: 512) supposed there was only one generation in the river Fulda, emerging from July to September. In winter the larvae are in diapause in second and third instar; we collected third instar larvae until the end of April and fourth instar larvae from April to September. However, Pinder (1983) found a population of third and fourth instar larvae that appeared in October and persisted through the winter, emerging in late April or early May.

FEEDING

Detritus, diatoms, pollen and spores of fungi have been found in the guts of larvae. Larvae were observed feeding on an algal/fungal/bacterial growth on a dead leaf. They are most probably selective grazers.

MICROHABITAT

The larvae are often found on stones, wood and plants (Lehmann, 1971; Becker, 1994; Klink, 2001; own observations), but also on the bottom. Pinder (1980, 1983) noted in a small chalk stream that *P. cultellatum* apparently occupies the same niche as *P. convictum*. The larvae were collected on sand and on soft sediments. In another small chalk stream he found only a few larvae on a bottom of coarse flint (Pinder & Farr, 1987). Tokeshi (1986) and Tokeshi & Pinder (1985; 1986) found *P. cultellatum* in small numbers on *Potamogeton*, *Myriophyllum spicatum* and *Ranunculus penicillatus*. We found the larvae mainly on plants and on wood in a tube of silt, especially near the water surface.

See also under Water type (Dimensions) and under Saprobity.

WATER TYPE

Current

The larvae are numerous in the potamal zone of the river Fulda (Lehmann, 1971) and are common in the whole German part of the Rhine (Caspers, 1980, 1991; Becker, 1994). It appears from recent investigations that in the Netherlands they live in many lowland brooks. When the current becomes stronger in winter, the larvae seem to survive in stretches or sites where the current is weaker (own observations, see also Dispersal). In streams with a moderate to fast current the species can be collected locally (Casas &

D

Valchez-Quero, 1989; Orendt, 2002; Michiels, 2004). Records from stagnant water are less numerous (see under Dimensions), but Brundin (1949) found the species often and abundantly in Swedish lakes.

Dimensions
Klink (2001) found the larvae in large quantities on wood in the rivers Danube and Tisza in Hungary, but in the Rhine they seem to be only locally common (Becker, 1994; see also Wilson & Wilson, 1984, as Pe 1), possibly because of a shortage of suitable substrate. In the Netherlands the species is without doubt most abundant in the middle courses of lowland brooks. The larvae live in large lakes (e.g. Brundin, 1949; Schmale, 1999) and canals (Tempelman, pers. comm.) and sometimes in small pools (van Kleef, unpublished; own data), but for unclear reasons the species is absent from most stagnant water bodies.

Permanence
Fritz (1981) reared 11 specimens from a temporary pool near the river Rhine, emerging in July and September. However, it is possible that these larvae were brought down by the river.

pH
Most records are from brooks and streams with a pH higher than 7.0, but in the Netherlands the species has also been found in four moorland pools with a pH of 5–6.

TROPHIC CONDITIONS AND SAPROBITY
Brundin (1949) collected *P. cultellatum* widely and sometimes abundantly in oligotrophic lakes in Sweden and not in eutrophic lakes. Most data from the Netherlands are from eutrophic, but unpolluted or only slightly polluted streams or stagnant waters. However, we also found the larvae in a moderately polluted lowland brook and the large numbers in the rivers Danube and Tisza indicate that the species is more or less tolerant of pollution.
When the larvae live mainly on plants, wood and stones they are less affected by oxygen shortage and toxic decomposition products than bottom dwellers. This may be the reason that in Dutch lowland brooks the species is rarely a bottom dweller, whereas in a shallow stream with a faster current Pinder (1980) collected the larvae mainly from the bottom. In contrast to other *Polypedilum* species the larvae are only slightly reddish.

SALINITY
There are no records from brackish water.

DISPERSAL
Pinder (1983) supposed that the larvae were transported in large numbers by drift and settled in the stretch of the stream under investigation. An irregular patchy occurrence of the larvae on plants in lowland streams is quite normal because of the dynamic character of these systems (Tokeshi & Townsend, 1987).

Polypedilum laetum (Meigen, 1818)

Polypedilum gr. *laetum* Lenz, 1954–62: 239–243 (pro parte)
Polypedilum laetum agg. Moller Pillot, 1984: 160–161 (pro parte)
nec *Polypedilum laetum* agg. Tolkamp, 1980: 151 et seq.

SYSTEMATICS AND IDENTIFICATION

Lenz (1954–62) included all larvae with equally long lateral mental teeth in his gr. *laetum*. This was borrowed by Moller Pillot (1984). Later, much material from the Netherlands appeared to belong to *P. convictum*, including that of Tolkamp (1980). There is no problem with identifying male adults or exuviae, but Langton (1991) stated that the exuviae material is very variable and may belong to more than one species.

DISTRIBUTION IN EUROPE AND THE NETHERLANDS

P. laetum has been collected in nearly the whole of Europe (Saether & Spies, 2004). In the Netherlands the species is almost entirely confined to hilly areas (Nijmegen, Southern Limburg) and has also been collected in the Keersop (province of Noord-Brabant) (Tempelman, in prep.).

LIFE CYCLE

Lehmann (1971) collected adults and pupae from June until September in the Fulda region and called *P. laetum* a univoltine summer species. Elsewhere, the emergence period is much longer and the species will have at least two generations. For instance, Michiels (1999) collected exuviae in the Salzach in Bavaria from early May until the middle of October.

MICROHABITAT

Lehmann (1971) found the larvae in sandy-silty sediment in the river Fulda, but also on slightly silty mosses and stones. Becker (1994) reared two males from stones. Cuppen (pers. comm.) collected the larvae on sand and gravel bottoms in fast-flowing brooks in Southern Limburg (the Netherlands). They build short, loose sand tubes (Lindegaard-Petersen, 1972).

WATER TYPE

P. laetum is mainly an inhabitant of running water. Lehmann (1971) collected the species only in the rhithral zone of the river Fulda; it is numerous in the Upper Rhine, but rarer in the lower parts of the river (Caspers, 1980; Klink & Moller Pillot, 1982). It is a common species in fast-flowing streams in Bavaria (Michiels, 1999; Orendt, 2002, 2002a). There are some records from lowland streams with a mean velocity of 30–60 cm/sec (Linding Å in Denmark, Lindegaard-Petersen, 1972; Keersop in the Netherlands, D. Tempelman, unpublished), but most records in the Netherlands are from faster-flowing brooks and streams, especially in Southern Limburg.
Brundin (1949) collected the species numerously in eutrophic lakes in Sweden, but elsewhere in Europe it is rare in lakes (e.g. Reiss, 1968; Lindegaard & Brodersen, 2000). Van Kleef (unpublished) collected the exuviae of one specimen in a Dutch moorland pool (det. P. Langton).

pH

As far as is known, *P. laetum* has been found mainly in water with a more or less high pH. In a Dutch moorland pool the pH was 6.06 (van Kleef, unpublished).

TROPHIC CONDITIONS AND SAPROBITY

Brundin (1949) collected *P. laetum* in large numbers in eutrophic lakes in Sweden and rarely in oligotrophic lakes. Klink (1985) found the exuviae rather abundantly in the organically polluted Grensmaas stretch of the river Meuse in Southern Limburg, the Netherlands, but not at the most polluted sites. Orendt (2002) called the species very tolerant of organic loading. On the other hand, the larvae have often been collected in unpolluted or only slightly polluted upper courses of rivers, for example in the Fulda (Lehmann, 1971) and the Elz in the Black Forest (Michiels, 2004).

Polypedilum leucopus (Meigen, 1830)

This taxon, mentioned in Pinder (1978: 138, as *P. leucopum*), is only a form of *P. nubeculosum* (see Ashe & Cranston, 1990: 305). The form has been collected in the Netherlands at many places (Kruseman, 1933).

Polypedilum nubeculosum (Meigen, 1804)

Polypedilum cf. *nubeculosum* Moller Pillot, 1984: 166
Polypedilum nubeculosum agg. Moller Pillot & Buskens, 1990: 12, 24, 50

IDENTIFICATION

The larvae in fourth instar can be identified quickly by the very dark gula (for the rest only present in *P. nubifer*; see Tempelman, in prep.). In third instar most specimens also have a dark gular patch, but in some cases the dark patch is absent. Larvae identified using Moller Pillot (1984: 166) as *P.* cf. *nubeculosum* (in later publications *nubeculosum* agg.) belong to this species; his group *nubeculosum* s.l. has no systematic value.

DISTRIBUTION IN EUROPE AND THE NETHERLANDS

The species is very common throughout Europe (Saether & Spies, 2004) and the Netherlands.

LIFE CYCLE

P. nubeculosum has two or three generations a year with a diapause in winter. Adults emerge from the end of April to early October (Kruseman, 1933; Mundie, 1957; Reiss, 1968; Lehmann, 1971; Lindegaard & Jónsson, 1987; Otto, 1991; Orendt, 1993). Near an urban water body we saw swarming males as early as the middle of March. In such cases a fourth generation cannot be excluded. According to Tourenq (1975) the adults emerge when the water temperature in spring reaches 8 °C. Sokolova (1968) also noted that the species needs relatively few degree-days for larval development.
The larvae overwinter in third and (mainly) in fourth instar; we found the last third instar larvae in April. Moore (1979) stated that the larvae hardly eat during short day length in winter.

OVIPOSITION

Based on an observation by Munsterhjelm (1920), Nolte (1993) stated that one egg mass contains 450 eggs. However, she found 510 and 537 eggs in two egg masses of an unidentified *Polypedilum* species. These egg masses were attached to wood at the water's edge.

MICROHABITAT

The larvae are typical bottom dwellers, but some specimens can be found on wood or stones, or sometimes on plants (even as a miner, e.g. Urban, 1975). In the bottom they usually make long tubes (Izvekova, 1980). They often live in sandy bottoms and also in organic silt, possibly depending on oxygen conditions (Otten, 1986; Peeters, 1988; own data). Int Panis et al. (1995) found the larvae at a mean depth of 1.5 cm in the sediment and their density was obviously correlated with oxygen conditions at the site. They seemed not to derive much benefit from higher oxygen contents in the water column. Koskenniemi & Sevola (1989) found the larvae abundantly in organic sediment in winter, some of them 3–7 cm deep in the sediment. As a rule, the larvae in the frozen part of the sediment did not survive.

DENSITIES

Titmus & Badcock (1981) stated that *P. nubeculosum* became more aggregated with increasing density. The first instar larvae may prefer to settle near other *Polypedilum* larvae rather than near *Chironomus* larvae. Varying densities have been reported from lakes, from less than 10 larvae/m^2 (mainly in summer) to between 400 and 4000 (mainly in autumn) (Lindegaard & Jónsson, 1983; Brown & Oldham, 1984; Ten Winkel, 1987).

FEEDING

Titmus & Badcock (1981) found mainly detritus, some diatoms and few other algae in the gut of *Polypedilum nubeculosum* larvae. In many cases bacteria seem to be the most important food (Moore, 1979a). Moore (1979) observed that the larvae selected algae of a well-defined small size, such as *Scenedesmus*, and therefore ate very little filamentous algae. However, some algae, such as *Scenedesmus*, were poorly digested or not digested at all. When available, diatoms or blue-green algae can form an important part of the diet. Especially when the larvae penetrate deeper in the sediment they eat more small animals (Izvekova, 1980). Observations by Lenz (1954) indicate that *Polypedilum* larvae can also feed on dead animals.

WATER TYPE

Current

The larvae live in stagnant and slow-flowing water (Langton, 1991). They are also found in the rhithral zone of streams, although in low numbers (Lehmann, 1971; Orendt, 2002a). The presence of food may be decisive (see below under Shade).

Shade

The larvae were absent in the shaded lowland brooks investigated by Tolkamp (1980). This may be caused by the scarcity of algae in combination with the dominance of coarse, slowly decaying organic material.

Dimensions

Steenbergen (1993) collected significantly fewer larvae in canals and ditches less than 10 m wide, but they were sometimes present in ditches less than 4 m wide (cf. Moller Pillot, 2003: 120). Remmert (1955) collected no larvae from small pools and also observed no adults. They can be numerous in large lakes (Mol et al., 1982; van der Velden et al., 1995) and appear to be the dominant chironomid in Lake IJssel in the Netherlands, at least up to 1750 m from the shore (Maenen, 1983).
In running water the larvae are also less abundant in narrow brooks and often numerous in streams and rivers. In lakes and rivers they are also found in deeper layers (sometimes in high densities) if these are well oxygenated (Kuijpers, pers. comm.; Brown &

Oldham, 1984), although such migration was not found by Neubert & Frank (1980) and Peeters (1988). Mol et al. (1982) collected the larvae rarely deeper than 6 or 8 m, in this case probably because of oxygen conditions.

Permanence
P. nubeculosum has been rarely collected in temporary water (Schleuter, 1985; Verdonschot et al., 1992; Moller Pillot, 2003; cf. Delettre, 1989), but most investigations of temporary non-acid pools refer to very small pools.

SOIL

Steenbergen (1993) collected the larvae (mostly in stagnant water), significantly less on clay and more on peat than on sand.

pH

The larvae of *P. nubeculosum* are scarce or absent in acid water bodies such as moorland pools, peat cuttings, sand pits, upper courses of streams, etc. (Buskens, 1983, 1987; Leuven et al., 1987; Werkgroep Hydrobiologie, 1993; Duursema, 1996; Moller Pillot, 2003; van Kleef, unpublished). However, they can be found in low numbers even at a pH of about 5 (Raddum & Saether, 1981; Delettre, 1989; Moller Pillot, 2003; van Kleef, unpublished). The species is more common in water at pH > 5 and they can also endure very alkaline conditions: in Hjarbæk Fjord, Denmark, the larvae were found in water with a pH of 10.5–11 (Lindegaard & Jónsson, 1983). Steenbergen (1993) stated no preference for pH > 7.5. It seems certain that pH itself has no influence on the larvae, but it will influence food quality and quantity.

TROPHIC CONDITIONS AND SAPROBITY

P. nubeculosum is a scarce inhabitant of oligotrophic lakes and is common to abundant in eutrophic lakes (Brundin, 1949; Saether, 1979; Lindegaard & Jonsson, 1983; 1987; Buskens & Verwijmeren, 1989; Langdon et al., 2006). The larvae are sometimes collected in large numbers in flowing water with moderate to severe organic pollution if the oxygen content is not very low (e.g. Otten, 1986; Peters et al., 1988).
Moore (1979a) noted an increase in the number of larvae when the oxygen content was declining and Grontmij | Aqua Sense (unpublished) sometimes found the larvae in nearly anoxic conditions. However, Int Panis et al. (1995) stated that the density was correlated with the oxygen content in the sediment because the larvae seemed not to benefit much from higher oxygen contents in the water column (see Microhabitat). Rossaro et al. (2007) stated that the haemoglobin content of the larva was very high, but they did not find the species in water poor in oxygen. Consistent with this, we observed that the larvae were absent in organic mud with bad oxygenation; in such cases larvae could sometimes be found on artificial substrate (own unpublished observations).
Bazerque et al. (1989) and Wilson & Ruse (2005) called the species intolerant of organic pollution. However, we think that the larvae have some preference for organically polluted water and even survive severe pollution, but that they are not able to survive for a long time in sediment without oxygen.

SALINITY

In Hjarbæk Fjord, Denmark, the larvae were absent from brackish water and appeared only after almost complete freshening (Lindegaard & Jónsson, 1983). In the Netherlands the species is very scarce at a chloride content above 1000 mg/l (Steenbergen, 1983; Krebs & Moller Pillot, in prep.). Tourenq (1975) sometimes found the larvae in oligohaline water with a chloride content of 1500 mg/l. This author referred to literature suggesting that the species can also develop in mesohaline water. Schreijer (1983) found one larva of *P.* gr. *nubeculosum* in water with a chloride content up to 2810 mg/l.

Polypedilum nubens (Edwards, 1929)

SYSTEMATICS

The species belongs to subgenus *Pentapedilum* Kieffer, 1912.

IDENTIFICATION

The adult male can be identified using Langton & Pinder (2007), but in some specimens the superior volsella has a lateral seta (Oyewo & Saether, in prep.), in which case the species can be identified as *P. uncinatum*. The exuviae can be identified without problems using Langton (1991). The larva of *P. nubens* is unknown.

DISTRIBUTION IN EUROPE

The species lives in Western, Southern and Central Europe (Saether & Spies, 2004). So far there are no records from the Netherlands or adjacent lowlands.

ECOLOGY

P. nubens is mainly known from lakes (Edwards, 1929; Douglas & Murray, 1980; Langton, 1991), but Michiels (2004) collected the exuviae in the lower parts of a fast-flowing brook in the Black Forest. According to Ruse (2002) the species lives mainly in water bodies with very low conductivity (< 200 µS/cm).

Polypedilum nubifer (Skuse, 1889)

Polypedilum pharao Tourenq, 1975: 196; Fittkau & Reiss, 1978: 435
Polypedilum aberrans Chernovskij, 1949: fig. 47; Michailova, 1988: 239–246

SYSTEMATICS

According to Michailova (1988) *P. aberrans* Chernovskij, 1949 is another species, based on mainly slight differences in cytology. This has not been accepted by Saether & Spies (2004).

DISTRIBUTION IN EUROPE AND THE NETHERLANDS

The species is commonly found in the Mediterranean region and in some countries in Central Europe (Saether & Spies, 2004). In the Netherlands there are two recent records (Wilhelm, unpublished; van den Hoek, unpublished). The species is thought to have extended its range in a northerly direction as a result of climate change.

WATER TYPE

Fittkau & Reiss (1978) considered the species to be characteristic of small temporary water bodies. At least one of the two Dutch records apply to a recently dug pool. When burrowed in a sandy bottom the larvae survive desiccation for up to 20 days at a water content of 30% (Tourenq, 1975).

ECOLOGY

The species is common and often numerous in rice fields. The larvae are probably not very harmful to developed rice plants, but damage the roots and leaves of rice seedlings

(Ferrarese, 1992; Wang, 2000). In Italian rice fields the larvae were absent in May and the first half of June (Ferrarese, 1992). The author does not mention whether this was the result of the larvae disappearing from the fields in winter. The larvae survive water temperatures of 35° C. (Tourenq, 1975).

SALINITY

The larvae survive polyhaline water (up to 20,000 mg chloride/l) for a long time and can emerge from water with 10,000 mg chloride/l (Tourenq, 1975).

Polypedilum pedestre (Meigen, 1830)

IDENTIFICATION

The larvae have often been identified as *P. pedestre* agg. (Moller Pillot, 1984). It can now be concluded that no other European species belongs to this aggregate. East European data under the name *P.* gr. *pedestre* based on identification using Chernovskij (1949) also refer to this species.

DISTRIBUTION IN EUROPE AND THE NETHERLANDS

P. pedestre has been collected in the whole of Europe (Saether & Spies, 2004). In the Netherlands the species is common in Southern Limburg and has been found in a number of brooks and streams in the eastern part of the country and in the province of Noord-Brabant (Moller Pillot & Buskens, 1990; Nijboer & Verdonschot, unpublished; Limnodata.nl). Some records from the western part of the country have to be verified.

LIFE CYCLE

Lehmann (19710 collected adults and pupae from May until September. Kawecka & Kownacki (1974) caught pupae in the river Raba in Poland in July and in September.

FEEDING

From May until September Kawecka & Kownacki (1974) found that the guts of larvae of *P.*gr. *pedestre* contained much mineral matter (45–80% of gut content). The remaining matter was mainly indeterminate organic matter and a low number of diatoms. Blue-green and green algae were nearly absent in the gut, although they were abundant in the river.

MICROHABITAT

The larvae live on hard substrates (stones, wood) and on the bottom. There are no more exact data about preference.

pH

The species has not been mentioned from acid streams. It is not known if the larvae can endure acidity.

WATER TYPE

Current and dimensions

Polypedilum pedestre is a species of brooks and streams, although there are a number of records from lakes (e.g. Sweden: Brundin, 1949; Lake Constance: Reiss, 1968; Bavaria: Orendt, 1993). The mention of the occurrence of this species in ponds in the 1961 translation of Chernovskij (1949) is an incorrect translation.

The larvae are mainly found at rather fast currents. Lehmann (1971) collected the species most often in de 'mittlere Salmoniden Region'. The Dutch records are mainly from fast-flowing brooks or from riffles with many stones. However, larvae have sometimes been collected in smaller brooks with moderate currents. It is present in the whole river Rhine (Caspers, 1980; 1991), but very scarce in the Dutch part (Klink & Moller Pillot, 1982; cf. Becker, 1994). It seems to be nearly absent from the Dutch part of the river Meuse (Klink & Moller Pillot, 1982; Smit, 1982; Peeters, 1988).

Permanence
Although there are no records from temporary water, Helešic et al. (1998) noted that *P. pedestre* is one of the few species that are able to survive dewatering of a periodically covered river bed at the wet bed.

SAPROBITY
The larvae of *P. pedestre* have often been found together with large quantities of *Chironomus* in polluted stretches of brooks and streams such as the Łyna in Polonia (Wielgosz, 1979), the Geul in the Dutch province of Limburg (Cuijpers & Damoiseaux, 1981) and the Keersop in the province of Noord-Brabant (own unpublished data). Nevertheless, in the Netherlands the larvae have also been found in brooks without anthropogenic pollution (L. Higler, pers. comm.; H. Cuppen, pers. comm.; Waterschap Zuiveringsschap Limburg, unpublished).

Polypedilum pullum (Zetterstedt, 1838)

Chironomus prolixitarsis Macan, 1949: 169 et seq.

SYSTEMATICS
The species belongs to subgenus *Tripodura* Townes, 1945.

IDENTIFICATION
Because of the interchanging of figures 4 and 5 in Pinder (1978: 136) adult males with marked wings have been identified as *P. pullum* instead of *P. scalaenum*. The hypopygium (and description?) of *P. pullum* by Hirvenoja (1962) applies to *P. bicrenatum*. Exuviae can be identified well using Langton (1991), but *P. pullum* and *P. scalaenum* are more or less similar. The larva of *P. pullum* is still unknown. The female has been described by Rodova (1978).

DISTRIBUTION IN EUROPE AND THE NETHERLANDS
P. pullum has been found nearly everywhere in Europe (Saether & Spies, 2004). Very few records are known from the Netherlands, mainly because the larva has not been described. Klink (1985a) collected adult males along the river IJssel near Kampen. Kuijpers et al. (1992) collected the exuviae in the water storage reservoirs in the Biesbosch and van Kleef (unpublished) collected the exuviae in two moorland pools (det. P. Langton). Some of the larvae collected in fast-flowing brooks may belong to this species (see under *P. bicrenatum*).

LIFE CYCLE
Macan (1949) collected emergent adults at the end of May (few) and in July. Lehmann (1971) found adults and pupae from July until September.

EGGS
Nolte (1993) reported 100 eggs in an egg mass.

MICROHABITAT

Lehmann (1971) collected larvae in silt on the bottom and on a firm substrate in the river Fulda. Becker (1994) reared two males from stones in the Rhine. Palomäki (1989) found the larvae locally on open sandy bottoms or detritus in Finnish lakes.

WATER TYPE

The larvae live in stagnant and flowing water. Fittkau & Reiss (1978) mentioned brooks, rivers, lakes and marshes. They are common in lakes in Scandinavia and in more or less montane lakes (e.g. Brundin, 1949; Macan, 1949; Langton, 1991; Orendt, 1993), but they can be collected in smaller pools in the Dutch lowland too (van Kleef, unpublished). The larvae were also present in the profundal zone of an oligotrophic lake in Finland (Paasivirta, 1974).
In streams and rivers they live at sites with moderate to rather fast currents (Lehmann, 1971; Caspers, 1991; Langton, 1991; Orendt, 2002a; Scheibe, 2002; Michiels, 2004). In the Netherlands Klink (1985a) found adult males of *P. pullum* in very low numbers between *P. scalaenum* along the river IJssel near Kampen. The larvae are also present in the lower reaches of the Rhine (Becker, 1994), but Wilson & Wilson (1984) must have (at least partly) mistaken *P. scalaenum* for *P. pullum*.

pH

P. pullum lives in streams with a high pH, but also in very acid conditions. Orendt (1999) recorded the presence in streams with a pH of 5.4–6.3. The species appeared in a Scottish lake after acidification between 1850 and 1950, and increased in numbers when the pH fell from 5.5 to 4.8 (Brodin & Gransberg, 1993). The exuviae of one specimen has been found in the Gerritsflesch (a Dutch moorland pool) at pH 4.0.

TROPHIC CONDITIONS AND SAPROBITY

According to Ruse (2002) *P. pullum* is a characteristic species of water with low conductivity. This is consistent with its widespread presence in oligotrophic lakes in Scandinavia (Brundin, 1949; Paasivirta, 1974) and the records from a tarn in England (Macan, 1949) and moorland pools in the Netherlands (van Kleef, unpublished). The species is probably more tolerant of eutrophication in streams and rivers, but some reports of such tolerance will apply to *P. scalaenum* (see Identification).

Polypedilum quadriguttatum Kieffer, 1921

SYSTEMATICS

The species belongs to subgenus *Tripodura* Townes, 1945.

IDENTIFICATION

Adult males can be identified using Langton & Pinder (2007). In Pinder (1978) the species can be identified as *P. bicrenatum* because figures 4 and 5 on p. 136 have been interchanged. Exuviae of *P. quadriguttatum* cannot be distinguished from *P. scalaenum* using Langton (1991); some differences are given in Langton & Visser (2003). The larvae are unknown. Reiss (pers. comm. to C. Becker) suggested that this species is only a form of *P. scalaenum*.

DISTRIBUTION IN EUROPE

P. quadriguttatum is an inhabitant of Western and Central Europe (Saether & Spies, 2004). The fact that the species is not yet known from the Netherlands may be because the larvae and pupae cannot be identified.

WATER TYPE

The locus typicus of the species is the river Oder, where the larvae live in sand (Lenz, 1954–62). Becker (1994) reared the species from the Rhine from stony, gravelly and muddy substrate. The larvae seem to live mainly in streams with moderate current (Pinder, 1974; Caspers, 1991; Michiels, 2004).

Polypedilum scalaenum (Schrank, 1803)

Polypedilum breviantennatum Chernovskij, 1949: 79–80; Ertlová, 1970: 296–297
nec *Polypedilum* gr. *scalaenum* Chernovskij, 1949: 79, fig. 48
nec *Chironomus (Polypedilum) scalaenus* Pagast, 1931: 227–228 (pro parte)

SYSTEMATICS

The species belongs to subgenus *Tripodura* Townes, 1945.

IDENTIFICATION

Pagast (1931) found two *Polypedilum* species in Lake Usma, Latvia. The larva described by him as *scalaenum* belongs to another species, but he also found the true *P. scalaenum* larva. The citation by Brundin (1949: 772) does not apply to *P. scalaenum*.
The male adult can be identified using Langton & Pinder (2007). In Pinder (1978) the wing membrane is incorrectly described as unmarked. Because figures 4 and 5 have been interchanged in the key by Pinder (1978), adult males of *P. scalaenum* and *P. pullum* have often been misidentified (as each other). The exuviae of these species are also very similar. For instance, Wilson & Wilson (1984) do not mention *P. scalaenum* in the river Rhine and identified all their exuviae as *P. pullum*. The female is described in Rodova (1978).
Identification of the larvae is problematic because one or more other species resemble *P. scalaenum* (see e.g. *P. apfelbecki* and *P. pullum*). For good figures of the mouth parts and antenna see Kiknadze et al. (1991). The larvae are fiery red in third and fourth instar.

DISTRIBUTION IN EUROPE AND THE NETHERLANDS

P. scalaenum has been found everywhere in Europe (Saether & Spies, 2004). In the Netherlands the species is common in the Pleistocene areas, in Southern Limburg and in the large rivers. It is very scarce elsewhere (Limnodata.nl; Nijboer & Verdonschot, unpublished).

LIFE CYCLE

Lehmann (1971) reported emergence from May until August in Germany; Becker (1994) collected many adults along the river Rhine in April, rarely in March. Tõlp (1958) and Otto (1991) found (at least) two generations a year. Orendt (1993: 110) found two emergence periods in summer: early July and the end of August. Konstantinov (1958) reported four generations a year in the river Amur in East-Siberia. The larvae hibernate in third and fourth instar; third instar larvae have been found until April.

OVIPOSITION

Nolte (1993) reported 127 eggs in one egg mass. However, this may be accidentally few because the numbers are often much more in other species of the genus. See for more information under *P. nubeculosum*.

MICROHABITAT

In brooks, rivers and lakes the larvae inhabit sandy soils without much organic material

(Lehmann, 1971; Srokosz, 1980; Tolkamp, 1980; Verdonschot & Lengkeek, 2006). They are also present in sand and gravel bottoms (van Urk & bij de Vaate, 1990; Peeters, 1988; Becker, 1994). Tolkamp (1980: 48, 192) stated a preference for sand with fine detritus and a strong preference for fine sand. The larvae can also be found incidentally on silted plant stems and wood (Lehmann, 1971). They are rare on bottoms with pebbles and cobbles and on stones in the littoral zone (Klink & Moller Pillot, 1982; Peeters, 1988) and the larvae settle only scarcely on artificial substrates (Ertlová, 1970). Otto (1991: 44) noted as much emergence from the *Phragmites* zone as from open water.

DENSITIES

Van Urk & bij de Vaate (1990) reported up to 10,000 larvae/m² in sand and gravel bottoms in rivers.

WATER TYPE

Current

The larvae are mainly inhabitants of brooks and the potamal zone of rivers, but also live in lakes. They always require the availability of an open sandy bottom (see Microhabitat). They are also found locally in faster running streams in hilly country (Braukmann, 1984; Orendt, 2002a; Waterschap Roer en Overmaas, unpublished) and in small tributaries of such streams (Dutch unpublished data).

Dimensions

P. scalaenum is a common inhabitant of large rivers (Lehmann, 1971; Caspers, 1980, 1991; Klink & Moller Pillot, 1982; Peeters, 1988; van Urk & bij de Vaate, 1990; Klink, 2002). Peeters found the larvae mainly 5–8 m deep and much less frequently in the littoral zone of the river. The larvae are scarce in the estuarine part of the Dutch rivers (Smit et al., 1994; van der Velden et al., 1995). In the Netherlands the species can be found in the middle courses of lowland brooks, but are more common in spring brooks and small upper courses of brooks (ten Cate & Schmidt, 1986; Verdonschot & Schot, 1987; own data).
The species is only reported locally from lakes (Pagast, 1931 pro parte; Brundin, 1949: rare; Reiss, 1968; Orendt, 1993). Pagast (1931) collected the larvae of this species only down to a depth of 3 m (not correctly interpreted by Brundin, 1949); however, Tõlp (1971) found them up to 27 m deep. In the Netherlands the species has been rarely collected in lakes and sometimes in canals, but never in fen lakes, pools or other small stagnant water bodies.

Permanence

The larvae are rare in temporary brooks (ten Cate & Schmidt, 1986).

SAPROBITY

Orendt (1993) collected the species only in a mesotrophic lake in Bavaria and supposed that it is characteristic for mesotrophic conditions. In the Netherlands the larvae are often found in relatively pure sand bottoms in brooks and rivers without anthropogenic pollution. However, they endure organic pollution very well if the current is faster and therefore the oxygen availability is good. For instance, Cuijpers & Damoiseaux (1981) found the species numerously in the rather heavily polluted lower reaches of the Geul. There are no records from sites with low oxygen content.

pH

In small brooks in the province of Overijssel in the Netherlands the larvae lived at pH 6.5–8.1 (ten Cate & Schmidt, 1986). Elsewhere, most records are from water bodies at pH > 7.

SALINITY

In Estonia the species (as *P. breviantennatum*) has been found in water with a salinity (not chlorinity!) of 0.4–8.5‰ (Tõlp, 1971). At 8.5‰ it was the only chironomid species present. In the Netherlands the species have never been found in brackish water.

Polypedilum sordens (van der Wulp, 1874)

Tendipedini gen.? l. *macrophtalma* Chernovskij, 1949: 84–87, fig. 60
Pentapedilum sordens Walshe, 1951: 70 et seq.; Dvořák, 1996: 27 et seq.
Pentapedilum tritum Burtt, 1940: 119 nec aliis

SYSTEMATICS

The larva of *P. sordens* is very aberrant from those of other species of the subgenus *Pentapedilum*. The species possibly deserves a separate status (see Saether & Sundal, 1999: 330). In second instar the second lateral teeth of the mentum are a little longer than the first laterals.

DISTRIBUTION IN EUROPE AND THE NETHERLANDS

The species has been found nearly everywhere in Europe (Saether & Spies, 2004); in the Netherlands it seems to be absent only from some brackish water regions.

LIFE CYCLE

The first emergence period of this species does not start until the middle of May and lasts until October (Berg, 1950; Shilova, 1976; Kouwets & Davids, 1984; Otto, 1991; Orendt, 1993). Schleuter (1985: 159–160) reported emergences from a pond in Germany until December. As a rule there are two generations, but Orendt (1993) and Schleuter (1985) consider a third or even fourth generation to be possible locally. Different authors stated a small spring generation and a much more numerous summer generation (e.g. van der Velde & Hiddink, 1987; Otto, 1991; Kondo & Hamashima, 1992; Orendt, 1993; Lindegaard & Brodersen, 2000). However, in De Zodden (the Netherlands) Mol et al. (1982a) found the larvae most abundant in spring.
The larvae winter in second, third and fourth instar on or within the tissues of plants (Berg, 1950; own data). Second instar larvae can be collected until early April.

D

OVIPOSITION

The eggs are laid under water and attached to the edge of plants (Kondo & Hamashima, 1992).

FEEDING

The larvae are filter feeders, feeding on what comes into the net: diatoms, algae (periphyton and plankton), bacteria and detritus (Burtt, 1940; Berg, 1950; Walshe, 1951). Kondo & Hamashima (1992) stated that the larvae are non-selective. Dvořák (1996) found mainly particulate organic matter (from epipelic origin or originating from phytoplankton) in the gut of the larvae, as well as diatoms and filamentous cyanophytes.

MICROHABITAT

The larvae are miners, but are hardly able to penetrate through the epidermis of living plant stems. They live as miners in soft plant tissues or as secondary miners in mines vacated by other species (Berg, 1950; van der Velde & Hiddink, 1987). They are found in leaves and more numerously in stems of *Stratiotes*, *Potamogeton*, *Typha* and *Nuphar lutea* (Berg, 1950; Walshe, 1951; Urban, 1975; van der Velde & Hiddink, 1987; Kondo

& Hamashima, 1992). Dvořák (1996) found the larvae mainly on the plant surface and Mol et al. (1982) collected some larvae on stones. Higler (1977) collected the larvae (as *P.* gr. *laetum*) only on submerged *Stratiotes* plants and not on the dense emersed vegetation near the shore of the broads. The larvae were also present on artificial plants (Higler, unpublished).
The larvae pupate within their tubes and when mature swim about near the water surface for a few hours preceding emergence (Berg, 1950). Because of their manner of living, the numbers of larvae collected are highly dependent on the sampling method.

WATER TYPE

Current
The larvae are scarcely found in lowland brooks (Bazerque et al., 1989; Limnodata.nl; own data). Braukmann (1984) did not collect the species at all in German brooks. The species was absent from the river Fulda (Lehmann, 1971). Some or all exuviae found in rather fast-flowing brooks and streams (Orendt, 2002a) may have originated from oxbow lakes along the stream.

Dimensions
Steenbergen (1993) collected the larvae most often in large canals and lakes and in significantly fewer numbers in ditches less than 4 m wide. However, Limnodata.nl contains a relatively large number of records from narrow ditches.

Permanence
There are hardly any records from temporary water (ten Cate & Schmidt, 1986; own data). In view of the life cycle of the species the chance of settlement in a temporary pool is very low.

pH

P. sordens has often been collected in water with a pH of 4 (Buskens, 1983, 1987; Leuven et al., 1987; Duursema, 1996; van Kleef, unpublished; own data). However, the species is common and often abundant in water at high pH: Steenbergen (1993) found the larvae significantly more often in water with pH > 8.0. Limnodata.nl gives a mean pH of 7.42.

TROPHIC CONDITIONS AND SAPROBITY

In lakes, ponds and ditches the species can be found from oligotrophic to eutrophic water (Brundin, 1949; Duursema, 1996). Orendt (1993) called the species eutraphent. Steenbergen (1993) collected slightly more larvae in water with high chlorophyll-a content, but significantly more in water with low content of orthophosphate. Mol et al. (1982a) stated that the larvae were more numerous in the more eutrophicated part of a gradient in 'De Zodden' near Maarsseveen.
Because they live on and in plant stems the larvae are less sensitive to accumulation of polluted material, but organic particles can disturb their filter system. Bazerque et al. (1989) collected the larvae rarely in polluted stretches of the river Somme and considered them to be species intolerant of organic pollution (code A), in agreement with Wilson & Ruse (2005). Without doubt the larvae prefer water with a stable oxygen regime (Steenbergen, 1993; Limnodata.nl), although they have been collected sometimes in oxygen poor conditions.

SALINITY

P. sordens has been rarely collected in water with more than 1000 mg Cl/l (Steenbergen, 1993). The species is most common in water with an intermediate conductivity: 200–

800 µS/cm (Ruse, 2002; Limnodata.nl). Krebs (1981, 1984) did not find the species in brackish water in the province of Zeeland, but this is because he did not investigate mining species.

Polypedilum tetracrenatum Hirvenoja, 1962

SYSTEMATICS

The species belongs to subgenus *Tripodura* Townes, 1945.

IDENTIFICATION

P. tetracrenatum was not included in the key by Pinder (1978), but has been added in the new key by Langton & Pinder (2007). Hirvenoja (1962) gives descriptions of the male and the female. The exuviae are keyed in Langton (1991). The larva is unknown, but Vårdal (2002) supposed, on the basis of genetic relationship, that the antennal segments 3 to 5 are not all reduced.

DISTRIBUTION IN EUROPE AND THE NETHERLANDS

P. tetracrenatum has been collected in only a few countries scattered over Europe (Saether & Spies, 2004). In the Netherlands there is only one record, from a moorland pool near Waalre (det. P. Langton).

WATER TYPE

The species is known from shallow standing water (Langton, 1991). Hirvenoja (1962) mentioned specimens from a lake in Finland.

Polypedilum tritum (Walker, 1856)

nec *Polypedilum tritum* Burtt, 1940: 119 (= *P. sordens*)
nec *Polypedilum tritum* Iovino & Miner, 1970: 203 et seq. (doubtful identification)
Polypedilum tritum auctt. (pro parte)
Polypedilum uncinatum Limnodata.nl (pro parte)

SYSTEMATICS AND IDENTIFICATION

See *P. uncinatum*.

DISTRIBUTION IN EUROPE AND THE NETHERLANDS

P. tritum has been reported from nearly the whole of Europe (Saether & Spies, 2004). There are no confirmed data from the Netherlands, although the species is probably regionally common, especially in the Holocene regions.

LIFE CYCLE

In the Borkener See in Germany (a permanent water body) adult males emerged from May until September, most numerously in May and sharply decreasing in numbers during the summer. Only one male has been verified using the new key by Oyewo & Saether (in prep.) (Dettinger-Klemm, pers. comm.).

MICROHABITAT

Without doubt many larvae live on or near the bottom, but larvae probably belonging to this species have been found several times on plants, even near the water surface (own data).

WATER TYPE

Oyewo & Saether (in prep.) report the occurrence of *P. tritum* in flowing water and in a lake, a pond and a bog. Dettinger-Klemm (pers. comm.) verified the identification of adult males from a temporary and a permanent water body. In the first case *P. tritum* and *P. uncinatum* were found together. There are many unconfirmed records of larvae from ditches and pools in different parts of the Netherlands. Humphries (1936) collected the larvae (of this species?) mainly in the sublittoral zone of Lake Windermere, rarely up to 20 m deep.

pH

Dettinger-Klemm (pers. comm.) verified the identification of two populations in water with a pH of 6.5–9. In contrast, Oyewo & Saether (in prep.) saw material from a bog and we have identified larvae from an acid moorland pool.

TROPHIC CONDITIONS AND SAPROBITY

Owing to identification problems there are very few confirmed records. It has been definitely established that *P. tritum* sometimes lives in very eutrophic water (Dettinger-Klemm, in litt.), but it appears that the larvae also (and probably more often) occur in less eutrophic conditions, as indicated, for example, by many unconfirmed records from ditches in the Holocene part of the Netherlands (D. Tempelman, pers. comm.). In one case, larvae of *P. uncinatum/tritum* were collected (in spring) in low numbers in organically polluted water with hardly any oxygen in summer (Prov. Waterstaat Zuid-Holland, unpublished).

SALINITY

The species is most probably able to live in slightly brackish water; see under *P. uncinatum*.

Polypedilum uncinatum (Goetghebuer, 1921)

Pentapedilum uncinatum Pinder 1978: 134
nec *Polypedilum uncinatum* Beattie 1978a
Polypedilum cf. uncinatum Moller Pillot, 1984: 166 pro parte
Polypedilum uncinatum agg. Moller Pillot & Buskens, 1990: 12, 24, 51 pro parte
Polypedilum tritum Dettinger-Klemm, 2003

SYSTEMATICS

P. uncinatum belongs to the subgenus *Pentapedilum*. This subgenus can be easily recognised in the adult stage by the macrotrichia on the wing membrane. The pupae and larvae have no conspicuous subgeneral characters.

IDENTIFICATION AND NOMENCLATURE

Adult males

The differences between *P. uncinatum* and *P. tritum* have been unclear for a long time. It appeared to be impossible to distinguish between these two species by means of the characters mentioned in Goetghebuer (1937–1954) and Pinder (1978). This question has been elaborated by Dettinger-Klemm (2003: 111–115). In Saether & Spies (2004) both species are united as *P. tritum*. In the meantime the problem has been solved by Oyewo & Saether (in prep.). The characters used in the past are indeed unreliable, but the males of both species can be distinguished because *P. tritum* has a tibial spur on the foreleg ending in a sharp point, while *P. uncinatum* has only a blunt process.

Another problem is that the adult male of *P. uncinatum* resembles some specimens of *P. nubens*. However, the latter species will be rare in the Netherlands since the exuviae are easy identifiable and have never been encountered.

Exuviae
Until now identification of exuviae has proved to be unreliable. Oyewo & Saether (in prep.) have keyed the pupae of *P. uncinatum* and *P. tritum*; it remains to be seen whether the difference given by these authors will hold after studying more material.

Larvae
Larvae identified using Moller Pillot (1984: 166) as *Polypedilum* cf. *uncinatum* (in Moller Pillot & Buskens, 1990: 24, 51, 78 *P. uncinatum* agg.) can belong to three or more different species. *P. tritum* is still absent in the key by Klink (2001) and most larvae of this species will be identified as *P. uncinatum*. Identification of the larvae is still not completely reliable because Oyewo and Saether did not have enough material and we have no reared or associated material of *P. tritum*. Nevertheless, it is very probable that the larvae can be identified by means of the characters mentioned in Tempelman (in prep.). Larvae identified using provisional keys by Moller Pillot and Tempelman (where *uncinatum* has been distinguished from cf. *tritum*) will usually have been named correctly.
It is certain that the species investigated by Dettinger-Klemm (2000, 2003) is *P. uncinatum*. The author has verified adult males with the tibial spur on the foreleg. The ratio between the second, third and fourth antennal segment in his fig. 50 on p. 114 is about 16:14:14 µm, which is characteristic for this species according to Tempelman (in prep.).

DISTRIBUTION IN EUROPE AND THE NETHERLANDS

Because of the impossibility of distinguishing between *P. uncinatum* and *P. tritum* so far little has been known about their distribution. The combined taxon *P. tritum/uncinatum* has been recorded throughout Europe, with the exception of some parts of the Mediterranean region (Saether & Spies, 2004). Oyewo & Saether (in prep.) saw specimens of *P. uncinatum* from many countries in Europe.
P. uncinatum is very common in the Pleistocene areas of the Netherlands and probably rather scarce in the Holocene parts of the country. According to Limnodata.nl, *Polypedilum uncinatum/tritum* has been recorded mainly in the Pleistocene and fen-peat regions.

LIFE CYCLE

The larvae have a diapause (oligopause) in winter in third and fourth instar, induced by short days of about 13 h (Dettinger-Klemm, 2000; 2003: 133, 248). Younger larvae do not survive winter conditions because these larvae have a lower lethal limit of 5 °C. Dettinger-Klemm (2000, 2003) called the species polyvoltine. He found two to three generations a year. In culture vessels, however, the duration of development lasted only about 30 days at 19 °C.
In the Netherlands the males can be observed swarming already early in April.

FEEDING

The guts of larvae of *P. tritum/uncinatum* were found to contain mainly fine particulate organic material and filamentous algae.

MICROHABITAT

Larvae were found on organic and anorganic bottoms, often on dead plant remains or on mosses and against the banks of ditches and pools. Larvae have also been found on living plants, but is not clear if they also live on plants near the water surface, as stated for larvae probably belonging to *P. tritum*.

SWARMING AND OVIPOSITION

We saw swarms of 50 or more males 20–100 cm above the water surface. Dettinger-Klemm (2000) reported adults swarming about 1 m above or beside pools. They were also swarming in cages of 125 x 85 x 80 cm and in this case fertilisation of the egg masses was 100%. In smaller aquaria only a small fraction of the eggs were fertilised. Egg masses were laid at dusk on the margins of a gauze laid on the water surface. Dettinger-Klemm (2003: 240) supposed that the females do not lay eggs on quasi-terrestrial soils. The number of eggs per egg mass (at 25 °C) was 50–180. However, the number of eggs is highly temperature dependent; at lower temperatures higher egg numbers are expected.

TEMPERATURE

In rearing the species Dettinger-Klemm (2000) reported a lower lethal limit for the eggs and small larvae of 5 °C and an upper lethal limit for larvae around 30 °C.

DENSITIES

High densities of larvae were found especially in temporary water, at least in one case nearly 200 larvae/dm^2 (unpublished own data; cf. Schleuter, 1985: 92; Delettre, 1989).

WATER TYPE

Oyewo & Saether (in prep.) report *P. uncinatum* from flowing and stagnant water. Dettinger-Klemm (2000, 2003) studied the species in temporary pools. In the Netherlands the species appears to be often numerous in pools, peat cuttings and alder carr and much less common in the upper courses of lowland brooks and in meadow pools. In larger brooks and rivers the larvae seem to be scarce or absent (see under *P. tritum*). The larvae are able to survive in mud when the water content is reduced to < 20% of saturation (Dettinger-Klemm, 2002, 2003). However, in very dry years a sharp decrease in numbers was noted (Moller Pillot, 2003).

pH

From material examined by Oyewo & Saether (in prep.) and our own data, it appears that *P. uncinatum* lives in water with very different pHs. However, we only collected the species numerously in acid and/or temporary water (peat cuttings, woodland pools and alder carr). The decaying plant material consisted mainly of dead grasses and leaves, but filamentous algae were sometimes present. In many cases the pH was lower than 4 (e.g. Waajen, 1982; Duursema, 1996; Moller Pillot, 2003). Buskens & Verwijmeren (1989) collected larvae of *P. uncinatum/tritum* mainly at low pH in deep sand pits.
The temporary pools of Dettinger-Klemm (2000, 2003) had a pH of 6 to 7.5. The preference for acid water seems to be restricted to permanent water bodies.

TROPHIC CONDITIONS AND SAPROBITY

P. uncinatum appears to live numerously in acid oligohumic as well as polyhumic water, but also in non-acid marshes and alder carr. This is an indication that factors like phytoplankton production do not play an important role. This corresponds with emergence early in spring (April). Most sites where the larvae have been found are characterised by a dominance of decomposing plant material. Waajen (1982) reported intensive decomposition and rather low oxygen contents in summer in most peat cuttings where the larvae were collected. In one pool the oxygen content remained below 1 mg/l during hot summer days.

Nevertheless, larvae probably belonging to *P. uncinatum* were collected in non-acid permanent water mainly in more or less mesotrophic conditions, although they may live in eutrophic and even brackish water (see below). Concurrence may possibly be a limiting factor in eutrophic permanent water.

SALINITY

Krebs (1981, 1984, 1985) sometimes collected *P. uncinatum/tritum* in slightly brackish water bodies. The adult males had been identified as *P. uncinatum*, but this could not be verified.

DISPERSAL

According to Dettinger-Klemm (2000) the adults show no effective dispersal behaviour. New pools did not seem to be colonised easily. The larvae were rarely found in drift samples (Moller Pillot, 2003).

Pseudochironomus prasinatus (Staeger, 1839)

SYSTEMATICS

Pseudochironomus is the only representative of the tribe Pseudochironomini in Europe.

IDENTIFICATION

P. prasinatus is the only species of the genus in Europe and can easily be identified from the adult male, pupa and larva. The female has been described and illustrated by Rodova (1978).

DISTRIBUTION IN EUROPE AND THE NETHERLANDS

P. prasinatus lives in the greater part of Europe, but seems to be absent from some parts of the Mediterranean area (Saether & Spies, 2004). Verified records of the species in the Netherlands are only known from the Pleistocene area in the eastern and southern part of the country.

LIFE CYCLE

From data summarised by Goddeeris (1989) we can state that as a rule *P. prasinatus* is a univoltine species, emerging during a long period in summer, sometimes starting as early as the end of May and ending in August or early September. A second generation is possible if the water temperature remains higher than 15 °C during the whole summer (and/or in eutrophic water) (Mundie, 1957; cf. Goddeeris, 1989; cf. Otto, 1991; Orendt, 1993). The larvae overwinter in second, third and fourth instar; in early fourth instar they enter diapause (Goddeeris, 1983, 1989). In the Netherlands third instar larvae could be found until April and most pupae were found in August.

FEEDING

According to Lenz (1954–62: 254) the larvae feed on algae and detritus. Bowen et al. (1998) also found wood fibres in the gut of another *Pseudochironomus* species in Canada.

MICROHABITAT

P. prasinatus is mainly a bottom dweller, but the larvae can also live numerously between mosses and algae on stones (Meuche, 1939; Reiss, 1968; Brodersen et al., 1998). Bijlmakers (1983) found the larvae most abundantly in a moorland pool on a bottom

with much organic material and living *Sphagnum*. However, Otto (1991: 44) collected the greatest numbers of emergent adults from sand/gravel without much detritus. Janecek (1995) reared them (mainly in winter) also from roots and rotten wood.
The Dutch records are either from acid water, and then often on organic bottoms, or in alkaline water, but then on mineral sand or on stones (Buskens, unpublished; own data). The larvae rarely climb on macrophytes (Soszka, 1975), but Otto found some between reed vegetation.

OVIPOSITION

Munsterhjelm (1920) stated that one female can lay up to six ropes of 69–83 eggs. These rope-shaped egg masses have a double layer of gelatine. (See also Nolte, 1993).

DENSITIES

The larvae are sometimes abundant. For instance, Brodersen et al. (1998) found a maximum of 244 larvae on a 0.25 m^2 stone surface. Brundin (1949) reported a maximum of 2000 larvae/m^2 . As a rule, densities are much lower. Humphries (1936), for example, found 55 larvae/m^2 at a depth of 3 m in Lake Windermere, and far lower densities in the deeper parts of the lake. Bijlmakers (1983) collected 100–500 larvae/m^2 in a shallow acid moorland pool in the Netherlands in summer.

WATER TYPE

Current and dimensions
The larvae are exclusively inhabitants of stagnant water (Lenz, 1954–62: 254; Dutch data). They are usually found only in the littoral and sublittoral zone of lakes, up to 9 m deep (Humphries, 1936; Brundin, 1949), but Johnson (1989) also collected them in the profundal zone of three lakes in Sweden. In the Netherlands few larvae are collected in very small pools, but Fittkau & Reiss (1978) mentioned small pools as a biotope.

Permanence
Because of their life cycle the occurrence of larvae in temporary water is nearly impossible.

pH

The larvae appear to be numerous in acid as well in more alkaline lakes and reservoirs in different European countries (e.g. Meuche, 1939; Mundie, 1957; Raddum & Saether, 1981; Otto, 1991). Raddum et al. (1984) noted an increased abundance of the species after liming and an increase in pH from 4.5–4.8 to 6.7–7.0.
Most Dutch records are from moorland pools with a pH of 3.5–6 (Buskens et al., 1986; Leuven et al., 1987; Limnodata.nl; van Kleef, unpublished). In deep sand pits in the Netherlands the species is not typical for very acid conditions and the larvae have also sometimes been collected in more alkaline water with a pH around 7 (Buskens & Verwijmeren, 1989; Buskens, unpublished data). In the Netherlands the larvae are rare in small alkaline ditches and pools, except for pools with a relatively good water quality (van Kleef, unpublished; own data).

TROPHIC CONDITIONS AND SAPROBITY

Some authors (e.g. Brodersen et al., 1998: 583, 590) consider *P. prasinatus* to be an indicator of oligotrophic or mesotrophic conditions. Saether (1979) did not mention the species, but American species of this genus display the same preference. However, the larvae have also been found in eutrophic lakes (e.g. Johnson, 1989; Otto, 1991; Dutch data).
Brundin (1949: 773) supposed that the larvae can live in the whole littoral and sublit-

toral zone of oligotrophic lakes and that in more eutrophic conditions the oxygen content is only sufficient on stones, etc. along the banks. However, Johnson (1989) found the larvae in the profundal zone of three Swedish lakes that were not oligotrophic (this author did not measure the oxygen content). The Dutch data on microhabitat and pH (see above) support the supposition of Brundin about the oxygen needs of the larvae; in any case the presence of the species indicates a low or tardy decomposition rate. The pupa of *P. prasinatus* has an unexpectedly small ring organ, indicating low tolerance to oxygen shortage (Rossaro et al., 2007).

SALINITY

According to Tõlp (1971) *P. prasinatus* has been found in Estonia in brackish waters with salinity 0.4–6.4‰. In the Netherlands the species is totally absent from brackish water.

Robackia demeijerei (Kruseman, 1933)

Tendipes (Parachironomus) demeijerei Kruseman, 1933: 195–196, fig. 58

SYSTEMATICS AND IDENTIFICATION

The genus *Robackia* was split off from *Parachironomus* by Saether (1977: 123). He gives a key for the adult males of the two European species. Langton (1991) keyed the exuviae of two European species without species names; in Langton & Visser (2003) *R. demeijerei* is included under this name. The genus (adult males) cannot be identified using Pinder (1978) or Langton & Pinder (2007), but is treated by Cranston et al. (1989). The larvae have been described and illustrated by Saether (1977) and Pankratova (1983: 205–207, fig. 166).

DISTRIBUTION IN EUROPE AND THE NETHERLANDS

R. demeijerei has been collected in large parts of Europe and also lives in the Nearctic. The species seems to be absent from the British Isles and (parts of?) Scandinavia (Saether, 1977; Saether & Spies, 2004). In the Netherlands there are only records from the large rivers. The species was not collected in the Netherlands between 1932 and 1995; after this year the numbers increased continuously (Klink, 2002). It is not known whether (and where) the species has survived the severe pollution of the Dutch rivers in the 1960s.

LIFE CYCLE

Pupae and adults have been collected (in low numbers) in May and August (Kruseman, 1933; R. Munts, unpublished data). In the Netherlands larvae in third and fourth instar have been collected as late as early October (M. Orbons, unpublished). Whitman & Clark (1984) found almost exclusively first and second instars in winter; they collected pupae in April and September (in Texas).

FEEDING

The larvae are without doubt predators of small animals living in the sandy bottom (Whitman & Clark, 1984).

MICROHABITAT

The larvae live only in sand bottoms, in flowing water or in the wave-break zone of lakes (Chernovskij, 1949; Pankratova, 1964; Whitman & Clark, 1984). The last authors found them in summer mainly at a depth of 0–10 cm, but in winter at a depth of 5–30

cm in the sand bottom. In April the larvae were evenly distributed from 0 to 20 cm depth. The larvae were much more numerous in riffles than in pools. In the river Waal in the Netherlands larvae can also be found in the bottom of the channel, where other chironomids are rare (Klink, 2002).
The larvae are psammorheobionts and belong to the interstitial invertebrate assemblage (Smit et al., 1994). They have a very thin and highly flexible body and live between the sand grains in rivers (Chernovskij, 1949). They can be overlooked if the sand is not carefully washed or if too large a mesh size is used (Chernovskij, 1949; Smit et al., 1994).

DENSITIES

Whitman & Clark (1984) collected from 1100 (in February) to 2700 (in October) larvae/m^3 sand, with a maximum of 3400 larvae/m^3 at optimal localities.

WATER TYPE

The larvae live in lakes (Saether, 1977) and in rivers (Chernovskij, 1949; Dutch data). In the Netherlands they have never been collected in small rivers, but Whitman & Clark (1984) found rather dense populations in small rivers in Texas. In lakes as well in rivers the larvae appear to live mainly at sites with wave action or currents (see under Microhabitat). They lived in the Volga before the regulation of the river and were still present for the first few years immediately after the regulation in 1958/1959 (Zinchenko, 1992). There are no records from fast-flowing streams.

pH

In the creeks in Texas investigated by Whitman & Clark (1984) the pH was about 6 in the water column to 5.5 in the sandy bottom. In the Dutch rivers the pH is always > 7.

TROPHIC CONDITIONS AND SAPROBITY

In lakes the larvae are characteristic of oligotrophic and mesotrophic conditions (Saether, 1979). In rivers the larvae tolerate some pollution. Pankratova (1964: 191) noted the occurrence of the larvae also in the polluted part of the river Oka in Russia. Zaćwilichowska (1970) found the larvae rarely in the polluted zone of the river San in Poland and numerously after recovery from pollution. In the Netherlands the species was not collected during the period when the large rivers were severely polluted (1960s and 1970s) (cf. Klink, 1986a). Today the larvae are (scarcely) collected there, although the rivers are still moderately polluted.
The most important precondition is probably the presence of a well oxygenated sandy bottom (see under Microhabitat). Whitman & Clark (1984) collected larvae in winter in the sandy bottom 20–30 cm below the surface, where they recorded an oxygen concentration of about 2.5 mg/l. The figures in the matrix table in Chapter 4 are only indicative. As always, the oxygen contents in the table apply to the water column!

DISPERSAL

The larvae were often found in drift samples in the Warta River in Poland (Grzybkowska, 1992).

Robackia pilicauda Saether, 1977

R. pilicauda has been collected in Norway, but not in other West-European countries. The larva and ecology are unknown. See for further information under *R. demeijerei*.

Saetheria Jackson, 1977

Tendipedinae genuinae N 9 Lipina Chernovskij, 1949: 61, fig. 22
Chironominae genuinae N 9 Lipina Pankratova, 1983: 171–172, fig. 137
Cryptochironomus borysthenicus Chernovskij, 1949: 62, fig. 23; Pankratova, 1983: 173–174, fig. 140

SYSTEMATICS AND DISTRIBUTION

There are probably three European species of *Saetheria*: *S. reissi*, *S. tylus* and an unnamed species described from larva by Nocentini (1985). *S. tylus* is only known from Eastern Europe and is possibly the same as *Cryptochironomus borysthenicus* Chernovskij. The species described by Nocentini is only known from Italy. The larvae collected in the Netherlands belong to the same species as 'Chironominae genuinae N 9 Lipina' (most probably *S. reissi*). We refer to this species as cf. *reissi*. *Saetheria reissi* has been recorded in large parts of Europe except for the Mediterranean area (Saether & Spies, 2004). Larvae of *S.* cf. *reissi* are rather common in East-European rivers and relatively scarce in Western Europe. In the Netherlands the larvae have been collected in five different lowland brooks in the eastern part of the country, most of them in the province of Limburg.

IDENTIFICATION

The adult male of *Saetheria tylus* has been illustrated by Cranston et al. (1989), that of *S. reissi* by Langton & Pinder (2007). The pupa of *Saetheria tylus* has been illustrated by Pinder & Reiss (1986), that of *S. reissi* by Langton (1991). The larvae of *S.* cf. *reissi* and spec. Nocentini have been keyed by Klink & Moller Pillot (2003). The larvae of cf. *reissi* differ from spec. Nocentini and the Nearctic species illustrated by Pinder & Reiss (1983) by having a five-segmented antenna, narrower ventromental plates and a narrower central tooth of the mentum (see figures).

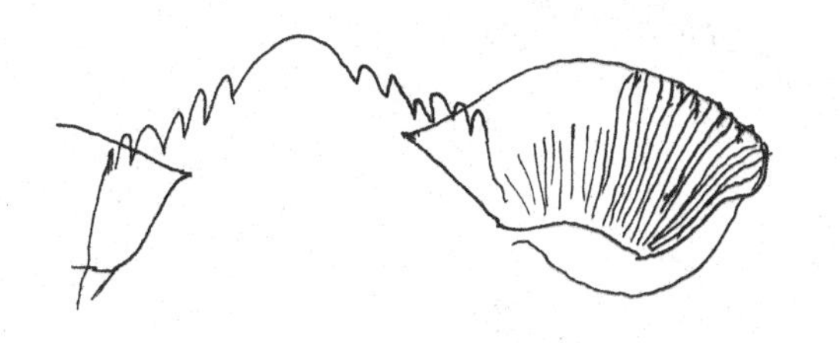

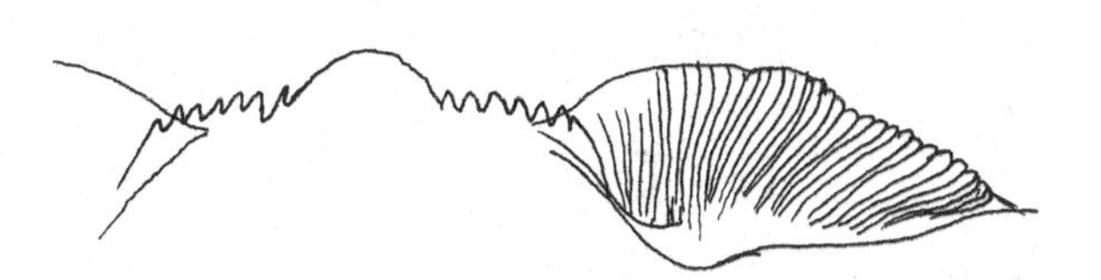

9 Saetheria reissi (top) and Saetheria spec. Nocentini: mentum (after Chernovskij, 1949 and Nocentini, 1985)

MICROHABITAT AND WATER TYPE

The larvae have been found in sandy bottoms in rivers (Pankratova, 1983; Pinder & Reiss, 1983; own data) and more rarely in lakes (Koskenniemi & Paasivirta, 1987). The Dutch records are from small rivers and lowland brooks.

Sergentia Kieffer, 1921

SYSTEMATICS AND NOMENCLATURE

Until recently most *Sergentia* specimens from Britain and the Netherlands were identified as *S. coracina* (see Pinder, 1978). After cytological investigation of the material (see Wülker et al., 1999) there appeared to be three species in Europe. Dutch pupae cannot be distinguished from *S. prima*. However, the larvae are aberrant from the type material described by Proviz & Proviz (1992). The Dutch species seems to be a new, undescribed species. We provisionally call this species S. near *prima*.

IDENTIFICATION

Adult males, pupae and larvae can be identified using Wülker et al. (1999). Adult males can also be identified using Langton & Pinder (2007). The whole genus *Sergentia* is absent from the key by Moller Pillot (1984). The mentum resembles that of *Tribelos* and *Graceus*. The larvae can be distinguished from the larvae of these genera by the mandible with 4 inner teeth. For figures see Pinder & Reiss, 1983: fig. 10.69; Kiknadze et al., 1991: fig. 43; Klink & Moller Pillot, 2003. However, the larvae of *Sergentia* near *prima* (see under Systematics) have a mandible with only 3 inner teeth. See under the species name.

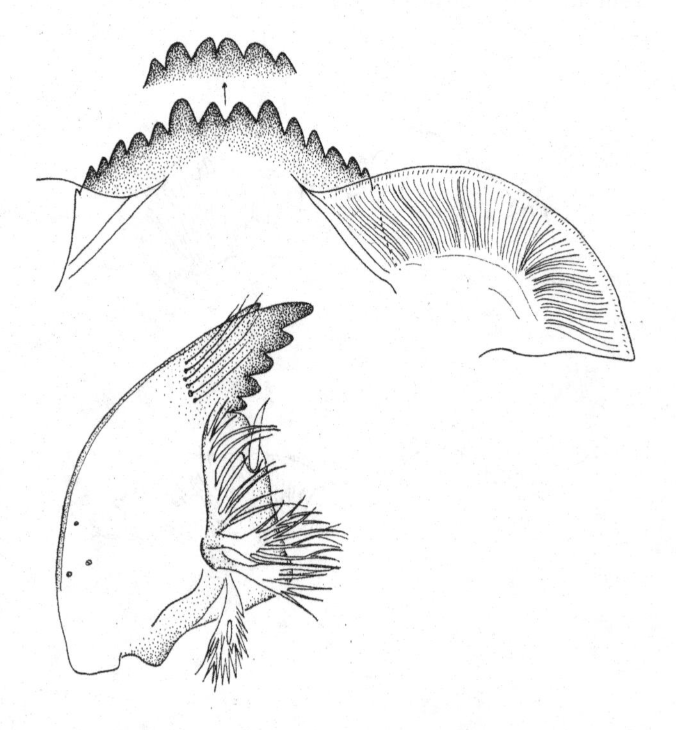

10 *Sergentia baueri: mentum and mandible (after Kiknadze et al., 1991)*

Sergentia baueri Wülker, Kiknadze, Kerkis & Nevers, 1999

Sergentia longiventris auct. pro parte
Sergentia psiloptera Langton, 1991: 296 pro parte

DISTRIBUTION IN EUROPE

S. baueri has been recorded in Germany, Finland and the British Isles (Saether & Spies, 2004; Langton & Pinder, 2007).

LIFE CYCLE, FEEDING

See *S. coracina.*

MICROHABITAT

The larvae in the Black Forest lived on mud and *Callitriche* or on a thin layer of mud on sand.

WATER TYPE

In contrast to *S. coracina* the larvae live in undeep water. In the Black Forest the species has been collected in pools, one with a considerable current and another eutrophied (Wülker et al., 1999).

Sergentia coracina (Zetterstedt, 1850)

Phaenopsectra coracina Saether, 1979: 67–70
Sergentia longiventris auct. pro parte

DISTRIBUTION IN EUROPE

S. coracina has been collected in most parts of Europe (Saether & Spies, 2004). However, the species is rare in the northwest European lowland, where there are few deep lakes. Brundin (1949: 652–653) called the species a glacial relict in this region. Most probably the species is absent from the Netherlands.

LIFE CYCLE

Kreamer & Harrison (1984) reported one generation a year in Canada, emerging in June. In the Kaliningrad region of Russia the adults emerge in September/October and partly in spring (Shcherbina, 1989). The larvae overwintered as prepupae and some in second and third instar.

EGGS

Shcherbina (1989) stated 184–364 eggs per female.

FEEDING

Walshe (1951) observed that the larvae had no filter-feeding mechanism. They stretch out of their tubes and feed from the surrounding mud. Bergquist (1987) supposed that *Sergentia* larvae living in the deeper part of forest lakes are more or less dependent on allochthonous material because in these lakes most algae decompose before reaching the profundal sediment.

MICROHABITAT

According to Walshe (1951) and Shcherbina (1989) the larvae are mud-dwellers, making tubes on the bottom of lakes. See also under Dispersal.

DENSITIES

Kreamer & Harrison (1984) reported 100–650 larvae/m^2 , but before emergence locally up to more than 6000 larvae/m^2 (see under Dispersal). Shcherbina (1989) found 880–1840 larvae/m^2 in summer and 700–800 larvae/m^2 in winter.

WATER TYPE

The larvae live in the profundal zone of lakes, but also in the littoral zone of arctic lakes (Humphries, 1936; Brundin, 1949: 645–646; Saether, 1979).

pH

The larvae increased in numbers in an acidified lake in Scotland at pH 4.8 (Brodin & Gransberg, 1993). Raddum & Saether (1981) found the species more numerous in acid lakes (pH < 5) than in lakes with a higher pH.

TROPHIC CONDITIONS AND SAPROBITY

Johnson (1989) called the genus indicative of oligomesotrophic systems. Saether (1979) indicated that the larvae can be found in the profundal zone of oligotrophic to slightly eutrophic lakes, but are found in the littoral zone only in oligotrophic and mesotrophic lakes. The larvae can endure low oxygen contents, especially in cold water, such as northern lakes (Brundin, 1949). Bergquist (1987: 229) mentioned that *Sergentia* larvae are more or less tolerant of low oxygen concentrations and thus better adapted to more humic and eutrophic conditions than true indicators of oligotrophic water. Langdon et al. (2006) found no obvious relation between the presence of *Sergentia* and changes in the trophic state of a lake.

DISPERSAL OF LARVAE

Kreamer & Harrison (1984) noted nocturnal movements of prepupae from the profundal zone towards the shore in spring and movements of larvae into deeper regions in the autumn. Mundie (1965) observed that the larvae (not only early instars) move from the lake bottom to the water mass in summer and as the summer progresses they are found in shallower water.

Sergentia prima Proviz & Proviz, 1997

Sergentia longiventris auct. pro parte

SYSTEMATICS

See also *Sergentia* near *prima* below.

DISTRIBUTION IN EUROPE

S. prima has been recorded from the northern part of Europe and from the British Isles (Saether & Spies, 2004; Langton & Pinder, 2007). At least the Scandinavian material belongs to this species, as stated by Wülker et al. (1999) who studied the cytology of the larvae. As far as is known the Dutch material belongs to *Sergentia* near *prima*.

Sergentia near prima

nec *Sergentia prima* Proviz & Proviz, 1997

SYSTEMATICS AND IDENTIFICATION

The adult and pupa of *Sergentia* near *prima* cannot be distinguished from those of *S. prima* (P. Langton, in litt.). The larvae (collected by A. Klink with associated pupae) correspond with those of *Phaenopsectra* sp. illustrated in fig. 10.58. B/E in Pinder & Reiss (1983). These larvae resemble those of *Tribelos* and *Graceus*. A short description, including the differences between these genera in Europe, is in preparation.

DISTRIBUTION IN EUROPE AND THE NETHERLANDS

The larvae have been collected in two canals in the Pleistocene part of the Netherlands and an oxbow lake in Belarus. The pupae and exuviae of '*S. prima*' collected in the Netherlands probably also belong to this species. They were collected in 8 samples at 3 locations in the fenland area of Utrecht/Noord-Holland (e.g. Limnodata.nl). It is possible that the British specimens of *S. prima* also belong to this species.

LIFE CYCLE, FEEDING

See *S. coracina*. In the Netherlands pupae have been collected in May and September.

MICROHABITAT

The Dutch populations most probably live on muddy bottoms.

WATER TYPE

The Dutch records are from three fen-peat lakes near Amsterdam (Het Hol, Naardermeer and Ankeveensche Plassen) and two canals. In Belarus the larva has been collected in an oxbow lake near the river Pripyat.

pH

In the Netherlands the species does not live in acid water. There are too few data to allow comparison with *S. coracina*.

Stenochironomus Kieffer, 1919

D

SYSTEMATICS

The European species belong to two different subgenera. *S. gibbus* belongs to *Stenochironomus* s.s. and *S. fascipennis* belongs to the subgenus *Petalopholeus* (Saether & Spies, 2004). The third species, *S. hibernicus* (= *S. tubanticus* Kruseman) cannot be placed in a subgenus because the pupa and larva are unknown. Borkent (1984) supposed that larvae of the subgenus *Stenochironomus* are wood mining, while those of *Petalopholeus* mine in the leaves of plants. Borkent (1984: 81) suggested that adults described as *S. gibbus* actually represent two or more species.

IDENTIFICATION

Borkent (1984) made a revision of the whole genus, with descriptions of all known adult males, females, pupae and larvae. The males of the three West-European species have been keyed by Langton & Pinder (2007). According to Borkent (1984) the long seta on the inferior volsella is not normal in *S. gibbus*, but the illustration is probably of an aberrant specimen. Rodova (1978) described and keyed the females of *S. gibbus* and *S. fascipennis*. These two species can also be identified from pupal exuviae (Langton,

1991). Langton also keyed the exuviae of a third species, possibly *S. hibernicus*. Kalugina (1959) described, illustrated and keyed the first instar larva of *S. gibbus* and also illustrated the larva in second instar. In first instar the mentum already looks like the later mentum, but there is only one central tooth and three pairs of lateral teeth. We found no principal differences between miners in wood and plant stems. Identification of the larvae to species level is still impossible.

DISTRIBUTION IN EUROPE AND THE NETHERLANDS

Three species of *Stenochironomus* have been recorded in North-Western Europe. *S. gibbus* lives almost throughout the whole of the European mainland, *S. fascipennis* is absent from the Mediterranean area and *S. hibernicus* is only known from a few countries, confined mostly to the northwestern part of Europe (Saether & Spies, 2004). In the Netherlands only *S. gibbus* and *S. hibernicus* (as *S. tubanticus*) have yet been collected (e.g. Kruseman, 1933; Kouwets & Davids, 1984). The larvae have been recorded from only about 40 localities; there are no records from the Holocene marine clay area.

LIFE CYCLE

Kruseman (1933) collected adults in the Netherlands from May to August; other authors (Lenz, 1954–62; Kouwets & Davids, 1984) only in July and August. There is possibly only one generation. Only a few larvae have been collected in winter, in third and fourth instar.

FEEDING

The larvae are true miners, feeding on the plant tissue (Lenz, 1954–62) or wood (Borkent, 1984; Bowen et al., 1988).

MICROHABITAT

Kalugina (1959, 1960) described the adaption of the larva of *S. gibbus* to living in wood. The first instar larva swims around, but after settling in decaying wood looks like a Buprestidae larva and is no longer able to creep or swim. Bowen et al. (1998) found *Stenochironomus* larvae most abundantly in the microbially softened surface layers of otherwise firm undecayed coarse woody debris, but the larvae were almost absent on the fresh substrates. According to Borkent (1984) *S. gibbus* is probably an exclusive wood miner, but he does not exclude the possibility that the species is ecologically plastic. Lenz (1954–62: 181–182) mentioned that the larvae of the three (!) West European species live in living stems and dead fragments of stems of *Phragmites*, *Typha* and *Scirpus*. In the Netherlands *Stenochironomus* larvae have also been found in stems of *Juncus effusus*. Meuche (1939) reported that *S. hibernicus* (as *tubanticus*) mines in stems of water plants.

WATER TYPE

According to Lenz (1954–62) the species is restricted to pools and the littoral zone of lakes. Fittkau & Reiss (1978) stated that *S. fascipennis* also lives in streams and rivers. According to Caspers (1991) *S. gibbus* lives in the upper parts of the Rhine in Germany. In the Netherlands about half of all records are from moorland pools, but the larvae have also been collected in fen lakes, in sand pits, in quite a number of lowland brooks and, rarely, in the river Meuse (R. Munts, pers. comm.). M. Siebert (pers. comm.) collected the larvae regularly in brooks in Germany, under the bark of branches. Braukmann (1984) only rarely found the larvae in lowland brooks. Klink (1986a, 1989) stated that a few centuries ago *Stenochironomus* larvae were rather abundant in the large rivers in the Netherlands.

pH

The larvae have often been collected in very acid moorland pools in the Netherlands (pH 4–5), but also in brooks and streams with a high pH. In one case exuviae were identified as *S. gibbus* at pH 4.6 (van Kleef, unpublished). Adult males of the same species were collected by Kouwets & Davids (1984) near Lake Maarsseveen at a pH of approx. 8.

Stictochironomus Kieffer, 1919

IDENTIFICATION

The male adults of the West European species can be identified using Langton & Pinder (2007). For *S. crassiforceps* see below. The females of some species have been described by Rodova (1978). Exuviae of all European species except *S. rosenschoeldi* can be identified using Langton (1991). No key has yet been drawn up for the larvae. The larvae of all species have been described (see under each species), but a more detailed study is necessary to make a complete key. We give here a provisional key to the Dutch species. As we have no reared or associated material of *S. pictulus*, the identity of these larvae is uncertain.

1a Nearly the whole head capsule pale. Second lateral teeth of the mentum strikingly smaller than third lateral teeth (see figure). Head length 0.4–0.45 mm — *S. maculipennis*

b Hind part of head capsule on upper and/or under side with pale brown or dark brown marking or spots. Second lateral mental teeth hardly smaller or as long as third teeth. Head length 0.45–0.55 mm — 2

2a Hind part of head capsule dorsally with tripartite longitudinal brownish marking, between which a pale V. Head length 0.5–0.55 mm — *S. pictulus*

b Hind part of the head dorsally without such marking; only an extended dark spot ventrally and laterally. Head length 0.45–0.5 mm — *S. sticticus*

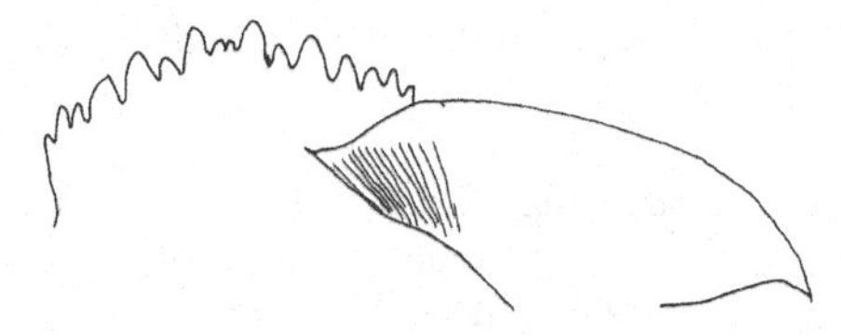

11 Stictochironomus maculipennis: mentum

DISTRIBUTION OF THE SPECIES

The distribution of the species will be treated in some detail for every species below. In the Netherlands and adjacent lowlands only three species can be expected because *S. crassiforceps* occurs only in northern, central and eastern regions and *S. rosenschoeldi* lives exclusively in the hypolimnion of lakes. Settlement of the latter species in deep sand pits is possible.

FEEDING, OVIPOSITION

See *S. sticticus*.

ECOLOGY

The ecology of the species appears to be very different. Attention must therefore be paid to the species name (see below).

Stictochironomus crassiforceps (Kieffer, 1922)

Allochironomus crassiforceps Goetghebuer, 1937–1954: 5, 65, fig. 200; Brundin, 1947: 59–60, fig. 91, 92
nec *Allochironomus* Chernovskij, 1949: 81, fig. 53

IDENTIFICATION

Shilova (1965b) argued that the adult male cannot be correctly identified using Goetghebuer (1937–1954). It is better to use the descriptions and figures by Brundin (1947) and Shilova (1965b). The female has been described by Shilova (1965b) and Rodova (1978). The pupa has been keyed by Langton (1991).
According to the descriptions and figures of Palmén (1962), Shilova (1965b) and Kiknadze et al. (1991) the larva differs hardly from that of *S. sticticus*. However, it resembles *S. maculipennis* a little more closely: the dark spot on the gula is only pale brown and the second lateral teeth of the mentum are somewhat smaller than the adjacent teeth. The mentum has a dark (sometimes only brown) line underneath the teeth as in *S. sticticus*; the pecten epipharyngis has 14–16 teeth.

DISTRIBUTION IN EUROPE

The species has been recorded in Central, Northern and Eastern Europe, but seems to be absent from the Netherlands, Belgium and other West European countries (Saether & Spies, 2004).

LIFE CYCLE

Around the Baltic Sea the species has one generation a year (Palmén, 1962; Shcherbina, 1989).

OVIPOSITION

Shilova (1965b) observed that flying females throw off solitary eggs onto the surface of lakes in a scattered pattern, even at 5–7 km from the shore.

ECOLOGY

The larvae live in stagnant water, lakes and pools (Palmén, 1962; Shilova, 1965b; Fittkau & Reiss, 1978). Shilova (1965b) mentioned a preference for sandy bottoms without much silt. Shcherbina (1989) collected the larvae mainly on sandy bottoms, but also on silt.

SALINITY

Palmén (1962) stated that *S. crassiforceps* is a common species in Finnish brackish lakes, but also lives in fresh water; see also Shilova (1965b). The occurrence of larvae of *Stictochironomus* gr. histrio in brackish waters in Estonia with up to 3000 mg Cl/l or more reported by Tõlp (1971) probably applies to this species.

Stictochironomus maculipennis (Meigen, 1818)

IDENTIFICATION

Identification of the adult male and the exuviae using Langton & Pinder (2007) and Langton (1991) respectively can present problems. The female is unknown.
Goetghebuer (1928) described the larva. The larva has also been separately reared and described by Gouin (1936). H. Cuppen and H. Vallenduuk collected pupae and associated larvae in a Dutch lowland brook. As mentioned by Moller Pillot (1984: 184), the species can be distinguished from *S. pictulus* (? see there), *S. sticticus* and *S. rosenschoeldi* by the relatively small head, the shorter second lateral teeth of the mentum and the absence of a dark spot on the gula. The number of teeth of the pecten epipharyngis appears to be unsuitable as a species character, contrary to what was stated by Moller Pillot (1984: 184). In *S. maculipennis* this number varies at least between 10 and 13.
Because of uncertainties about the differences between the larvae of *S. maculipennis* and *S. pictulus* (possibly in the literature also between the adult males, see under *S. pictulus*), some of the ecological data given below can apply to the latter species.

DISTRIBUTION IN EUROPE AND THE NETHERLANDS

The species has been recorded in many European countries and will be present nearly everywhere in Europe (Saether & Spies, 2004). In the Netherlands the species has been collected in brooks in the eastern and southern part of the country and in a lake in the coastal dune region.

MICROHABITAT

Larvae most probably belonging to this species were collected in lakes by Rearadevall & Prat (1989) on silty bottoms of limestone and clay (cf. Prat, 1978). Schmale (1999) found the larvae in dune lakes with much silt. Tolkamp (1980) collected the larvae in two Dutch lowland brooks and obtained different results: in the Ratumsebeek the larvae lived nearly exclusively on bare sand; in the Snijdersveerbeek they were found more often on sand with detritus. In Dutch lowland streams Verdonschot & Lengkeek (2006) found more *Stictochironomus* larvae on detritus than on sandy bottoms. This may apply to *S. maculipennis* and/or *S. pictulus.*
The species has been recorded as mining in stems of macrophytes (e.g. Fittkau & Reiss, 1978), but this could not be verified.

D

WATER TYPE

Current

The species is rather common in lowland brooks, at least in the eastern part of the Netherlands (own data). It seems to prefer more or less fast-flowing lowland brooks: 50% of the 17 records in Limnodata.nl are from brooks with a current velocity > 45 cm/sec. Orendt (2002a) reported the occurrence of the species in at least one fast-flowing stream in Bavaria. Caspers (1991) and Becker (1994) recorded the presence of the species in the Rhine. Lehmann (1971) collected no *Stictochironomus* at all in the river Fulda.
The species is common in Spanish lakes and has also been collected in at least one dune lake in the Netherlands (det. P. Langton, unpublished). The figures for occurrence in table 4 of Chapter 4 apply to the Dutch situation, where the species lives mainly in running water.

Dimensions

The larvae appear to be more common in rather small brooks (often less than 5 m wide), but can be found also in larger brooks, streams and rivers. Almost the only reports of occurrence in stagnant water are of its frequent presence in Spanish lakes. The larvae can also be found in the profundal zone (Rieradevall & Prat, 1989).

Permanence
There is at least one record from a lowland brook that dries up in very dry years.

pH

As far as is known, the Dutch records from lowland brooks were in water with a pH between 7 and 8 (see e.g. limnodata.nl). The dune lake probably had a slightly higher pH.

TROPHIC CONDITIONS AND SAPROBITY

The larvae have been collected in about 20 lowland brooks in the Netherlands, often without any pollution, but also in very eutrophic brooks with more than 1 mg orthophosphateP/l or more than 2 mg ammoniumN/l. A relatively large proportion of these brooks are shaded by surrounding woods. Low oxygen contents have never been recorded in these brooks.
According to Prat (1978), *S. maculipennis* is a characteristic inhabitant of mesotrophic and moderately eutrophic Spanish reservoirs. In some of these lakes the sediment can be black, indicating severe oxygen depletion. The species probably does not require an oxygen content as high as *S. sticticus*, but the differences between the Spanish and Dutch data cannot easily be explained because the Dutch larvae live mainly on sandy bottoms in flowing water, which are relatively well oxygenated.

Stictochironomus pictulus (Meigen, 1830)

IDENTIFICATION

The male hypopygia and the exuviae differ very little from those of *S. maculipennis* (see Prat, 1978a; Langton, 1991). The exuviae have been keyed and described by Langton (1991) and illustrated by Pinder & Reiss (1986). The larva has been described briefly by Bryce (1960), but this description does not indicate any difference from the larva of *S. sticticus*. It does not mention the dark spot on the head capsule or the number of teeth of the pecten epipharyngis. Tentatively, we take the larvae with the characteristic marking on the dorsal side of the head and living in a number of lowland brooks to be *S. pictulus* (see the provisional key above).

DISTRIBUTION IN EUROPE AND THE NETHERLANDS

The species has been found scarcely, but widely distributed in the European mainland (Saether & Spies, 2004). In the Netherlands Kruseman (1933) collected the adult males in different parts of the country, but only one unpublished record of the species has been made since then. The larvae mentioned above were collected in the province of Noord-Brabant.

MICROHABITAT, WATER TYPE

Judging from the literature, *S. pictulus* appears to inhabit flowing and stagnant water, among which more or less small streams. Fittkau & Reiss (1978) and Langton (1991) noted that the species lives in streams and lakes. Lenz (1954–62) mentioned the river Oder and Lago Maggiore. Orendt (1993) collected the species in a mesotrophic and a eutrophic lake in Bavaria. Srokosz (1980) found *S. pictulus* in the river Nida in Poland in sand and muddy sand, but sometimes also between the vegetation. Orendt (2002a) reported its presence in two streams in Bavaria. The records by Kruseman (1933) indicate that the larvae live in smaller streams and/or small lakes or pools.
From this information we think that the larvae with dark marking on the dorsal side of the head capsule belong to this species. We collected or saw specimens from a large

river, from lowland brooks and a small pool in the province of Noord-Brabant and from a pool in Belgium (leg. Int Panis, 1992).

OXYGEN

Int Panis et al. (1996) found an extremely high haemoglobin concentration (30.89 µg Hb/mg body weight) in larvae probably belonging to *S. pictulus* (named cf. *maculipennis*), suggesting an optimal ability to respond to oxygen shortage. According to investigations by Int Panis et al. (1995), the larvae seem to be less dependent on the oxygen content within the sediment, probably because they migrate between the sediment surface and deeper layers. In the pupa of *S. pictulus* Rossaro et al. (2007) measured a small ring organ which is characteristic of species living in oxygen rich environments. Consistent with this, they found this species only at high oxygen concentrations.

Stictochironomus rosenschoeldi (Zetterstedt, 1838)

IDENTIFICATION

The adult male has been keyed by Langton & Pinder (2007). In Rodova (1978) the female is distinguished from *S. sticticus* only by colour characters. The exuviae have not been keyed by Langton (1991). Kiknadze et al. (1991) made a description of the larva, accompanied by very good figures. The larva resembles *S. sticticus*. According to Lenz (1954–62) the pecten epipharyngis has 9–11 teeth. Kiknadze et al. (1991) stated 18 teeth.

DISTRIBUTION IN EUROPE AND THE NETHERLANDS

S. rosenschoeldi has been recorded in several West European countries, including Germany and the United Kingdom (Saether & Spies, 2004). Brundin (1949: 652–653) called the species a glacial relict. The species has not been found in the Netherlands and is probably absent there (see under Water type).

MICROHABITAT

Brundin (1949: 643) found *S. rosenschoeldi* on gyttja bottoms in the profundal zone of oligotrophic lakes.

WATER TYPE

The larvae live only in lakes (Brundin, 1949; Fittkau & Reiss, 1978). In Central Europe the larvae occur only in the hypolimnion, in more northern regions they also live in the littoral zone (Brundin, 1949: 779; Bergquist, 1987).

pH

The larvae often live in very acid lakes (pH 4.5–6), as stated by Raddum & Saether (1981), but can also be numerous in lakes with a pH of around 7 (Brundin, 1949).

TROPHIC CONDITIONS AND SAPROBITY

Brundin (1949) supposed that *S. rosenschoeldi* is a typical inhabitant of cold oligotrophic lakes and requires a high oxygen content. Everywhere in Europe except the northern regions the larvae live only in the hypolimnion of these lakes. Saether (1979) considered the species to be characteristic of the profundal zone of oligotrophic to mesotrophic lakes. Bergquist (1987) reported a temporal decrease in larval numbers in Lake Siksjön in central Sweden after some input of nutrients, which could have lowered oxygen availability.

Stictochironomus sticticus (Fabricius, 1781)

Stictochironomus histrio Saether, 1979: 69; Smit et al., 1996: 487–510; Langton, 1991: 290, fig. 1170, 118a–d
? nec *Stictochironomus* gr. *histrio* Tõlp, 1971: 96

IDENTIFICATION

For identification of males and pupal exuviae see under the genus. Goetghebuer (1937–1954: 56) contains a key to three varieties of this species. The female has been described and illustrated by Rodova (1978). The larva can be distinguished from *S. maculipennis* and *S. pictulus* (see above).

DISTRIBUTION IN EUROPE AND THE NETHERLANDS

The species has been recorded in many European countries and it will be present nearly everywhere in Europe (Saether & Spies, 2004). The species is widely dispersed over the Netherlands.

LIFE CYCLE

The species has two generations a year, emerging from early May until September (Otto, 1991; van de Bund, 1994). The spring generation is usually largest (Kouwets & Davids, 1984; Ten Winkel & Davids, 1987), but in one year Otto (1991) collected a greater number in summer. Often some of the larvae remain in second and third instar during summer; for example, at some sites Smit et al. (1996) collected no large larvae in mid summer (July). According to ten Winkel (1987) *S. sticticus* is predominantly a univoltine species that can be partly or totally bivoltine if food conditions and temperature are favourable (see also Heinis, 1993: 36). In the Danish lake Stigsholm the species seemed to be univoltine (Lindegaard & Brodersen, 2000: 320). The larvae winter in third and fourth instar, rarely in second instar (Ten Winkel & Davids, 1987; ten Winkel, 1987).

FEEDING

The larvae collect their food by deposit feeding (Heinis & Crommentuijn, 1989). In the guts of larvae of different *Stictochironomus* species in third and fourth instar we found many sand grains and most of the remaining material consisted of detritus, suggesting that they are less selective grazers than most other Chironomini. This may be why the mouth parts are often more worn down than in other genera. However, according to Ali et al. (1987) the larva of the related *S. affinis* (in India) constructs tubes from sand and other substrate particles. At one end of the tube it makes a sticky net in which food particles are trapped.

MICROHABITAT

The larvae of *S. sticticus* live in sandy bottoms (Palomäki, 1989; van de Bund, 1994; Smit et al., 1994, 1996) and rarely between vegetation or on stones (Mol et al., 1982; Buskens, unpublished). Kashirskaya (1989) found the larvae only on silty sand, not on silt with detritus or on silty clay. Smit et al. (1992) noted that the larvae occur in sediments with higher silt contents than *Lipiniella araenicola*. Steenbergen (1993) sometimes collected larvae in water bodies with a peaty bottom. From the literature it is not clear whether or not this species lives on organic sediment in the profundal zone of lakes or in streams. The younger larvae in particular are found at a depth of 0–2 cm in the sand; fourth instar larvae often 2–5 cm deep, in winter sometimes down to a depth of 10 cm (Ten Winkel & Davids, 1987; Heinis, 1993). According to Ten Winkel (1987) and van de Bund (1994) older larvae burrow deeper to escape predation by bream.

OVIPOSITION

In late summer small larvae of *S. sticticus* appeared at the highest sites on the Haringvliet mud flats and subsequently appeared gradually at lower sites. This suggests that oviposition took place at one of the exposed sites on the flats and that the larvae later migrated to other areas (Smit et al., 1996). In captivity the females of *S. affinis* deposited their egg masses on or very close to the water, apparently exhibiting a preference for floating objects (Ali et al., 1987). See also *S. crassiforceps*.

DENSITIES

Van de Bund (1994) found densities of young larvae up to 14,000/m² in May and June and up to 5000/m² in autumn, and densities of fourth instar larvae of up to 4000/m² in summer. Ten Winkel & Davids (1987) give a mean density of 1132 larvae/m² in August for the sandy bottom in the shallow part of Lake Maarsseveen. The abundance of second and third instar larvae in summer was primarily affected by water mite predation, that of fourth instar larvae by bream predation. Van der Velden et al. (1995) found a mean density of 1367 larvae/m² after freshening of Lake Volkerak-Zoommeer. Much lower densities are found in other lakes, especially in smaller water bodies.

WATER TYPE

Current

Fittkau & Reiss (1978) reported the occurrence of the species in rivers, but not in smaller streams. However, records from rivers are scarce. Caspers (1991) collected the species in the Rhine in Germany, but only in the middle stretch from Bingen to Bonn. We found no records of this species in brooks or small streams. It seems to be unlikely that the larvae named *S. ? histrio* by Braukmann (1984) and collected in small German brooks and streams belong to this species (see under *S. pictulus*). In the Netherlands there are no records from the large rivers, but *S. sticticus* is a common species in the enclosed Rhine-Meuse delta, for example in the Haringvliet (Smit et al., 1994). The larvae are common in lakes.

Dimensions

The larvae live mainly in the littoral zone of large lakes, but are sometimes collected in small lakes and large pools less than 50 m wide with a well oxygenated (usually sandy) bottom. They appear to be absent from the deeper zones of lakes (rarely deeper than 6–8 m). Larvae (probably of this species) have been collected even in the (very large, undeep) IJsselmeer, although scarcely (Maenen, 1983). Nearly all *Stictochironomus* larvae collected in Dutch sand pits probably belong to this species (Buskens, unpublished data). In flowing water the larvae appear to be restricted to large rivers, where they are scarce (see under Current).

D

pH

Palomäki (1989) collected 231 larvae/m² at pH 6.0 in the Finnish lake Alajärvi. Buskens (unpublished data) found the larvae as a rule not in (very) acid water, but once at pH 4.6. Most Dutch records are from lakes and estuaries with a pH of 6.5–9. The species may have only a weak preference for less acid conditions (e.g. *S. rosenschoeldi*).

TROPHIC CONDITIONS AND SAPROBITY

Saether (1979) mentioned the occurrence of the species in more or less oligotrophic to moderately eutrophic lakes. Johnson (1989) considered the whole genus indicative of oligomesotrophic systems in Swedish lakes. Ruse (2002) stated that *S. sticticus* is an indicator of water with low conductivity. Steenbergen (1993) collected the larvae of *Stictochironomus* in the province of Noord-Holland, mainly in water with less than

0.05 mg orthophosphateP/l and less than 0.1 mg ammoniumN/l.
The larvae are not tolerant of hypoxia: according to Heinis & Crommentuijn (1989, 1992) they are not able to feed at oxygen concentrations as low as 2.5 mg /l. When living somewhat deeper in the substrate they depend on ventilation to get more oxygenated water and may also migrate between the sediment surface and the deeper sediment layers (see *S. pictulus*). Consistent with these investigations the larvae are not found in deeper layers of Lake Maarsseveen, where hypoxic or even anoxic conditions can exist during stratification periods (Mol et al., 1982; Ten Winkel, 1987).

SALINITY
Brodersen et al. (1998: 589) noted the occurrence of *Stictochironomus* (without species name) in an undeep slightly brackish Danish lake (possibly > 500 mg Cl/l) and supposed that the larvae are rather tolerant of increased salinity. However, in Lake Volkerak-Zoommeer in the Netherlands the larvae occurred only some years after freshening of the lake, when the typical inhabitants of brackish water had already totally disappeared (van der Velden et al., 1995). Steenbergen (1993) collected the genus only in water with less than 300 mg Cl/l. Ruse (2002) stated that *S. sticticus* is an indicator of water with low conductivity. Therefore, we think that *S. sticticus* does not live in brackish water and that the records by Tõlp (1971) in Estonia apply to *S. crassiforceps*. (Tõlp identified the larvae of 'gr. *histrio*' using Chernovskij, 1949).

Tribelos intextum (Walker, 1856)

Endochironomus intextus Pinder, 1978: 122, fig. 59A, 152A

IDENTIFICATION
Until recently the larvae of *Tribelos intextum* have not been distinguished from those of *Graceus ambiguus* and some *Sergentia* species. A key to the larvae of *Tribelos* and similar genera is in preparation. See under *Sergentia*.

DISTRIBUTION IN EUROPE AND THE NETHERLANDS
Tribelos intextum has been recorded in large parts of Europe, but is possibly absent from some countries in Eastern and Southern Europe (Saether & Spies, 2004). The species is not very common in the Netherlands. More than a hundred localities are known in the fen-peat and Pleistocene regions, but there are no records from clay (Moller Pillot & Buskens, 1990; Steenbergen, 1993; Nijboer & Verdonschot, unpublished; Limnodata.nl).

LIFE CYCLE
Goddeeris (1983, 1986) reported only one emergence period in the Belgian Ardennes, in May–June. From May until the end of August he found no larvae in fourth instar. In the Netherlands adults are collected in May and August (Kruseman, 1933; Kouwets & Davids, 1984). In June and October Mol et al. found only juvenile larvae. The emergence of the second generation in August is most probably dependent on temperature and food (cf. also Kouwets & Davids, 1984). Brundin (1949) reported emergence in Southern Sweden in May–June and near Abisko at the end of July.
From October until April we collected larvae in second, third and fourth instar.

MICROHABITAT
Van der Velde & Hiddink (1987) found the larvae in folded leaves of *Nuphar lutea*. Lenz (1954–62) reared the species from coarse plant material. Usually, however, the larvae

are collected on sandy bottom with greater or lesser quantities of detritus or on peat, locally on stones and rarely on living plants (e.g. Mol et al., 1982; Bijlmakers, 1983). The latter author found the larvae only outside the vegetation zone.

DENSITIES

In lake Heiligensee near Berlin Neubert & Frank (1980) found the highest numbers of *T. intextum* in September and early October, up to 1673 larvae/m^2. In this lake the species was one of the most dominant chironomids. In most other cases much lower numbers of larvae have been collected; for example Goddeeris (1983) never collected more than 100 larvae/m^2 in Belgian fish ponds.

WATER TYPE

Current

The larvae are inhabitants of stagnant water and are rarely collected in slowly flowing water (e.g. Tõlp, 1956; Limnodata.nl).

Dimensions

The larvae are collected mainly in lakes and larger pools and only scarcely in small pools and ditches (Steenbergen, 1993; own data). The larvae live in the littoral zone, mainly at a depth of 0–3 m (Neubert & Frank, 1980), rarely somewhat deeper (Lenz, 1954–62).

Permanence

The species has never been found in temporary water.

SOIL

The larvae live in water bodies on peat or sand, never or very rarely on clay (Steenbergen, 1993; own data).

pH

T. intextum has been collected only rarely at $pH > 8$ and relatively often in slightly acid water (Koskenniemi & Paasivirta, 1987; Steenbergen, 1993; Limnodata.nl). Raddum et al. (1984) found the larvae in low numbers in a Norwegian lake at pH 4.5–4.8. Leuven et al. (1987) and van Kleef (unpublished) collected the larvae scarcely at pHs between 5 and 6 and very rarely at $pH < 4$.

TROPHIC CONDITIONS AND SAPROBITY

Brundin (1949) collected the species numerously in several oligotrophic lakes in Sweden. Steenbergen (1993) found the larvae only in 13 water bodies in the Dutch province of Noord-Holland, with an orthophosphate content lower than 0.15 mgP/l and chlorophyll-a usually less than 20μg/l. Elsewhere the species lives sometimes in eutrophic water (e.g. Neubert & Frank, 1980; Limnodata.nl), although most records are from more or less mesotrophic conditions (Buskens, unpublished; van Kleef, unpublished; Limnodata.nl; own data). It is not known why the species is absent from so many lakes (e.g. Orendt, 1993).

SALINITY

The species has not been reported from brackish water.

Xenochironomus xenolabis (Kieffer, 1916)

IDENTIFICATION

The adult male, pupa and larva are easy to identify. The female has been described and keyed by Saether (1977).

DISTRIBUTION IN EUROPE AND THE NETHERLANDS

The species has been recorded throughout Europe (Saether & Spies, 2004). In the Netherlands the species has been collected mainly in the Holocene part of the country, but also lives elsewhere, mainly in canals and small rivers (Nijboer & Verdonschot, unpublished; Brabantse Delta, unpublished.; Limnodata.nl).

LIFE CYCLE

Lehmann (1971) collected adults from June to August, but in the Netherlands the first exuviae have been collected at the end of April. Becker (1994) collected adults until October. Most probably two generations can occur, at least locally.
In winter we collected many larvae in third instar, but probably some of the larvae were in second or fourth instar. Third instar larvae can be present until early May.

MICROHABITAT

The larvae live exclusively in sponges, where they make short superficial tunnels (Pagast, 1934). In the river Meuse the larvae are found mainly on the undersides of stones, probably because this affords them shelter against strong currents and wave action (Smit, 1982; Peeters, 1988).

FEEDING

Pagast (1934) supposed that the larvae live as parasites within the sponges. However, Shilova (1974) argued that they filter the water with the help of the numerous bristles on labrum and maxillae.

WATER TYPE

Current
Lehmann (1971) collected the species along the lower reaches of the river Fulda. In the river Rhine the species is a common inhabitant of colonies of sponges, occurring in the whole German stretch of the river (Caspers, 1980, 1991; Becker, 1994). In the Netherlands the larvae have been collected mainly in stagnant water and under stones in the large rivers (see Microhabitat), but also in small rivers with slow currents.

Dimensions
Steenbergen (1993) collected the larvae only in water bodies more than 10 m wide, but the larvae can sometimes be found in rather narrow ditches (cf. Limnodata.nl). There are no records from brooks less than 8 m wide.

TROPHIC CONDITIONS AND SAPROBITY

Bazerque et al. (1989) collected the exuviae only in the less polluted stretches of the lowland river Somme in France (independent of current velocity). They considered the species slightly tolerant (code B), in agreement with Wilson & Ruse (2005). Smit (1982) collected the larvae less often in the more polluted part of the river Meuse. Steenbergen (1993) and Limnodata.nl contain few records at high phosphate or low oxygen contents.
According to Sládeček (1973) sponges do not live in α-mesosaprobic water.

SALINITY

Steenbergen (1993) collected the larvae scarcely in water with more than 300 mg l/l. We have few other records from oligohaline water.

Zavreliella marmorata (van der Wulp, 1858)

IDENTIFICATION

The male adult has been keyed and illustrated by Langton & Pinder (2007). In most regions only the females are encountered (see under Distribution); they are described and illustrated in Rodova (1978). The description of the female by Goetghebuer (1937–1954: 68, fig. 18, Taf. II fig. 42) is very incomplete and his figures are not very accurate.

DISTRIBUTION IN EUROPE AND THE NETHERLANDS

Z. marmorata lives in large parts of Europe, but possibly not in Ireland, Spain and Portugal (Saether & Spies, 2004). In central Europe *Z. marmorata* is obligately parthenogenetic; in southern Europe it is bisexual (Pinder & Reiss, 1983).
In the Netherlands the species has been collected at more than 200 localities, a high percentage of which are in the fen-peat areas with varying levels of groundwater seepage from the higher Pleistocene areas. There are no records from the province of Zeeland and Southern Limburg.

LIFE CYCLE

The pupae and exuviae have been collected in the Netherlands from May to August. Most probably there are usually two generations a year. In winter some larvae have been collected in second instar.

MICROHABITAT

Walshe (1951) collected larvae from among dead leaves and weeds in a pond. Higler (1977) collected them on *Stratiotes* plants, with a clear preference for emergent plants near the margins.

FEEDING

Walshe (1951) observed how the larvae extend their head out of the anterior orifice of their case and feed on particles of algae and detritus in the vicinity. This food is also mentioned by Zavrel (1926).

WATER TYPE

The larvae live in lakes, canals and ditches of different dimensions, rarely in slow-flowing ditches and small rivers. As mentioned under Distribution the larvae seem to prefer places with seepage of ground water (see also Klink, 2008).

SOIL

The larvae are often encountered on peat and sand and rarely on clay (Steenbergen, 1993; own data).

pH

There are no records from acid water. In 10% of the records in Limnodata.nl the pH of the water is lower than 7.1.

TROPHIC CONDITIONS AND SAPROBITY

Although the species has been stated to be an inhabitant of eutrophic water (Brundin, 1949; Verstegen, 1985), the larvae are mainly collected in water with low phosphate, ammonium and chlorophyll-a contents (Steenbergen, 1993; Limnodata.nl). A lowered oxygen content appears not to present a problem (see also Brundin, 1949), but the species has not been found in severely anthropogenically polluted water. Limnodata.nl states that the species is scarce in water with 1 mg ammoniumN/l. Wilson & Ruse (2005) considered the species intolerant of organic pollution.

SALINITY

The species has been found rarely in brackish environments. Steenbergen (1993) sometimes collected the larvae in slightly brackish water with a chloride content lower than 1000 mg/l.

4 TABLES OF BIOLOGICAL AND ECOLOGICAL PROPERTIES OF CHIRONOMINI LARVAE

(in principle for the Dutch situation)

The values in the tables are based on data from the literature and the numerous unpublished studies in the Netherlands referred to in Chapter 1.

blank	unknown
?	not sure

EXPLANATION OF TABLES 2–4

For each factor a species is awarded a score out of ten indicating the relative chance of encountering the species in a one square metre quadrant. This method was introduced by Zelinka & Marvan (1961) and used, among others, by Sládeček (1973) and Moog (1995). The values are based on the Dutch climate and Dutch circumstances. For a discussion of the usefulness of the tables outside the Netherlands, see the relevant section in the Introduction (Chapter 1). Figures lower than 0.5 have not been used. When the occurrence of a species is very exceptional this is recorded as 0.

As a rule, values are not given for a species when no data were available for the whole range in question. In such cases, please consult the species text in Chapter 3. Relatively few data were available for the genus *Chironomus* because most species have been identified very infrequently in the past and the identifications were rarely cytologically verified.

The figures in the tables apply only to water which meets the other requirements of the species. The figures for species that live only in brooks and small streams, for example, do not apply to pools or rivers. *Dicrotendipes nervosus* lives as much in α-mesosaprobic water as in β-mesosaprobic water (see table 2), but this is not true in water which contains no oxygen at night or in water without plants or a hard substrate, where the species rarely live, as can be seen in the right part of the table resp. in table 1. In this case oxygen (or the substrate) appears to be the problem, not the polluting matter. Bear in mind that as a rule a species can endure a negative influence better when all other factors are optimal.

TABLE 1: GENERAL ECOLOGY

gener = number of generations/year (see the comments in Chapter 2)
adult = flying period of adults (and egg deposition) in months
developm = duration of larval development in months (if more generations: in summer)
hibern = hibernation: larval instar; as far as is known all Chironomini have a diapause in winter.
eggs = number of eggs in one egg mass; bold figures are from two or more data (see the comments in Chapter 2)
food = only the most important food for third and fourth instar is given; younger larvae eat more detritus and other fine materials:
AN = animals; AL = algae; FP = fine particulate organic material;
FD = fine detritus; CD = coarse detritus; D = detritus; PL = plant tissue
habitat = preferred microhabitat of the larvae: bo = near or in the bottom (SA = sand; D, CD, FD see under food); pl = on plants;
hs = hard substrate

Values between brackets are found more rarely

TABLE 1	gener	adult	developm	hibern	eggs	food	habitat
Chironomus acutiventris	2(–3?)	5–9	2?			FP/AL ?	SA+FD
Chironomus annularius	(1–)2–4	4–9	1–2	(3–)4	**1000**	FP/AL	FD
Chironomus anthracinus	1	5 (7–9)	12 (24)	(3–)4	**700**	AL/FP	FD
Chironomus aprilinus	2(–3?)	4–10	3	3–4	**970**		bo
Chironomus cingulatus	2–3	4–9	2		**1050**	D/AL	FD
Chironomus dorsalis	4(–6?)	3–10	1	4	**800**		CD, FD
Chironomus luridus	3?	4–10		3–4	**980**		CD
Chironomus melanescens	3–4(–5?)	4–10	1	(3?–)4		CD?	CD
Chironomus muratensis	2	5–9	2	4		AL/FP	SA
Chironomus nuditarsis	3–4	4–11	1½				FD
Chironomus nudiventris	2(–3?)	5–9	2?	4		AL/FP	SA+FD
Chironomus pallidivittatus	2(–3?)	5–9	1½			CD,FP?	CD
Chironomus piger	3–5	4–10	1	(3–)4	**700**		D
Chironomus pilicornis	1	4	12	4	650	AL, D?	D
Chironomus plumosus	(1–)2–3	4–8	1½	(3–)4	**2000**	AL/FP	FD
Chironomus pseudothummi	3	4–8	1½	3–4	580		bo
Chironomus riparius	4–7	3–11	1	(1–)3–4	**500**	FP	FD
Chironomus salinarius	2(–4??)	5–9	1–3?	3–4	**550**	FP	bo
Chironomus tentans	2	4–9	2	(2?–)3–4	**2100**	FP	D
Cladopelma gr. goetghebueri	2	5–9?	2?	(2–)3		AL/FP?	bo
Cladopelma gr. viridulum	2	(4–)5–10	2½	2–4	400	FP(?)	bo
Cryptochironomus defectus	2	(4–)5–9	2	2–3(–4)	**300**	AN, FP	bo
Cryptochironomus redekei	2	(4–)5–9	2	4		AN	bo
Cryptotendipes	2	5–9	2	2(–3)		AL/FP	bo
Demicryptochironomus	1 (2?)	5–9	12	3–4		AN	bo

TABLE 1 (continued)	gener	adult	developm	hibern	eggs	food	habitat
Dicrotendipes lobiger	2	5–9	2	3			pl
Dicrotendipes nervosus	2–3	5–9	2	2–3(–4)	350	AL/FP	pl, hs
Dicrotendipes notatus	2(–3?)	5–9	2?	3(–4)		CD??	bo
Dicrotendipes pulsus	2(–3?)	4–9	2–3	(2–)4		AL/FP	
Einfeldia carbonaria	1–2	5–8	2 (12 ?)	3–4	1120?	AL ?	bo
Einfeldia dissidens	1–2	5–9	2 (12 ?)	3–4		AL ?	bo
Einfeldia pagana	1–?2	5–8	(2?)12	3–4	**500**	FP	bo
Endochironomus albipennis	2(–3?)	(4–)5–10	2½	(2–)3–4	> 500	AL/FP	pl
Endochironomus gr. dispar	2	4–9	2½?	3–4		AL/(FP)	pl, bo
Endochironomus tendens	2	5–9	2½?	3–4	1000	AL/FP	pl
Glyptotendipes barbipes	2–3?	(4–)5–10	2	(2–)3–4			bo, hs
Glyptotendipes cauliginellus	2	5–9	2	(?2–)3–4	1000	AL/FP	pl
Glyptotendipes pallens agg.	2–3	(4–)5–10	2	2–4	1500	AL/FP	pl, hs
Glyptotendipes paripes	2–3	(4–)5–9	2	3–4	1400	AL	bo
Glyptotendipes scirpi	2?	4–9	2?				pl
Glyptotendipes signatus	2	6–9	2	2–3		FP	pl, hs
Harnischia	2	(4–)5–9	2	(2?–)3		AN	bo
Kiefferulus tendipediformis	2?	5–8	2?	2–4	± 450?	FP?	bo
Lipiniella araenicola	2	5–8	2	4		AL	bo
Microchironomus deribae	2	5–9	2–3	3(–4)			bo
Microchironomus tener	2	5–9	2				bo
Microtendipes pedellus gr.	2–3	(3–)4–9	2	(3–)4	**550**	AL/FP	bo
Pagastiella orophila	1(–2?)	5–8	12 (2?)	(3–)4		AL	bo
Parachironomus arcuatus	(2–)3	(4–)5–9	1½	2–3(–4)	500?	(AL)	pl, hs
Paracladopelma (all species)	(2–)3	4–10	2	2–3(–4)		AN	bo
Paratendipes albimanus	1(–2?)	5–9	12 (3?)	1–3	**430**?	FP	bo
Phaenopsectra cf. flavipes	2(–3?)	4–10		2–3(–4)	150	FP	pl
Polypedilum bicrenatum	(1–)2	6–9	2	2–3		FP	bo
Polypedilum convictum	2(–3?)	5–10	2	2(–3?)			bo
Polypedilum cultellatum	2–3	(4–)5–9	2	2–3(–4)		AL	pl, hs
Polypedilum pedestre	2?	5–9	2?			FP	hs, bo
Polypedilum nubeculosum	2–3(–4?)	(3–)4–9	1½–2	3–4	450	AL/FP	bo
Polypedilum scalaenum	2–3(–4?)	4–8	1½	3–4	127		bo
Polypedilum sordens	2(–3?)	(5–)6–10	1½	2–4		AL/(FP)	pl
Polypedilum uncinatum	2–3	4–10	2	3–4	**130**		bo
Pseudochironomus prasinatus	1(–2)	5–8(–9)	12 (2)	2–4	**75**	AL/FP	bo
Robackia demeijerei	2	5–9	3	1–2(–4?)		AN	bo
Stenochironomus	1?	5–9	12?	3–4		PL	pl
Stictochironomus sticticus	1–2	5–9	3 (12)	(2–)3–4		FP	bo
Tribelos intextum	(1–)2	5–9	3, 12	2–4			pl, bo
Xenochironomus xenolabis	1–2	(4–)5–9	2?, 12	3		AL/FP	sponges
Zavreliella marmorata	2?	5–8	2?	2–?	200	AL/FP	pl

TABLE 2: SAPROBITY AND OXYGEN

(see sections 2.12 and 2.13)

We adopt the definition of saprobity by Sládeček (1973: 28): the amount and intensity of decomposition of organic matter. This author (ibid., p. 41) mainly follows the system of Caspers (1966), which is derived largely from conditions in stagnant water. However, many other workers use this system mainly for flowing water, sometimes even only for flowing water (e.g. Moog, 1995: I-22). The main difference between stagnant and flowing waters in this context lies in the availability of oxygen. In this book we try to give the demands of each species with respect to the most important factors found in (undeep) stagnant as well in flowing water. For this reason we keep saprobity and oxygen content separate in an attempt to show whether a species tolerates (or even prefers) a given amount of organic pollution and in what way the oxygen content determines the occurrence of the larvae.

The oxygen contents were measured in the water column, but estimation of saprobity has to take more than just the water column into account. An important part of the decomposition takes place on the bottom. Our polysaprobic level is the beta-polysaprobic level of Caspers (1966), Sládeček (1973) and Moog (1995: I-26), in which anoxybiosis is the rule in the bottom sediments. Ice cover in winter has not been taken into account because it is rarely significant in the Netherlands. When ice and snow cover continues for long periods in winter, a lower saprobity can result in anoxia, especially in shallow water bodies.

ol = oligosaprobic
B = β-mesosaprobic
A = α-mesosaprobic
p = polysaprobic
stab = stable oxygen regime: always above 50% saturation
unst = unstable: minimum between 10% and 50% saturation
low = sometimes (but not longer than a few hours) less than 5% saturation
rott = rotting: in summer almost daily less than 5% saturation for hours

TABLE 2	saprobity							oxygen			
	ol	ol/B	B	B/A	A	A/p	p	stab	unst	low	rott
Chironomus annularius	0	1	2	2	2	2	1	1	3	4	2
Chironomus bernensis	0	1	2	3	2	2	0	3	5	2	0
Chironomus commutatus	0	0	2	3	2	2	1	3	4	2	1
C. luridus in permanent water	0	0	2	2	2	2	2	2	3	3	2
Chironomus nuditarsis	0	0	2	4	2	1	1	1	3	4	2
Chironomus nudiventris	0	0	2	3	3	2	0?	7	3	0?	0
Chironmus obtusidens	0	0	1	3	3	2	1	2	5	2	1
Chironomus pallidivittatus	0	0	1	2	3	2	2	2	3	3	2
Chironomus plumosus	0	1	1	2	2	2	2	2	3	3	2
C. riparius in brooks with pH >6	0	0	1	1	2	3	3	1	2	3	4
Chironomus tentans	0	1	1	2	2	2	2				
Cladopelma gr. goetghebueri	0	1	3	3	2	1	0	6	3	1	0
Cladopelma gr. viridulum	0	1	2	3	3	1	0	5	3	2	0
Cryptochironomus defectus	0	2	3	2	2	1	0	4	4	2	0
Cryptochironomus redekei	0	1	3	3	2	1	0	4	4	2	0
Cryptochironomus supplicans	0?	2	2	2	2	1	1	3	3	2	2
Demeijerea rufipes	0	2	6	2	0	0	0	4	5	1	0
Demicryptochironomus	1	3	3	3	0	0	0	6	4	0	0

TABLE 2 (continued)	saprobity							oxygen			
	ol	ol/B	B	B/A	A	A/p	p	stab	unst	low	rott
Dicrotendipes lobiger	2	3	2	2	1	0	0	5	3	1	1
Dicrotendipes nervosus	0.5	1	2	2	2	2	0.5	3	4	2	1
Dicrotendipes notatus	0.5	2	2	2	2	1.5	0	3	4	2	1
Dicrotendipes pulsus	2	3	3	2	0	0	0	6	3	1	0
Einfeldia carbonaria	0	0	2	3	2	2	1	6	3	1	0
Einfeldia dissidens	0	0	3	3	2	1	1	3	3	3	1?
Einfeldia pagana	0	2	3	2	2	1	0	4	3	2	1?
Endochironomus albipennis	0.5	1	4	3	1	0.5	0	4	3	2	1
Endochironomus gr. dispar	0.5	2	4	2	1	0.5	0	5	3	1.5	0.5
Endochironomus tendens	2	2	3	2	1	0	0	5	3	1.5	0.5
Fleuria lacustris	0	0	2	3	2	2	1	3	3	2	2
Glyptotendipes barbipes	0	1	1	2	4	3	2	2	4	3	1
Glyptotendipes cauliginellus	0	2	3	2	2	1	0	4	4	2	0
Glyptotendipes pallens agg.	0	1	2	2	2	2	1	3	3	2	2
Glyptotendipes paripes	1	2	2	2	2	1	0	3	4	3	0
Glyptotendipes scirpi	0	2	4	3	1	0	0	4	4	2	0
Glyptotendipes signatus	0	1	2	3	3	1	0	4	4	2	0
Harnischia	1	2	4	2	1	0	0	6	4	0	0
Kiefferulus tendipediformis	0	1	3	4	1	1	0	6	2	1	1
Lauterborniella agrayloides	2?	3	3	2	0?	0	0	10	0	0	0
Lipiniella araenicola	0	2	6	2	0	0	0	8	2	0	0
Microchironomus deribae	0	1	2	2	2	2	1	3	3	2	2?
Microchironomus tener	0	2	4	2	2	0	0	6	3	1	0
Microtendipes chloris agg.	0	2	4	2	1	1	0	5	3	1	1
Microtendipes pedellus agg.	0	2	5	3	0	0	0	7	2	1	0
Pagastiella orophila	4	5	1	0	0	0	0	7	3	0	0
Parachironomus arcuatus	0.5	1	1.5	2.5	2	2	0.5	1.5	4	3	1.5
Parachironomus biannulatus	0.5	3	4	2	0.5	0	0	5	4	1	0
Parachironomus frequens	0	1	3	3	2	1	0	5	4	1	0
Paracladopelma nigritulum	3	3	3	0.5	0.5	0	0	7	2.5	0.5	0
Paratendipes albimanus	0.5	1	2	3	2	1	0.5	5	4.5	0.5	0
Phaenopsectra cf. flavipes	2	2	2.5	2	1	0.5	0	5	4	1	0
Polypedilum albicorne	2	3	3	2?	0	0	0	7	3	0	0
Polypedilum bicrenatum	1	2	4	2	1	0	0	6	4	0	0
Polypedilum convictum	2	4	3	1	0	0	0	8	2	0	0
Polypedilum cultellatum	1	2	5	2	0	0	0	7	3	0	0
Polypedilum laetum	1	2	3	2	2	0	0	9	1	0	0
Polypedilum uncinatum								2	3	3	2
Polypedilum nubeculosum	0.5	1	2.5	2.5	2	1	0.5	4	4	1.5	0.5
Polypedilum pedestre	0	1	2	2	2	2	1	5	5	0	0
Polypedilum scalaenum	1	2.5	3	2	1.5	0	0	7	3	0	0
Polypedilum sordens	2	3	3	1	0.5	0.5	0	7	2	1	0
Pseudochironomus prasinatus	3	3	3	1	0	0	0	9	1	0	0
Robackia demeijerei	2	3	3	2	0	0	0	8	2	0	0
Stictochironomus maculipennis	1	3	3	2	0	0	0	8	2	0	0
Stictochironomus sticticus	1	4	3	1	0	0	0	8	2	0	0
Tribelos intextum	2	3	3	2	0	0	0	7	3	0	0
Xenochironomus xenolabis	0	3	5	2	0	0	0	7	3	0	0
Zavreliella marmorata	0	2	5	2	1	0	0	7	2	1	0

TABLE 3: pH AND CHLORINITY

(see sections 2.10 and 2.16)

The pH and chlorinity mentioned are for the water column. Both can be much higher in the bottom. In the tables we use the chlorinity, not the total salt content (1 ‰ salinity = 0.54 g Cl/l).

TABLE 3	pH					chlorinity (g Cl/l)				
	< 4.5	5	6	7	>7.5	<0.3	0.3–1	1–3	3–10	> 10
Chironomus annularius	0	0.5	1.5	4	4	4	4	1.5	0.5	0
Chironomus aprilinus	0	0	0	0	10	0	1	4	4.5	0.5
Chironomus bernensis	0	0	1	4	5	9	1	0	0	0
Chironomus cingulatus	0.5	1.5	2	3	3	7	2	1?	0	0
Chironomus commutatus	0	0?	1	4	5	9	1	0	0	0
Chironomus dorsalis	1	1.5	2.5	2.5	2.5	6	2	1	1	0
Chironomus luridus						8	1.5	0.5	0	0
Chironomus muratensis	0	0	0	2	8					
Chironomus nuditarsis	0	0.5	1.5	4	4	8	1.5	0.5	0	0
Chironomus nudiventris	0	0	0	3	7	9	1	0	0	0
Chironomus obtusidens	0	0?	2	4	4	10	0	0	0	0
Chironomus pallidivittatus	0	0	1	2	7	4	4	2	0	0
Chironomus piger						5	3	1.5	0.5	0
Chironomus plumosus	0	0.5	1.5	4	4	7	1.5	1	0.5	0
Chironomus pseudothummi	3	3	3	1?	0?	10	0	0	0	0
Chironomus riparius	0.5	1.5	2	3	3	8	1.5	0.5	0	0
Chironomus salinarius	0	0	0	0	10	0	0.5	1.5	4	4
Chironomus tentans	0	1	1	3	5					
Cladopelma gr. goetghebueri	2	2	2	2	2	9.5	0.5	0?	0	0
Cladopelma gr. viridulum	1	2	2.5	2.5	2	9	1	0	0	0
Cryptochironomus defectus	1	1	2	3	3					
Cryptochironomus obreptans	1	1	2	3	3					
Cryptochironomus redekei	0	1	1	3	5	5	4	1	0	0
Demeijerea rufipes	0	0	0	3	7	6	4	0	0	0
Demicryptochironomus	0.5	0.5	2	3	4	10	0	0	0	0
Dicrotendipes lobiger	0.5	0.5	2	6	1	9	1	0	0	0
Dicrotendipes nervosus	0	0.5	1	3.5	5	5	3	1.5	0.5	0
Dicrotendipes notatus	0	0.5	0.5	7	2	6	3	1	0	0
Dicrotendipes pallidicornis	0	0	0	0	10	0	2	4	4	0
Dicrotendipes pulsus	2	3	2	1.5	1.5	9	1	0?	0	0
Einfeldia carbonaria	0	0	0	1	9	8	2?	0	0	0
Einfeldia dissidens	0	0	1	3	6		?	0	0	0
Einfeldia pagana	0	0	0.5	1.5	8	7	2	1	0	0
Endochironomus albipennis	0.5	0.5	1	4	4	5	4	1	0	0
Endochironomus gr. dispar	0.5	2	3	3	1.5	9	0.5	0.5	0	0
Endochironomus tendens	2	2	2	2	2	8	1.5	0.5	0	0

TABLE 3 (continued)	pH					chlorinity (g Cl/l)				
	< 4.5	5	6	7	>7.5	<0.3	0.3–1	1–3	3–10	> 10
Fleuria lacustris	0	0	0	1	9	7	3	0	0	0
Glyptotendipes barbipes	0	0	0.5	2.5	7	1	3	5	1	0
Glyptotendipes cauliginellus	0	0.5	1.5	4	4	7	2	1	0	0
Glyptotendipes ospeli	0	0	1	5	4	8	2	0	0	0
Glyptotendipes pallens agg.	0	0.5	1.5	4	4	7	2.5	0.5	0	0
Glyptotendipes paripes	2	2	2	2	2	8	1.5	0.5	0	0
Glyptotendipes scirpi	0	0	2	4	4	9	1	0	0	0
Glyptotendipes signatus	0	0	0.5	3.5	6	10	0?	0	0	0
Harnischia	0	0	2	4	4	8	1	1	0	0
Kiefferulus tendipediformis	0	0	0	4	6	7	2	1	0	0
Lauterborniella agrayloides	0	0	1	4	5	10	0	0	0	0
Lipiniella araenicola	0	0	0	2	8	9	1	0	0	0
Microchironomus deribae	0	0	0	0	10	0	1	4	4	1
Microchironomus tener	0	0	0.5	2.5	7	7	3	0	0	0
Microtendipes pedellus gr.	0.5	1.5	2	3	3	8	1.5	0.5	0	0
Pagastiella orophila	1	3	3	2	1	10	0	0	0	0
Parachironomus arcuatus	0	0.5	1	3.5	5	6	2.5	1	0.5	0
Parachironomus biannulatus	0	0	0	5	5	9	1	0	0	0
Parachironomus frequens	0	0	0.5	1.5	8	8	2	0	0	0
Paratendipes albimanus	0	0.5	1.5	5	3	9	0.5	0.5	0	0
Paratendipes nudisquama	3	3	1.5	1.5	1	10	0	0	0	0
Phaenopsectra cf. flavipes	1	1.5	2	3	2.5	9	0.5	0.5	0	0
Polypedilum albicorne	?	1	3	4	2	10	0	0	0	0
Polypedilum bicrenatum	0	0.5	1.5	4	4	8.5	1	0.5	0	0
Polypedilum convictum	0	0?	1?	4	5	10	0	0	0	0
Polypedilum cultellatum	0	1	2	4	3	10	0?	0	0	0
Polypedilum uncinatum	3	3	1	2	1					
Polypedilum nubeculosum	0	1	2	3.5	3.5	7	2	1	0	0
Polypedilum pedestre	0	0	2?	4	4	10	0	0	0	0
Polypedilum scalaenum	0	0	0.5	4.5	5	9	0.5?	0.5?	0?	0
Polypedilum sordens	0.5	1	1	3.5	4	8	1.5	0.5	0	0
Pseudochironomus prasinatus	3	4	2	0.5	0.5	9.5	0.5	0	0	0
Stictochironomus maculipennis	0	0	1?	4	5	10	0	0	0	0
Stictochironomus sticticus	0	0.5	1.5	4	4	9	1?	0	0	0
Tribelos intextum	1	2	2	3	2	10	0	0	0	0
Xenochironomus xenolabis	0	0	0	1?	9	8	2	0	0	0
Zavreliella marmorata	0	0	0	2	8	9	1	0	0	0

T

TABLE 4: CURRENT AND PERMANENCE

(see section 2.9)

Current

The given velocities apply only to stagnant water, brooks and small streams. They are thought to have been measured 10 cm below the surface (if not near the bottom) in the middle of the stream. Some authors (e.g. Rossaro & Mietto, 1998) mention the current at the spot where the larva lives and these current values are much lower for bottom dwellers. In principle this is a better record, but working with these current velocities is not so useful in practice and we have too little information about these velocities.

In using our interpretation of current, species like *Polypedilum cultellatum* or *Phaenopsectra flavipes* that live on plants near the water surface prefer a slower current than many bottom dwellers, even though (or even because) they actually experience higher velocities. The table shows that these species hardly live in streams with a fast current. The figures in the tables do not reflect the situation in large rivers; the current (as defined above) in rivers is much faster and some larvae live at other sites, such as the underside of stones.

The figures apply to the Netherlands and adjacent lowlands. This means that species living in Scandinavian lakes, but rarely in lakes in and around the Netherlands, can have a low figure in the current class < 10 cm/sec.

Permanence

The figures given for permanence mainly apply to water bodies that dry up in summer, but in principle not to pools or ditches, which are sometimes in contact with nearby permanent water. In the latter case, species characteristic of permanent water may be carried by stream to places where no water was previously present. Neither do the figures apply to temporary exposure of parts of an estuary or river during low tide or a low stage of the river. A water body is considered dry only when the bottom is dry.

TABLE 4	current (cm/s)					permanence (dry weeks/y.)				
	< 10	10-25	25-50	50-75	> 75	> 12	6–12	< 6	rarely	not
Chironomus acutiventris	3	5	2	0	0	0	0	0	0	10
Chironomus annularius	8	1.5	0.5	0	0	0	1	1	3	5
Chironomus bernensis	3	4	2	1	0	0	0	0?	1	9
Chironomus cingulatus	9	1	0	0	0	0	0?	1?	2?	7
Chironomus commutatus	8	1.5	0.5	0	0					
Chironomus dorsalis	8	2	0	0	0	2	2.5	2.5	2	1
Chironomus luridus	7	2	1	0	0	1.5	2.5	2.5	2	1.5
Chironomus muratensis	8	2	0	0	0	0	0	0	0	10
Chironomus nuditarsis	8	1.5	0.5	0	0	0	0	1	1	8
Chironomus nudiventris	8	2	0	0	0	0	0	0	0	10
Chironomus obtusidens	4	4	2	0	0	0	1	3	3	3
Chironomus pallidivittatus	8	2	0	0	0	0	1	2	3	4
Chironomus piger	8	1.5	0.5	0	0	2.5	2.5	2	1.5	1.5
Chironomus plumosus	8	2	0.5	0.5	0	0	0	0.5	0.5	9
Chironomus pseudothummi	6	3	1	0	0	2	2	2	2	2
Chironomus riparius	2	3	3	1.5	0.5	2	2.5	2.5	1.5	1.5
Chironomus tentans	8	2	0	0	0					
Cladopelma gr. goetghebueri	8.5	1	0.5	0	0	0	0.5	0.5	1	8
Cladopelma gr. viridulum	9	1	0	0	0	0	0	0.5	0.5	9
Cryptochironomus defectus	2	3	3	1.5	0.5	0	0	1	2	7

TABLE 4 (continued)	current (cm/s)					permanence (dry weeks/y.)				
	‹ 10	10-25	25-50	50-75	› 75	› 12	6–12	‹ 6	rarely	not
Cryptochiron. denticulatus	0	1	4	4	1	0	0	0	0	10
Cryptochironomus obreptans	8	2	0	0	0	0	0	1	2	7
Cryptochironomus psittacinus	8	1	1	0	0	0	0	0	0	10
Cryptochironomus redekei	10	0	0	0	0	0	0	0	0	10
Cryptochironomus rostratus	0	1	2	4	3	0	0	0	0	10
Cryptotendipes holsatus	4	4	2	0	0	0	0	0	0	10
Demeijerea rufipes	8	1.5	0.5	0	0	0	0	0	0	10
Demicryptochironomus	3	5	1.5	0.5	0	0	0	0	0	10
Dicrotendipes lobiger	8	2	0	0	0	0	0	0.5	0.5	9
Dicrotendipes nervosus	4	3	2	1	0	0	0	0.5	0.5	9
Dicrotendipes notatus	4	4	2	0	0	0	0.5	1	2.5	6
Dicrotendipes pulsus	9.5	0.5	0	0	0	0	0.5	0.5	1	8
Einfeldia carbonaria	9	1	0	0	0	0	0	0	0	10
Einfeldia dissidens	9	1	0	0	0	0	0	1	1	8
Einfeldia pagana	8	1	1	0	0	0	0	0	1	9
Endochironomus albipennis	9	1	0	0	0	0	0	0	1	9
Endochironomus gr. dispar	9	0.5	0.5	0	0	0	1	1	2	6
Endochironomus tendens	8	2	0	0	0	0	0	0.5	1.5	8
Fleuria lacustris	10	0	0	0	0	0	0	0	0	10
Glyptotendipes barbipes	8	2	0	0	0	0	0.5	1.5	2	6
Glyptotendipes cauliginellus	9.5	0.5	0	0	0	0	0	0	1	9
Glyptotendipes pallens agg.	8	1.5	0.5	0	0	0	0.5	1.5	2	6
Glyptotendipes paripes	9.5	0.5	0	0	0	0	0.5	1.5	2	6
Glyptotendipes scirpi	9.5	0.5	0	0	0	0	0	0	1	9
Glyptotendipes signatus	9	1	0	0	0	0	0	0.5	0.5	9
Harnischia curtilamellata	6	1	2	1	0	0	0	0	0	10
Harnischia fuscimana	3	3	3.5	0.5	0	0	0	0	0	10
Kiefferulus tendipediformis	6	3	1	0	0	0	0	0.5	0.5	9
Lauterborniella agrayloides	10	0	0	0	0	0	1	2	3	4
Lipiniella araenicola	8	2	0	0	0	0	0	0	0	10
Microchironomus tener	9	1	0	0	0	0	0	0	0	10
Microtendipes chloris agg.	7	2	0.5	0.5	0	0	0	0	0.5	9.5
Microtendipes pedellus agg.	1	5	2	1.5	0.5	0	0	0	0.5	9.5
Microtendipes rydalensis	0	2	3	4	1					
Pagastiella orophila	10	0	0	0	0	0	0	0	0	10
Parachironomus arcuatus	5	3.5	1	0.5	0	0.5	0.5	1	1	7
Parachironomus biannulatus	8	2	0	0	0	0	0	0	1	9
Parachironomus frequens	1	5	3	1	0	0	0	0.5	0.5	9
Paracladopelma nigritulum	2	5	2.5	0.5	0	0	0	0.5	1.5	8
Paralauterborniella	2	5	3	0	0	0	0	0	1?	9
Paratendipes albimanus	2	5	2	0.5	0.5	0	0	0.5	0.5	9
Phaenopsectra cf. flavipes	6	2.5	1	0.5	0	0.5	0.5	1	1	7
Polypedilum albicorne	2?	2?	2	2	2	0	0	0	1	9
Polypedilum bicrenatum	3	4	2	0.5	0.5	0	0	0?	1	9
Polypedilum convictum	0	2	2	3	3	0	0	0?	1	9
Polypedilum cultellatum	2	3	3	1	1	0	0	0	1	9
Polypedilum laetum	0	0	3	3	4					
Polypedilum uncinatum	7	2	0.5	0.5	0	1	2	3	2	2
Polypedilum nubeculosum	4	3.5	1.5	0.5	0.5	0	1	1	2	6
Polypedilum pedestre	1	1	2	3	3	0	0	0	0	10
Polypedilum scalaenum	2	2	3	2	1	0	0	1	1	8
Polypedilum sordens	8	1.5	0.5	0	0	0	0	0	1	9
Pseudochironomus prasinatus	10	0	0	0	0	0	0	0	0	10
Robackia demeijerei	2	4	3	1	0	0	0	0	0	10
Stictochironomus maculipennis	1	2	4	2	1	0	0	0	1	9
Stictochironomus sticticus	9	1	0	0	0	0	0	0	0	10
Tribelos intextum	9	1	0	0	0	0	0	0	0	10
Xenochironomus xenolabis	8	2	0	0	0	0	0	0	0	10
Zavreliella marmorata	9	1	0	0	0	0	0	0	1	9

REFERENCES

Adam, J.I. & O.A. Saether, 1999. Revision of the genus *Nilothauma* Kieffer, 1921 (Diptera: Chironomidae). – Ent. scand. Suppl. 56: 1-107.

Ali, A. & R.D. Baggs, 1982. Seasonal changes of chironomid populations in a shallow natural lake and in a man-made water cooling reservoir in Central Florida. – Mosquito News 42: 76-85.

Ali, A., P.K. Chaudhuri & D.K. Guha, 1987. Description of *Stictochironomus affinis* (Johannsen) (Diptera: Chironomidae), with notes on its behaviour. – Florida Ent. 70: 259-267.

Ali, A. & R.C. Fowler, 1983. Prevalence and dispersal of pestiferous Chironomidae in a lake front city of Central Florida. – Mosquito News 43: 55-59.

Armitage, P.D., 1968. Some notes on the food of the chironomid larvae of a shallow woodland lake in South Finland. – Ann. Zool. Fenn. 5: 6-13.

Armitage, P.D., P.S.Cranston & L.C.V. Pinder, 1995. The Chironomidae – Biology and ecology of non-biting midges. – Chapman & Hall (London).

Ashe, P., 1983. A catalogue of chironomid genera and subgenera of the world including synonyms (Diptera: Chironomidae). – Ent. scand. Suppl. 17: 1-68.

Ashe, P. & P.S. Cranston, 1990. Family Chironomidae. – In: Soos, A. & L. Papp (eds.): Catalogue of Palaearctic Diptera 2. Akadémiai Kiadó, Budapest: 113-355.

Baker, J.H. & L.A. Bradnam, 1976. The role of bacteria in the nutrition of aquatic detritivores. – Oecologia (Berlin) 24: 95-104.

Baker, R.L. & S. L. Ball, 1995. Microhabitat selection by larval *Chironomus tentans* (Diptera: Chironomidae): effects of predators, food, cover and light. – Freshw. Biol. 34: 101-106.

Bakhtina, V.I., 1980. Life cycles and production of mass species of chironomids in fattening ponds. – Acta Universitatis Carolinae – Biologica 1978: 13-20.

Balushkina, E.V., 1987. Functional importance of the larvae of chironomids in continental water bodies. – Trudy zool. inst. Akad. Nauk SSSR 142: 1-179. (In Russian).

Barnes, L.E., 1983. The colonization of ball-clay ponds by macro-invertebrates and macrophytes. – Freshw. Biol. 13: 561-578.

Baz', L.G., 1959. Biologiya i morfologiya predstaviteljej roda *Microtendipes*, obitayushchih v vodoprovodnom kanale Uchinskogo vodokhranilishche. – Trudy vses. gidrobiol. obshch. 9: 74-84.

Bazerque, M.F., H. Laville & Y. Brouquet, 1989. Biological quality assessment in two rivers in the northern plain of France (Picardie) with special reference to chironomid and diatom indices. – Acta Biol. Debr. Oecol. Hung. 3: 29-39.

Beattie, D.M., 1978. Chironomid populations in the Tjeukemeer. – Thesis Leiden University. 150 pp.

Beattie, D.M., 1978a. Life-cycle and changes in carbohydrates, proteins and lipids of *Pentapedilum uncinatum* Goet. (Diptera; Chironomidae). – Freshw. Biol. 8: 109-113.

Beattie, D.M., H.L. Golterman & J. Vijverberg, 1978. An introduction to the limnology of the Friesian lakes. – Hydrobiologia 58: 49-64.

Beattie, D.M., 1982. Distribution and production of the larval chironomid populations in Tjeukemeer. - Hydrobiologia 95: 287-306.

Beck, E.C. & W.M. Beck, 1969. Chironomidae (Diptera) of Florida. III. The Harnischia complex (Chironominae). – Bull. Florida State Mus. Biol. Sci. 13: 277-313.

Becker, C., 1995. Ein Beitrag zur Zuckmückenfauna des Rheins (Diptera: Chironomidae). – Thesis Bonn. Aachen: Shaker Verlag. 265 pp.

Beedham, G.E., 1966. A chironomid (Dipt.) larva associated with the lamellibranchiate mollusc, *Anodonta cygnea* L. –Entomologist's mon. Mag. 101: 142-143.

Bell, H.L., 1970. Effects of pH on the life cycle of the midge *Tanytarsus dissimilis*. – Can. Ent. 102: 636-639.

Benthem-Jutting, T. van, 1938. A freshwater pulmonate (*Physa fontinalis* [L.]) inhabited by the larva of a non-biting midge (*Tendipes* [*Parachironomus*] *varus* Gthg.). – Arch. Hydrobiol. 32: 693-699.

Berezina, N.A., 1998. Description of *Omisus caledonicus* (Diptera, Chironomidae) larva. – Ent. Rev. 78: 403-405. (Translation from Russian article).

Berg, C.O., 1950. Biology of certain Chironomidae reared from Potamogeton. – Ecol. Monographs 20: 83-101.

Bergquist, B., 1987. Effects of peatland drainage and fertilization on the abundance and biomass of *Stictochironomus rosenschoeldi* (Zett.) and *Sergentia longiventris* (Kieff.) (Diptera: Chironomidae) in Lake Siksjön, Central Sweden. – Ent. scand. Suppl. 29: 225-231.

Bijlmakers, L., 1983. De verspreiding en oecologie van chironomidelarven (Chironomidae: Diptera) in twee vennen in de omgeving van Oisterwijk (N.Br.). – Versl. K.U. Nijmegen. 118 pp. + bijl.

Biró, K, 1988. Kleiner Bestimmungsschlüssel für Zuckmückenlarven (Diptera: Chironomidae). – Wasser und Abwasser Suppl. Bd. 1/88: 1-329.

Biró, K., 2000. Chironomidae (Insecta, Diptera) from Hungary 2. New records of *Lipiniella moderata* Kalugina, 1970. – Spixiana 23: 157-158.

Biró, K. & A. Klink, 2005. Chironomidae (Insecta, Diptera) from Hungary 3. The pupa of *Paratendipes nubilus* (Meigen). – Acta zool. hung. 51: 181-185.

Bjelke, U., I.M. Bohman & J. Herrmann, 2005. Temporal niches of shredders in lake littorals with possible implications on ecosystem functioning. – Aq. Ecol. 39: 41-53.

Borkent, A., 1984. The systematics and phylogeny of *Stenochironomus* complex (*Xestochironomus*, *Harrisius*, and *Stenochironomus*) (Diptera: Chironomidae). – Mem. Ent. Soc. Can. 128: 1-269.

Bowen, K.L., N.K. Kaushik & A.M. Gordon, 1998. Macroinvertebrate communities and biofilm chlorophyll on woody debris in two Canadian oligotrphic lakes. – Arch. Hydrobiol. 141: 257-281.

Bowker, D.W., M.T. Wareham & M.A. Learner, 1983. The selection and ingestion of epilithic algae by *Nais elinguis* (Oligochaeta: Naididae). – Hydrobiologia 98: 171-178.

Braukmann, U., 1984. Biologischer Beitrag zu einer allgemeinen regionalen Bachtypologie. – Thesis Giessen Univ. 473 pp.

Brock, T.C.M., 1984. Aspects of the decomposition of *Nymphoides peltata* (Gmel.) O.Kuntze (Menyanthaceae). – Aq. Botany 19: 131-156.

Brodersen, K.P., P.C. Dall & C. Lindegaard, 1998. The fauna in the upper stony littoral of Danish lakes: macroinvertebrates as trophic indicators. – Freshw. Biol. 39: 577-592.

Brodin, Y-W. & M. Gransberg, 1993. Responses of insects, especially Chironomidae (Diptera), and mites to 130 years of acidification in a Scottish lake. – Hydrobiologia 250: 201-212.

Brooks, S.J., P.G. Langdon & O. Heiri, 2007. The identification and use of Palaearctic Chironomidae larvae in Palaeoecology. – Quaternary Res. Ass. Techn. Guide 10.

Brown, A.E. & R.S. Oldham, 1984. Chironomidae (Diptera) of Rutland Water. – Arch. Hydrobiol. Suppl. 69: 199-227.

Brown, A.E., R.S. Oldham & A. Warlow, 1980. Chironomid larvae and pupae in the diet of Brown trout (*Salmo trutta*) and Rainbow trout (*Salmo gairdneri*) in Rutland Water, Leicestershire. – In: Murray, D.A. (ed.): Chironomidae: Ecology, Systematics, Cytology and Physiology. Oxford, Pergamon Press: 323-329.

Brundin, L., 1947. Zur Kenntnis der schwedischen Chironomiden. – Ark. f. Zool. 39A: 1-95.

Brundin, L., 1949. Chironomiden und andere Bodentiere der südschwedischen Urgebirgsseen. – Rep. Inst. Freshw. Res. Drottningholm 30: 1-914.

Brundin, L., 1956. Die bodenfaunistischen Seetypen und ihre Anwendbarkeit auf die Südhalbkugel. Zugleich eine Theorie der produktionsbiologischen Bedeutung der glazialen Erosion. – Rep. Inst. Freshw. Res. Drottningholm 37: 186-191.

Bryce, D., 1960. Studies on the larvae of the British Chironomidae (Diptera) with keys to the Chironominae and Tanypodinae. – Trans. Soc. Br. Entom. 14: 19-62.

Bund, W. van de, 1994. Food web relations of littoral macro- and meiobenthos. – Thesis Amsterdam University. 106 pp.

Burtt, E.T., 1940. A filter-feeding mechanism in a larva of the Chironomidae (Diptera: Nematocera). – Proc. R. Ent. Soc. Lond. (A) 15: 113-121.

Buskens, R., 1983. De makrofauna, in het bijzonder de chironomiden, en de vegetatie van een vijftigtal geeutrofieerde, zure of laag-alkaliene stilstaande wateren op de Nederlandse zandgronden. – Lab. Aq. Oecol. Kath. Univ. Nijmegen, Rapport 159: 1-75 + 17 tables.

Buskens, R.F.M., 1987. The chironomid assemblages in shallow lentic waters differing in acidity, buffering capacity and trophic level in the Netherlands. – Ent. scand. Suppl. 29: 217-224.

Buskens, R.F.M., 1989. Monitoring of chironomid larvae and exuviae in the Beuven, a soft water pool in the Netherlands, and comparisons with palaeolimnological data. – Acta Biol. Debr. Suppl. Oecol. Hung. 3: 41-50.

Buskens, R.F.M., 1989a. Beuven: Herstel van een ecosysteem. – Vakgroep Aq. Oecol. en Biogeol. Kath. Univ. Nijmegen. 154 pp.

Buskens, R.F.M., R.S.E.W. Leuven, J.A. van der Velden & G. van der Velde, 1986. The spatial and temporal distribution of Chironomid larvae in four lentic soft waters differing in acidity. – Proc. 3rd Eur. Congr. Ent., Amsterdam 1: 75-78.

Buskens, R.F.M. & G.A.M. Verwijmeren, 1989. The chironomid communities of deep sand pits in the Netherlands. – Acta Biol. Debrecina Suppl. Oecologica Hungarica 1989 (3): 51-60.

Cannings, R.A. & G.G.E. Scudder, 1978. The littoral Chironomidae (Diptera) of saline lakes in central British Columbia. – Can. J. Zool. 56: 1144-1155.

Casas, J.J. & A. Vilchez-Quero, 1989. A faunistic study of the lotic chironomids (Diptera) of the Sierra Nevada (S.E. of Spain): changes in the structure and composition of the populations between spring and summer. – Acta Biol. Debr. Oecol. Hung. 3: 83-93.

Caspers, H., 1966. Stoffwechseldynamische Gesichtspunkte zur Definition der Saprobitätsstu-fen. – Verh. Internat. Verein. Limnol. 16: 801-808.

R

Caspers, H., 1972. Experimentelle Untersuchungen über den Einfluss von Schwefelwasserstoff auf limnische und marine Bodentiere.- Verh. Internat. Verein. Limnol. 18: 946-954.

Caspers, N. 1980. Die Makrozoobenthos-Gesellschaf-ten des Rheins bei Bonn. – Decheniana (Bonn) 133: 93-106.

Caspers, N., 1991. The actual biocoenotic zonation of the river Rhine exemplified by the chironomid midges (Insecta, Diptera). – Verh. Internat. Verein. Limnol. 24: 1829-1834.

Cate, L. ten & G. Schmidt, 1986. Makrofauna-levens-gemeenschappen in beekbovenlopen. – Versl. Rijksinst. Natuurbeheer, Leersum: 1-163.

Charles, W.N., K. East, D. Brown, M.C. Gray & T.D. Murray, 1974. Production studies on the larvae of four species of Chironomidae in the mud at Loch Leven in Scotland during 1970-71. – Ent. Tidskr. 95 Suppl.: 34-41.

Chernovskij, A.A., 1938. Vertikalnoe raspredelenie zhivotnykh v tolshche ila nekotorykh ozer okrestnostej Leningrada. – Zool. Zh. 17: 1030-1054.

Chernovskij, A.A., 1949. Opredelitelj lichinok komarov semejstva Tendipedidae. – Opredel. po faune SSSR 31: 1-186.

Collé, C., 1983. Methodologisch onderzoek aan makrofaunabemonsteringstechnieken in verschillende vegetatiestructuren van poldersloten in Demmerik. – Rapport Rijksinst. Natuurbeheer, Leersum. 49 pp.

Contreras-Lichtenberg, R., 1986. Revision der in der Westpaläarktis verbreiten Arten des Genus *Dicrotendipes* Kieffer, 1913 (Diptera, Nematocera, Chironomidae). – Ann. Naturhist. Mus. Wien 88/89 B: 663-726.

Contreras-Lichtenberg, R., 1989. Beitrag zur Kenntnis der Weibchen in der Westpaläarktis verbreiteter Arten des Genus *Dicrotendipes* Kieffer, 1913 (Diptera, Nematocera, Chironomidae). – Acta Biol. Debr. Oecol. Hung. 2: 173-179.

Contreras-Lichtenberg, R., 1996. Contribution to the knowledge of female west palaearctic *Glyptotendipes* Kieff. (Diptera, Nematocera, Chironomidae). – Hydrobiologia 318: 17-23.

Contreras-Lichtenberg, R., 1999. Revision der west-palaearktischen Arten des Genus *Glyptotendipes* Kieffer, 1913 (Insecta: Diptera, Nematocera, Chironomidae), Teil 1: Subgenus *Phytotendipes* Goetghebuer, 1937. – Ann. Naturhist. Mus. Wien 101 B: 359-403.

Contreras-Lichtenberg, R., 2001. Revision der west-palaearktischen Arten des Genus *Glyptotendipes* Kieffer, 1913 (Insecta: Diptera, Nematocera, Chironomidae), Teil 2: Sg. *Glyptotendipes* s.str. Kieffer, 1913 und Sg. *Trichotendipes* Heyn, 1993. – Ann. Naturhist. Mus. Wien 103 B: 417-451.

Cranston, P.S., M.E. Dillon, L.C.V. Pinder & F. Reiss, 1989. The adult males of Chironominae (Diptera: Chironominae) of the Holarctic region. Keys and diagnoses. – Ent. scand. Suppl. 34: 353-502.

Crawford, P.J. & D.M. Rosenberg, 1984. Breakdown of conifer needle debris in a new northern reservoir, Southern Indian Lake, Manitoba. – Can. J. Fish. Aquat. Sci. 41: 649-658.

Credland, P.F., 1973. A new method for establishing a permanent laboratory culture of *Chironomus riparius* Meigen (Diptera: Chironomidae). – Freshw. Biol. 3: 45-51.

Cuppen, H.P.J.J., 1980. De macrofauna in een aantal droogvallende en permanente stilstaande wateren in het ruilverkavelingsgebied Brummen – Voorst. – Regionale Milieuraad Oost-Veluwe. Apeldoorn. 79 pp. + app.

Cuijpers, P. & M. Damoiseaux, 1981. De Geul. Biologische beoordeling van de waterkwaliteit, met behulp van diverse systemen. – Rep. Natuurh. Genootsch. Limburg, Maastricht: 1-120 + app. (unpubl.).

Dam, H. van & R.F.M. Buskens, 1993. Ecology and management of moorland pools: balancing acidification and eutrophication. – Hydrobiologia 265: 225-263.

Danks, H.V., 1971. Life history and biology of *Einfeldia synchrona* (Diptera: Chironomidae). – Can. Ent. 103: 1597-1606.

Danks, H.V., 1971a. Overwintering of some north temperate and arctic Chironomidae. II. Chironomid biology. – Can. Ent. 103: 1875-1910.

Danks, H.V. & J.W. Jones, 1978. Further observations on winter cocoons in Chironomidae (Diptera). - Can. Ent. 110: 667-669.

Davies, L.J. &H.A. Hawkes, 1981. Some effects of organic pollution on the distribution and seasonal incidence of Chironomidae in riffles in the River Cole. – Freshw. Biol. 11: 549-559.

Dejoux, C., 1971. Contribution à l'étude des premiers états des chironomides du Tchad (Insectes, Diptères) (cinquième note). Description de *Chironomus (Cryptochironomus) deribae*, Freeman, 1957 et *Polypedilum (Polypedilum) fuscipienne* Kieffer, 1921. – Cahier O.R.S.T.O.M., sér. Hydrobiologie 5: 87-100.

Delettre, Y.R., 1989. Influence de la durée et de l'intensité de l'assèchement sur l'abondance et la phénologie des Chironomides (Diptera) d'une mare semi-permanente peu profonde. – Arch. Hydrobiol. 114: 383-399.

Delettre, Y., P. Tréhen & P. Grootaert, 1992. Space heterogeneity, space use and short-range dispersal in Diptera: a case study. – Landscape Ecol. 6: 175-181.

Dettinger-Klemm, P.-M. A., 2000. Temporäre Stillgewässer – Charakteristika, Ökologie und Bedeutung für den Naturschutz. – In: NUA (ed.): Gewässer ohne Wasser? NUA Seminarbericht 5: 17-42. Recklinghausen.

Dettinger-Klemm, P.-M. A., 2000a. Influence of temperature and photoperiod on development in three species of Chironomidae (Diptera) – Chironomus dorsalis Meigen, 1818, Polypedilum uncinatum (Goetghebuer, 1921) and Paralimnophyes hydrophilus (Goetghebuer, 1921) – living in temporary ponds. – In: Hoffrichter, O. (ed.): Late 20th Century Research on Chironomidae: 295-308. Aachen, Shaker Verlag.

Dettinger-Klemm, P.-M. A., 2002. Drought-tolerance and the impact of the photoperiod on growth and emergence in Polypedilum tritum (Walker, 1856) (= Polypedilum uncinatum Goetghebuer, 1921 syn. nov.). – Dtsch. Ges. Limnol., Tagungsber. 2001. 6 pp.

Dettinger-Klemm, P.-M. A., 2003. Chironomids (Diptera, Nematocera) of temporary pools – an ecological case-study. – Thesis Marburg Univ. 371 pp.

Dettinger-Klemm, P.-M. A. & H.W. Bohle, 1996. Überlebensstrategien und Faunistik von Chironomiden (Chironomidae, Diptera) temporärer Tümpel. – Limnologica 28: 403-421.

Douglas, D.J. & D.A. Murray, 1980. A checklist of the Chironomidae (Diptera) of the Killarney Valley catchment area Ireland. – In: Murray, D.A. (ed.): Chironomidae: Ecology, Systematics, Cytology and Physiology. Oxford, Pergamon Press: 123-129.

Drake, C.M., 1982. Seasonal dynamics of Chironomidae (Diptera) on the Bulrush *Schoenoplectus lacustris* in a chalk stream. – Freshw. Biol. 12: 225-240.

Driver, E.A., 1977. Chironomid communities in small prairie ponds: some characteristics and controls. – Freshw. Biol. 7: 121-133.

Duursema, G., 1996. Vennen in Drenthe. Een onderzoek naar ecologie en natuur op basis van macrofauna. – Assen, Zuiveringsschap Drenthe. 140 pp.

Dvořák, J., 1996. An example of relationships between macrophytes, macroinvertebrates and their food resources in a shallow eutrophic lake. – Hydrobiologia 339: 27-36.

Dvořák, J. & E.P.H. Best, 1982; Macroinvertebrate communities associated with the macrophytes of Lake Vechten: structural and functional relationships. – Hydrobiologia 95: 115-126.

Edwards, F.W., 1929. British non-biting midges (Diptera, Chironomidae). – Trans. ent. Soc. Lond. 77: 279- 430.

Ellenbroek, G.A. & J.L.J. Hendriks, 1972. Vergelijkend hydrobiologisch onderzoek in de Kroonbeek en de Teelebeek, een schone en een verontreinigde laaglandbeek. – K.U. Nijmegen, Zoöl. Lab. afd. Dieroecol., rapp. 55: 1-94.

Entz, B., 1965. Untersuchungen an Larven von *Chironomus plumosus* Meig. im Benthos des Balatonsees in den Jahren 1964-1965. – Annal. Biol. Tihany 32: 129-139.

Ertlová, E., 1970. Chironomidae (Diptera) aus Donauaufwuchs. - Biológia (Bratislava) 25: 291-300.

Ertlová, E., 1974. Einige Erkenntnisse über Chironomiden (Diptera, Chironomidae) aus Bryozoen. – Biológia (Bratislava) 29: 869-876.

Ferrarese, U., 1992. Chironomids of Italian rice fields. – Neth. J. Aquat. Ecol. 26: 341-346.

Fischer, J., 1969. Zur Fortpflanzungsbiologie von *Chironomus nuditarsis* Str. – Rev. Suisse Zool. 76: 23-55.

Fittkau, E.-J. & F. Reiss, 1978. Chironomidae. – In: Illies, J. (ed.): Limnofauna europaea. 2. Aufl. Stuttgart: 404-440.

Frank, C., 1987. A comparative study of chironomid (Diptera) emergence data from 14 lakes in the urban region of West Berlin. – Ent. scand. Suppl. 29: 211-216.

Fritz , H.-G., 1981. Über die Mückenfauna eines temporären Stechmückenbrutgewässers des Natur-schutzgebietes "Kühkopf Knoblochsaue". – Hessische Faun. Briefe 1: 38-49.

Fritz, H.-G., 1982. Ökologische und systematische Untersuchungen an Diptera/ Nematocera (Insecta) in Überschwemmungsgebieten des nördlichen Oberrheins. Ein Beitrag zur Ökologie grosser Flussauen. – Thesis Darmstadt.

Gaevskaya, N.S., 1969. The role of higher aquatic plants in the nutrition of the animals of freshwater basins. – Translation by D.G. Maitland-Muller. – Mann, K.A. (ed.): Natn. Lending Libr. Sci. Technol., Boston Spa, Yorkshire. 629 pp.

Geiger, H.J., H.M. Ryser & A. Scholl, 1978. Bestimmungsschlüssel für die Larven von 18 Zuckmückenarten der Gattung *Chironomus* Meig. (Diptera, Chironomidae). – Mitt. Naturf. Ges. Bern NF 35: 89-106.

Gendron, J.M. & H. Laville, 1992. Diel emergence patterns of drifting chironomid (Diptera) pupal exuviae in the Aude river Eastern Pyrenees, France). – Neth. J. Aquat. Ecol. 26: 273-279.

Goddeeris, B.R., 1983. Het soortspecifieke patroon in de jaarcyclus van de Chironomidae (Diptera) in twee visvijvers te Mirwart (Ardennen). – Thesis Kath. Univ. Leuven. 177 pp. + bijl.

Goddeeris, B.R., 1986. Diapause in Chironomidae (Diptera) in two ponds in the Belgian Ardennes. – Proceedings of the 3rd European Congress of Entomology, Amsterdam: 1: 174.

Goddeeris, B.R., 1989. Life cycle characteristics in *Pseudochironomus prasinatus* (Staeger, 1839) (Diptera: Chironomidae). – Bulletin Koninklijk Belgisch Instituut voor Natuurwetenschappen, Entomologie 59: 165-171.

Goddeeris, B.R., A.C. Vermeulen, E. De Geest, H. Jacobs, B. Baert & F. Ollevier, 2001. Diapause induction in the third and fourth instar of *Chironomus riparius* (Diptera) from Belgian lowland brooks. – Arch. Hydrobiol. 150: 307-327.

Goedkoop, W. & R.K. Johnson, 1992. Modelling the importance of sediment bacterial carbon for profundal macroinvertebrates along a lake nutrient gradient. – Neth. J. Aquat. Ecol. 26: 477-483.

Goetghebuer, M., 1912. Etudes sur les Chironomides de Belgique. – Mém. Acad. r. Belg. Cl. Sci. 3: 1-26.

Goetghebuer, M., 1928. Diptères (Nématocères). Chironomidae. III. Chironomariae. – Faune de France 18: 1-174.

Goetghebuer, M. 1936. Les Cératopogonides et les Chironomides de Belgique au point de vue hydrobiologique. – Bull. Ann. Soc. ent. Belg. 76: 313-326.

Goetghebuer, M., 1937-1954. Tendipedidae (Chironomidae). b) Subfamilie Tendipedinae (Chironominae). A. Die Imagines. – In: Lindner, E. (ed.): Die Fliegen der palaearktischen Region 13c: 1-138.

Gouin, F., 1936. Métamorphoses de quelques Chironomides d'Alsace et de Lorraine. – Revue fr. Ent. 3: 151-173.

Griffiths, D., 1973. The structure of an acid moorland pond community. – J. Anim. Ecol. 42: 263-283.

Gripekoven, H., 1913. Minierende Tendipediden. – Arch. Hydrobiol. Suppl. 2: 129-230.

Grodhaus, G., 1980. Aestivating chironomid larvae associated with vernal pools. - In: Murray, D.A. (ed.): Chironomidae: Ecology, Systematics, Cytology and Physiology. Oxford, Pergamon Press:315-322.

Groenendijk, D., J.F. Postma, M.H.S. Kraak & W. Admiraal, 1998. Seasonal dynamics and larval drift of *Chironomus riparius* (Diptera) in a metal contaminated lowland river. – Aq. Ecol. 32: 341-351.

Grzybkowska, M., 1992. Diel drift of Chironomidae in a large lowland river (Central Poland). – Neth. J. Aquat. Ecol. 26: 355-360.

Guibé, J., 1942. Chironomes parasites de Mollusques Gastropodes. *Chironomus varus lymnaei* Guibé espèce jointive de *Chironomus varus varus* Goetgh. – Bull. biol. Fr. Belg. 76: 283-297.

Haas, H., 1956. Der Einflusz der O_2-Spannung der Wassers auf die Entwicklung der Tubuli und Analpapillen von Chironomus thummi. – Biol. Zentralblatt 75: 712-732.

Haas, H. & K. Strenzke, 1957. Experimentelle Untersuchungen über den Einflusz der ionalen Zusammensetzung des Mediums auf die Entwicklung der Analpapillen von Chironomus thummi. – Biol. Zentralblatt 76: 513-528.

Hall, R.E., 1951. Comparative observations on the chironomid fauna of a chalk stream and a system of acid streams. – J. Soc. Br. Ent. 3: 253-262.

Hamburger, K., P.C. Dall & C. Lindegaard, 1995. Effects of oxygen deficiency on survival and glycogen content of *Chironomus anthracinus* (Diptera, Chironomidae) under laboratory and field conditions. – Hydrobiologia 297: 187-200.

Hammen, H. van der, 1992. De macrofauna van Noord-Holland. – Haarlem, Prov. Noord-Holland, Dienst Ruimte en Groen. 256 pp.

Hawtin, E., 1998. Chironomid communities in relation to physical habitat. – In: Bretschko, G. & J. Helešic (eds.): Advances in river bottom ecology: 175-184. Backhuys, Leiden.

Heinis, F., 1993. : Oxygen as a factor controlling occurrence and distribution of chironomid larvae. – Thesis Amsterdam Univ. 155 pp.

Heinis, F. & T. Crommentuijn, 1989. The natural habitat of the deposit feeding chironomid larvae *Stictochironomus histrio* (Fabricius) and *Chironomus anthracinus* Zett. in relation to their responses to changing oxygen concentrations. – Acta Biol. Debr. Oecol. Hung. 3: 135-140.

Heinis, F. & T. Crommentuijn, 1992. Behavioural responses to changing oxygen concentrations of deposit feeding chironomid larvae (Diptera) of littoral and profundal habitats. – Arch. Hydrobiol. 124: 173-185.

Heinis, F. & W.R. Swain, 1996. Behavioral responses to changing oxygen concentrations in relation to distribution patterns of selected chironomid larvae. – In: Velthuis, H.H.W. (ed.), Proc. 3rd Eur. Congr. Ent. Nederl. Entomol. Ver., Amst.: 87-90.

Heinis, F., K.R. Timmermans & W.R. Swain, 1990. Short-term sublethal effects of cadmium on the filter-feeding chironomid larva *Glyptotendipes pallens* (Meigen) (Diptera). – Aq. Toxicology 16: 73-86.

Helešic, J., F. Kubiček & S. Zahrádková, 1998. The impact of regulated flow and altered temperature regime on river bed macroinvertebrates. – In: Bretschko, G. & J. Helešic (eds.): Advances in river bottom ecology: 225-243. Backhuys, Leiden.

Henrikson, L., J.B. Olofsson & H.G. Oscarson, 1982. The impact of acidification on Chironomidae (Diptera) as indicated by subfossil stratification. – Hydrobiologia 86: 223-229.

Heyn, M.W., 1992. A review of the systematic position of the North American species of the genus *Glyptotendipes*. – Neth. J. Aq. Ecol. 26: 129-137.

Higler, L.W.G., 1977. Macrofauna-cenoses on *Stratiotes* plants in Dutch broads. – Verh. R.I.N. 11: 1-86.

Hilsenhoff, W.L., 1966. The biology of *Chironomus plumosus* (Diptera:

Chironomidae) in Lake Winnebago, Wisconsin. – Ann. Ent. Soc. Amer. 59: 465-473.

Hirvenoja, M., 1962. Zur Kenntnis der Gattung *Polypedilum* Kieff. (Dipt., Chironomidae). – Ann. Ent. Fenn. 28: 127-136.

Hirvenoja, M., 1998. Delimitation of the pupae of the European species of the subgenus *Camptochironomus* (Diptera, Chironomidae). – Entomol. Fenn. 8: 215-218.

Hirvenoja, M. & P. Michailova, 1991. The karyotype, morphology and ecology of *Glyptotendipes aequalis* Kieffer (Diptera, Chironomidae). – Entomol. Fenn. 2: 87-96.

Hodkinson, I.D. & K.A. Williams, 1980. Tube formation and distribution of *Chironomus plumosus* L. (Diptera: Chironomidae) in a eutrophic woodland pond. – In: Murray, D.A. (ed.): Chironomidae: Ecology, Systematics, Cytology and Physiology. Oxford, Pergamon Press: 331-337.

Holzer, M., 1980. Die Belebung der Gewässer von Sandkiesanschwemmungen unterhalb des aktiven Stromes des Flusses March in der Obermährischen Talsenkung. – Acta Univ. Palackianae Olomuc. Fac. Rerum Nat. 67: 107-129 (in Czech).

Humphries, C.F., 1936. An investigation of the profundal and sublittoral fauna of Windermere. – J. Anim. Ecol. 5: 29-52.

Hynes, H.B.N., 1970. The ecology of running waters. – Liverpool Uiniv. Press. 555 pp.

Hynes, H.B.N. & N.K. Kaushik, 1969. The relationship between dissolved nutrient salts and protein production in submerged autumnal leaves. – Verh. internat. Verein. Limnol. 17: 95-lo3.

Iersel, P. van, 1977. Zuurstofritmiek in de Boven-Slinge bij Winterswijk. – LH Wageningen, Natuurbeheer rapp. 298. 78 pp. + app.

Int Panis, L., B. Goddeeris & R. Verheyen, 1995. On the relationship between the oxygen microstratification in a pond and the spatial distribution of the benthic chironomid fauna. – In: Cranston, P.S. (ed.): Chironomids. From genes to ecosystems. CSIRO Publ., East Melbourne, Australia: 323-328.

Int Panis, L., B. Goddeeris & R. Verheyen, 1996. On the relationship between vertical microdistribution and adaptations to oxygen stress in littoral Chironomidae (Diptera). – Hydrobiologia 318: 61-67.

Iovino, A.J. & F.D. Miner, 1970. Seasonal abundance and emergence of Chironomidae of Beaver Reservoir, Arkansas (Insecta: Diptera). – J. Kans. Ent. Soc. 43: 197-216.

Istomina, A.G., M.T. Siirin, N.V. Polukonova & I.I. Kiknadze, 2000. *Chironomus sokolovae* sp. n. iz gruppy *obtusidens* (Diptera, Chironomidae). – Zool. Zh. 79: 928-938.

Izvekova, E.I., 1980. Pitanie. – Trudy vses. gidrobiol. obshch., zool. inst. AN SSSR 23: 72-101.

Izvekova, E.I., 2000. On the new substrates for phytophilous chironomid larvae during and after summer increase of the water level in reservoir. - In: Hoffrichter, E.O. (ed.): Late 20th century research on Chironomidae. Shaker verlag, Aachen: 309-312.

Izvekova, E.I., A.A. Kuzminych & S.G. Nikolaev, 1996. Chironomidy nekotorych malych rek bassejna reki Oki i vozmozjnostj ispolzovaniya ich lichinok v kachestve indikotorov zagryazneniya. – In: Shobanov, N.A. & T.D. Zinchenko (eds.): Ekologiya, evolyutsiya i sistematika chironomid. Tolyatti/ Borok: 132-137. (In Russian)

Izvekova, E. & A.A. Lvova-Katchanova, 1972. Sedimentation of suspended matter by *Dreissena polymorpha* Pallas and its subsequent utilization by Chironomidae larvae. – Polsk. Archiw. Hydrobiol. 19: 203-210.

Janecek, B.F.U., 1995. *Tanytarsus niger* Andersen (Diptera: Chironomidae) and the chironomid community in Gebhartsteich, a carp pond in northern Austria. – In: Cranston, P. (ed.): Chironomids, from genes to ecosystems. CSIRO Publications, East Melbourne: 281-296.

Janković, M., 1971. Anzahl der Generationen der Art *Chironomus plumosus* in den Karpfenteichen Serbiens. – Limnologica (Berlin) 8: 203-210.

Janse, J. & D. Monnikendam, 1982. Macrophytenen macrofaunagemeenschappen in vaarten. – Basis-rapport proj. E.K.O.O. 2: 1-74 + app.

Janzen, L., 2003. Typisierung und Bewertung von Fliessgewässern mit Hilfe der Chironomidae (Zuckmücken)-Fauna anhand des AQEM Datensatzes. – Diplomarbeit Univ. Duisburg-Essen Abt. Hydrobiol. 1-117.

Jernelov, A., M. Magell & A. Svensson, 1981. Adaptation to an acid environment in Chironomus riparius from Smoking Hills N.W.T., Canada. – Holarct. Ecol. 4: 116-119.

Johnson, R.K., 1984. Distribution of *Chironomus plumosus* and *C. anthracinus* with respect to sediment parameters in mesotrophic Lake Erken. – Verh. Internat. Verein. Limnol. 22: 750-758.

Johnson, R.K., 1986. Life histories and coexistence of *Chironomus plumosus* (L.), *C. anthracinus* Zett. (Diptera: Chironomidae) and *Pontoporeia affinis* (Crustacea: Amphipoda) in mesotrophic Lake Erken. – Thesis Uppsala Univ.

Johnson, R.K., 1989. Classification of profundal chironomid communities in oligotrophic/humic lakes of Sweden using environmental data. – Acta Biol. Debr. Oecol. Hung. 3: 167-175.

Johnson, R.K. & B. Pejler, 1987. Life histories and coexistence of the two profundal *Chironomus* species in Lake Erken, Sweden. – Ent. scand. Suppl. 29: 233-238.

Jónasson, P.M., 1972. Ecology and production of the profundal benthos in relation to phy-

R

toplankton in Lake Esrom. – Oikos Suppl. 14: 1-148.
Jónasson, P.M. & J. Kristiansen, 1967. Primary and secondary production in Lake Esrom. Growth of *Chironomus anthracinus* in relation to seasonal cycles of phytoplankton and dissolved oxygen. – Int. Revue ges. Hydrobiol. 52: 163-217.
Jónsson, E., 1987. Flight periods of aquatic insects at Lake Esrom, Denmark. – Arch. Hydrobiol. 110: 259-274.
Kajak, Z., 1963. The effect of experimentally induced variations in the abundance of *Tendipes plumosus* L. larvae on intraspecific and interspecific relations. – Ekologia Polska (A) 11: 355-367.
Kajak, Z., 1987. Determinants of maximum biomass of benthic Chironomidae (Diptera). – Ent. scand. Suppl. 29: 303-308.
Kajak, Z., K. Dusoge, A. Hillbricht-Ilkowska, E. Pieczynski, A. Prejs, I. Spodniewska & T. Weglenska, 1972. Influence of the artificially increased fish stock on the lake biocenosis. – Verh. Internat. Verein. Limnol. 18: 228-235.
Kalugina, N.S., 1958. On the habitats and feeding of larvae of Glyptotendipes glaucus Mg. (Diptera, Chironomidae) from the Ucha water reservoir. – Zool. Zh. 37: 1045-1057. (in Russian).
Kalugina, N.S., 1959. O nekotorykh vozrastnykh izmeneniyakh v stroenii i biologii lichinok chironomid (Diptera Chironomidae). – Trudy vses. gidrobiol. obshch. Akad. Nauk SSSR 9: 85-107.
Kalugina, N.S., 1960. Die ontogenetischen Veränderungen in der Morphologie der Chironomidenlarven. – Verh. XI. Internat. Kongr. Ent. 1: 182-184.
Kalugina, N.S., 1961. Taxonomy and development of *Endochironomus albipennis* Mg., *E. tendens* F. and *E. impar* Walk. (Diptera, Tendipedidae). – Ent. Obozr. 40: 900-919. (in Russian)
Kalugina, N.S., 1963. Systematics and development of Glyptotendipes glaucus Mg. and G. gripekoveni Kieff. (Diptera, Chironomidae). – Ent. Obozr. 42: 889-908. (in Russian).
Kalugina, N.S., 1963a. Mesta obitaniya lichinok i smena pokolenii u semi vidov *Glyptotendipes* Kieff. i *Endochironomus* Kieff. (Diptera, Chironomidae) iz Uchinskogo vodokhranilishcha. – Mozh. Vodokhr. Izd. Mosk. Univ.: 173-212.
Kalugina, N.S., 1970. A new species of the genus *Lipiniella* (Diptera, Chironomidae). – Zool. Zh. 49: 1034-1038.
Kalugina, N.S., 1971. Chironomidae in piscicultural ponds (Diptera, Chironomidae). – Limnologica 8: 211-213.
Kalugina, N.S., 1972. Midges (Diptera, Chironomidae) of fish-ponds. – Proc. 13 Int. Congr. Ent. Moscow 1968, 3: 460-461.
Kalugina, N.S., 1979. Morpho-ekologicheskie gruppy roda *Glyptotendipes* Kieff. (Diptera, Chironomidae). – In: Narchuk, E.P. (ed.): Ecological and morphological principles of Diptera systematics (Insecta): 33-35. Izd. Zool. Inst. Ac. Sc. SSSR.
Kashirskaya, E.V., 1989. Chironomid larvae of the Volgograd water reservoir (fauna, ecology, role in communities). – Acta Biol. Debr. Oecol. Hung. 3: 199-207.
Kawecka, B. & A. Kownacki, 1974. Food conditions of Chironomidae in the River Raba. – Ent. Tidskr. 95 Suppl.: 120-128.
Ketelaars, H.A.M., A.M.J.P. Kuijpers & L.W.C.A. van Breemen, 1992. Temporal and spatial distribution of chironomid larvae and oligochaetes in two Dutch storage reservoirs. – Neth. J. Aquat. Ecol. 26: 361-370.
Keyl, H.-G., 1962. Chromosomenevolution bei *Chironomus* II. Chromosomenumbauten und phylogenetische Beziehungen der Arten. – Chromosoma (Berl.) 13: 464-514.
Kiknadze, I., I. Kerkis, A. Shilova & M. Filippova, 1989. A review of the species of the genus *Lipiniella* Shilova(Diptera). I. *L. arenicola* Shil. and *L. moderata* Kalug. - Acta Biol. Debr. Oecol. Hung. 2: 115-128.
Kiknadze, I.I., A.I. Shilova, I.E. Kerkis, N.A. Shobanov, N.I. Zelentsov, L.P. Grebenyuk, A.G. Istomina & V.A. Prasolov, 1991. – Kariotipy i morfologiya lichinok tribby Chironomini. Atlas. – Novosibirsk: Nauka, Sib. otdelenie. 115 pp.
Klaren, P., 1987. De levenscyclus en auoecologische aspekten van *Lipiniella arenicola* Shilova (Diptera: Chironomidae) in het Haringvliet. – Lab. Aq. Oecol. KU Nijmegen / RWS-RIZA. 85 pp. + app.
Klink, A.G., 1983. Makro-Evertebraten in en langs 2 zandputten in de randmeren (Gooimeer, Huizen en Veluwemeer, Harderwijk). – Rapp. Hydrobiol. Adviesbur. Klink. 16 pp.
Klink, A., 1985. Hydrobiologie van de Grensmaas. Huidig funktioneren, potenties en bedreigingen. – Rapp. Meded. Hydrobiol. Adviesbur. Klink 15: 1-38, app. pp.1-111.
Klink, A., 1985a. Een inventarisatie van volwassen Chironomidae bij Kampen (IJssel). - Rapp. Meded. Hydrobiol. Adviesbur. Klink 21: 1-5 + 2 app.
Klink, A., 1986. Geschiedenis van de verzuring in Nederland. – Rapp. Meded. Hydrobiol. Adviesbur. Klink 27: 1-43, app. 1-10.
Klink, A., 1986a. Palaeolimnologisch onderzoek naar de geschiedenis van kopafwijkingen bij muggelarven in de grote Nederlandse rivieren. - Rapp. Meded. Hydrobiol. Adviesbur. Klink 25: 1-15, app. 1-21.
Klink, A., 1989. The lower Rhine: Palaeoecological analysis. – In: Petts, G.E. (ed.): Historical change of large alluvial rivers: Western Europe: 183-201.
Klink, A., 1991. Maas 1986 – 1990. Evaluatie van

5 jaar hydrobiologisch onderzoek van makro-evertebraten. - Rapp. Meded. Hydrobiol. Adviesbur. Klink 39: 1-38 + app.

Klink, A., 1994. Makro-evertebraten in relatie tot bodemvormingsprocessen in de Nieuwe Merwede, Hollandsch Diep en Dordtsche Biesbosch. – Rapp. Meded. Hydrobiol. Adviesbur. Klink 49: 1-70 + app.

Klink, A., 2001. Determinatiesleutel voor de larven van de in Nederland voorkomende soorten *Polypedilum*. - Rapp. Meded. Hydrobiol. Adviesbur. Klink 74: 1-10.

Klink, A., 2002. Zandsuppletie in kribvakken in de Waal. Effecten op de macrofauna 2. Een jaar na baggeren en suppleren. - Rapp. Meded. Hydrobiol. Adviesbur. Klink 78: 1-27.

Klink, A., 2008. Monitoring aquatische macrofauna in de Kaliwaal en de Leeuwense Waard (2007). - Rapp. Meded. Hydrobiol. Adviesbur. Klink 98: 1-21 + app.

Klink, A.G. & H. Moller Pillot, 1982. Onderzoek aan de makro-evertebraten in de grote Nederlandse rivieren. – Wageningen/Tilburg (private publ.). 57 pp.

Klink, A.G. & H.K.M. Moller Pillot, 2003. Chironomidae larvae. Key to the higher taxa and species of the lowlands of Northwestern Europe. - CD-ROM, Expert center for Taxonomic Information, Amsterdam. (out of trade)

Klink, A., J. Mulder, M. Wilhelm & M. Jansen, 1995. Ecologische ontwikkelingen in de wateren van Blauwe Kamer 1989 – 1995. Doorzicht afgenomen en inzicht toegenomen. - Rapp. Meded. Hydrobiol. Adviesbur. Klink 58: 1-79 + app.

Klötzli, A.M., 1974. Revision der Gattung *Chironomus* Meig. V. *Chironomus nuditarsis* Keyl. Morphologische Beschreibung und Vergleich mit ähnlichen Arten. – Arch. Hydrobiol. 74: 68-81.

Koehn, T. & C. Frank, 1980. Effect of thermal pollution on the chironomid fauna in an urban channel. – In: Murray, D.A. (ed.): Chironomidae: Ecology, Systematics, Cytology and Physiology. Oxford, Pergamon Press: 187-194.

Kolosova, N.N. & I.S.M. Lyachov, 1957. The larva *Einfeldia* of the group *carbonaria* Mg. f.l. *reducta* Tshern. (Diptera, Tendipedidae), and its biology. – Zool. Zh. 7: 1101-1104.

Kondo, S. & S. Hamashima, 1992. Habitat preferences of four chironomid species associated with aquatic macrophytes in an irrigation reservoir. – Neth. J. Aquat. Ecol. 26: 371-377.

Konstantinov, A.S., 1957. On the taxonomy of the mosquito larvae of the genus *Chironomus* Meig. – Zool. Zh. 36: 885-893. (In Russian).

Konstantinov, A.S., 1961. Feeding in some predatory chironomid larvae. – Vopr. Ichthiol. 1 (3) 20: 570-582. (Russian with english summary).

Koskenniemi, E., 1992. The role of chironomids (Diptera) in the profundal macrozoobenthos in Finnish reservoirs. – Neth. J. Aquat. Ecol. 26: 503-508.

Koskenniemi, E. & L. Paasivirta, 1987. The chironomid (Diptera) fauna in a Finnish reservoir during its first four years. – Ent. scand. Suppl. 29: 239-246.

Koskenniemi, E. & P. Sevola, 1989. Winter regulation effects on littoral chironomids in Hungarian reservoirs. – Acta Biol. Debr. Oecol. Hung. 3: 215-218.

Koskinen, R., 1969. Larval growth in *Chironomus salinarius* Kieff. (Diptera, Chironomidae) in Western Norway. – Ann. zool. fenn. 6: 266-268.

Kouwets, F.A.C. & C. Davids, 1984. The occurrence of chironomid imagines in an area near Utrecht (the Netherlands) and their relations to water mite larvae. – Arch. Hydrobiol. 99: 296-317.

Kownacki, A., 1989. Taxocenes of Chironomidae as an indicator for assessing the pollution of rivers and streams. – Acta Biol. Debr. Oecol. Hung. 3: 219-230.

Kreamer, G. & A.D. Harrison, 1984. Seasonal and diurnal migration of larval *Sergentia coracina* (Diptera: Chironomidae) in Lake Matamek, Eastern Quebec. – Verh. Internat. Verein. Limnol. 22: 388-394.

Krebs, B.P.M., 1978. Waarnemingen aan soortensamenstelling en populatiedynamiek van chironomidenlarven (Diptera, Chironomidae) in een brakke sloot. - Delta Inst. Hydrobiol. Onderz., Rapp. en Versl. 1978-6. 1-29.

Krebs, B.P.M., 1979. *Microchironomus deribae* (Freeman, 1957) (Diptera, Chironomidae) in the Delta region of the Netherlands. – Hydrobiological Bulletin (Amsterdam) 13: 144-151.

Krebs, B.P.M., 1981. Aquatische macrofauna van binnendijkse wateren in het Deltagebied. I. Zuid-Beveland. – Delta Inst. Hydrobiol. Onderz., Rapp. & Versl. 1981-8: 1-158.

Krebs, B.P.M., 1982. Chironomid communities of brackish inland waters. – Chironomus 2 (3): 19-23.

Krebs 1984. Aquatische macrofauna van binnendijkse wateren in het Deltagebied. II. Zeeuws-Vlaanderen, oostelijk deel. – Delta Inst. Hydrobiol. Onderz., Rapp. en Versl. 1984-2: 1-124.

Krebs, B.P.M., 1985. Aquatische macrofauna van binnendijkse wateren in het Delta-gebied. III. Noord-Beveland, Tholen en St. Philipsland. – Delta Inst. Hydrobiol. Onderz., Rapp. en versl. 1985-9: 1-58.

Krebs, B.P.M., 1988. Some records of two rare chironomid species in the Netherlands. – Spixiana Suppl. 14: 29-33.

Krebs, B.P.M., 1990. Aquatische macrofauna van binnendijkse wateren in het Delta-gebied.

R

IV: Schouwen-Duiveland. – Delta Inst. Hydrobiol. Onderz., Rapp. en versl. 1990-07: 1-124.

Krebs, B.P.M. & H.K.M. Moller Pillot, in prep. Influence of some environmental factors on the abundance of Chironomidae in a predominantly brackish water area.

Kreuzer, R., 1940. Limnologisch-ökologische Untersuchungen an holsteinischen Kleingewässern. – Arch. Hydrobiol. Suppl. 10: 359-572.

Krieger-Wolff, E. & W. Wülker, 1971. Chironomiden (Diptera) aus der Umgebung von Freiburg i. Br. (mit besonderer Berücksichtigung der Gattung *Chironomus*). – Beitr. naturk. Forschung Südw.Dd. 30: 133-145.

Kruglova, V.M., 1940. Novye lichinki khironomid (triba Chironomariae) iz zapadnoi Sibiri. – Trudy Biol. inst. Tomski Gosudarstv. Univ. 7: 219-227.

Kruseman, G., 1933. Tendipedidae Neerlandicae. Pars 1. Genus Tendipes cum generibus finitimis. – Tijdschr. Entomol. 76: 119-216.

Kruseman, G., 1934. Welche Arten von Chironomus s.l. sind Brackwassertiere? – Verh. Int. Ver. Limnol. 6: 163-165.

Kurazhskovskaya, T.N., 1969. Stroennie slyunnykh zhelez lichinok khironomid. – Trudy Inst. Biol. vnutr. Vod AN SSSR 19: 185-195.

Kurazhskovskaya, T.N., 1971. On the biology of *Glyptotendipes varipes*. – Limnologica 8: 219-220.

Kuijpers, A.M.J.P., H.A.M. Ketelaars & L.W.C.A. van Breemen, 1992. Chironomid pupal exuviae and larvae of two storage reservoirs in the Netherlands. – Neth. J. Aquat. Ecol. 26: 379-383.

Langdon, P.G., Z. Ruiz, K.P. Brodersen & I.D.L. Foster, 2006. Assessing lake eutrophication using chironomids: understanding the nature of community response in different lake types. – Freshw. Biol. 51: 562-577.

Langton, P.H., 1991. A key to pupal exuviae of West Palaearctic Chironomidae. – P.H. Langton, Huntingdon (private publ.). 386 pp.

Langton, P.H. & L.C.V. Pinder, 2007. Keys to the adult male Chironomidae of Britain and Ireland. – Freshw. Biol. Ass. Sc. Publ. 64: 239 + 168 pp., 276 figs.

Langton, P.H. & H. Visser, 2003. Chironomidae exuviae – a key to pupal exuviae of the West Palaearctic Region. – CD-ROM, Expert center for Taxonomic Information, Amsterdam. (out of trade)

Laville, H., 1971. Recherches sur les Chironomides (Diptera) lacustres du massif de Néouvielle (Hautes-Pyrénées). – Ann. Limnol. 7: 173-332.

Lehmann, J., 1969. Zur Ökologie und Verbreitung dreier für Schleswig-Holstein neuer Chironomiden-arten (Diptera, Nematocera). – Faun.-ökol. Mitt. 3: 262-268.

Lehmann, J. 1970. Revision der europäischen Arten (Imagines) der Gattung *Parachironomus* Lenz (Diptera, Chironomidae). – Hydrobiologia 33: 129-158.

Lehmann, J., 1971. Die Chironomiden der Fulda. – Arch. Hydrobiol. Suppl. 37: 466-555.

Lellák, J., 1968. Positive Phototaxis der Chironomiden-Larvulae als regulierender Faktor ihrer Verteilung in stehenden Gewässern. – Ann. Zool. Fenn. 5: 84-87.

Lenz, F., 1951. Neue Beobachtungen zur Biologie der Jugendstadien der Tendipedidengattung *Parachironomus* Lenz. – Zool. Anz. 147: 95-111.

Lenz, F., 1954. Beitrag zur Kenntnis der Ernährungsweise der Tendipedidenlarven. – Zool. Anz. 153: 197-204.

Lenz, F., 1954-62. Tendipedinae (Chironominae). b) Subfamilie Tendipedinae. B. Die Metamorphose der Tendipedinae. – In: Lindner, E. (ed.): Die Fliegen der palaearktischen region 13c: 139-260.

Lenz, F., 1959. Zur Metamorphose und Ökologie der Tendipediden-Gattung *Paracladopelma*. –Arch. Hydrobiol. 55: 429-449.

Lenz, F., 1959a. Die Metamorphose der Gattung *Cryptotendipes* Lenz. – Dt. ent. Z. 6: 238-250.

Lenz, F., 1960. Die Metamorphosestadien der Tendipedidengattung *Demicryptochironomus* Lenz. – Abh. naturw. Ver. Bremen 35: 450-463.

Lenz, F. 1960a. Die Tendipediden-Gattung *Cryptocladopelma* Lenz in oberitalienischen Gewässern. – Mem. Ist. Ital. Idrobiol. 12: 165-184.

Leuven, R.S.E.W., J.A. van der Velden, J.A.M. Vanhemelrijk & G. van der Velde, 1987. Impact of acidification on chironomid communities in poorly buffered waters in the Netherlands. – Ent. scand. Suppl. 29: 269-280.

Lindeberg, B., 1958. A new trap for collecting emerging insects from small rock-pools, with some examples of the results obtained. – Suom. Hyönt. Aikak. 24: 186-191.

Lindeberg, B., 1959. *Chironomus lugubris* Zett. (Dipt. Chironomidae) from Tvärminne, S.W. Finland. – Ann. Ent. Fenn. 25: 224-227.

Lindeberg, B. & T. Wiederholm, 1979. Notes on the taxonomy of European species of *Chironomus* (Diptera: Chironomidae). – Ent. scand. Suppl. 10: 99-116.

Lindegaard, C., 1995. Chironomidae (Diptera) of European cold springs and factors influencing their distribution. – J. Kansas Ent. Soc. 68 (2) suppl.: 108-131.

Lindegaard, C. & E. Jónsson, 1983. Succession of Chironomidae (Diptera) in Hjarbæk Fjord, Denmark, during a period with change from brackish water to freshwater. – Mem. Am. ent. Soc., Philadelphia 34: 169-185.

Lindegaard, C. & E. Jónsson, 1987. Abundance,

population dynamics and high production of Chironomidae (Diptera) in Hjarbæk Fjord, Denmark, during a period of eutrophication. – Ent. scand. Suppl. 29: 293-302.

Lindegaard-Petersen, C., 1972. An ecological investigation of the Chironomidae (Diptera) from a Danish lowland stream (Linding Å). Arch. Hydrobiol. 69: 465-507.

Luferov, V.P., 1972. The role of light in the populating of water bodies by epibiotic chironomid larvae. – Proc. 13 Int. Congr. Ent. Moscow 1968, 3: 469-470.

Macan, T.T., 1949. Survey of a moorland fishpond. – J. Anim. Ecol. 18: 160-186.

Mackey, A.P., 1976. Quantitative studies on the Chironomidae (Diptera) of the Rivers Thames and Kennet. II. The Thames flint zone. – Arch. Hydrobiol. 78: 310-318.

Mackey, A.P., 1979. Trophic dependencies of some larval Chironomidae (Diptera) and fish species in the River Thames. – Hydrobiologia 62: 241-247.

Maenen, M.M.J., 1983. Inleidend onderzoek naar de verspreiding van Chironomidae in het IJsselmeergebied April tot en met September 1982. – Rep. Lab. Aquat. Oecol. Kath. Univ. Nijmegen, 144: 1-50.

Marlier, G., 1951. La biologie d'un ruisseau de plaine, le Smohain. – Verh. Kon. Belg. Inst. Natuurw. 114: 1-98. Brussels.

Martin, J., A. Blinov, K. Alieva & K. Hirabayashi, 2007. A molecular phylogenetic investigation of the genera closely related to *Chironomus* Meigen (Diptera: Chironomidae). – In: Andersen, T. (ed.): Contributions to the systematics and ecology of aquatic Diptera – a tribute to Ole A. Saether: 193-203. Caddis Pr., Columbus.

Mason, C.F. & R.J. Bryant, 1975. Periphyton production and grazing by chironomids in Alderfen Broad, Norfolk. – Freshw. Biol. 5: 271-277.

Matěna, J., 1986. Übersicht der bisher in Böhmen gefundenen Arten der Gattung *Chironomus* Meig. (Diptera, Chironomidae) mit Bemerkungen zu ihrer Ekologie. – Dipterol. bohemosl. 4: 35-37. (In Czechian).

Matěna, J., 1989. Seasonal dynamics of a *Chironomus plumosus* (L.) (Diptera, Chironomidae) population from a fish pond in Soutern Bohemia. – Int. Revue ges. Hydrobiol. 74: 599-610.

Matěna, J., 1990. Succession of *Chironomus* Meigen species (Diptera, Chironomidae) in newly filled ponds. – Int. Revue ges. Hydrobiol. 75: 45-57.

Matěna, J. & J. Frouz, 2000. Distribution and ecology of *Chironomus* species in the Czech Republic (Diptera, Chironomidae). - In: Hoffrichter, E.O. (ed.): Late 20th century research on Chironomidae. Shaker Verlag, Aachen: 543-548.

McLachlan, A.J., 1974. The development of chironomid communities in a new temperate impoundment. – Ent. Tidskr. Suppl. 95: 162-171.

McLachlan, A.J., 1976. Factors restricting the range of *Glyptotendipes paripes* Edwards (Diptera: Chirono-midae) in a bog lake. – J. Anim. Ecol. 45: 105-113.

McLachlan, A.J., 1977. Some effects of tube shape on the feeding of *Chironomus plumosus* L. (Diptera: Chironomidae). – J. Anim. Ecol. 46: 139-146.

McLachlan, A.J. & Cantrell, M.A., 1976. Sediment development and its influence on the distribution and tube structure of *Chironomus plumosus* L. (Chironomidae, Diptera) in a new impoundment. – Freshw. Biol. 6: 437-443.

McLachlan, A.J. & S.M. Mclachlan, 1975. The physical environment and bottom fauna of a bog lake. – Arch. Hydrobiol. 76: 198-217.

McLachlan, A.J., L.J. Pearce & J.A. Smith, 1979. Feeding interactions and cycling of peat in a bog lake. – J. Anim. Ecol. 48: 851-861.

Menzie, C.A., 1980. The chironomid (Insecta: Diptera) and other fauna of a *Myriophyllum spicatum* L. plant bed in the Lower Hudson River. – Estuaries 3: 38-54.

Merks, A.G.A. & J.W. Rijstenbil, 1981. Saline seepage and vertical distribution of oxygen in a brackish ditch. – Hydrobiol. Bull. 15: 111-121.

Meuche, A., 1939. Die Fauna im Algenbewuchs. Nach Untersuchungen im Litoral ostholsteinischer Seen. – Arch. Hydrobiol. 31: 501-507.

Michailova, P.V., 1987. Comparative karyological studies of three species of the genus *Glyptotendipes* Kieff. (Diptera, Chironomidae) from Hungary and Bulgaria and *Glyptotendipes salinus* sp. n. from Bulgaria. – Folia Biol. (Kraków) 35: 43-56.

Michailova, P.V., 1988. A review of the genus *Polypedilum* Kieffer. The cytotaxonomy of *Polypedilum aberrans* Tshernovskij (Diptera, Chironomidae). – Spixiana Suppl. 14: 239-246.

Michailova, P.V., 1989. The polytene chromosomes and their significance to the systematics of the family Chironomidae, Diptera. – Acta Zool. Fenn. 186: 1-107.

Michailova, P. & R. Contreras-Lichtenberg, 1995. Contribution to the knowledge of *Glyptotendipes pallens* (Meigen, 1804) and *Glyptotendipes glaucus* (Meigen, 1818). – Ann. Naturhist. Mus. Wien 97 B: 395-410.

Michiels, S., 1999. Die Chironomidae (Diptera) der unteren Salzach. – Lauterbornia 36: 45-53.

Michiels, S., 2004. Die Zuckmücken (Diptera: Chironomidae) der Elz – ein Beitrag zur Limnofauna des Schwarzwaldes. – Mitt. bad. Landesver. Naturkunde u. Naturschutz N.F. 18: 111-128.

Mol, A.W.M., M. Schreijer & P. Vertegaal, 1982. De makrofauna van de Maarsseveense plassen. – Rapp. Rijksinst. Natuurbeheer: 1-134, 1-187.

Mol, A.W.M., M. Schreijer & P. Vertegaal,

R

1982a. De trofiegradiënt in de Maarsseveense Zodden. – Rapp. Rijksinst. Natuurbeheer: 1-179.
Moldován, J., 1987. Description of a multispecies *Chironomus* community (Diptera: Chironomidae) at an experimental sewage-treatment plant. – Ent. scand. Suppl. 29: 381-386.
Moller Pillot, H.K.M., 1971. Faunistische beoordeling van de verontreiniging in laaglandbeken. – Tilburg. 285 pp.
Moller Pillot, H.K.M., 1984. De larven der Nederlandse Chironomidae (Diptera). Inleiding, Tanypodinae en Chironomini. – Nederl. faun. Meded. 1A. 277 pp.
Moller Pillot, H., 2003. Hoe waterdieren zich handhaven in een dynamische wereld. – St. Noordbr. Landsch., Haaren. 182 pp.
Moller Pillot, H.K.M. & R.F.M. Buskens, 1990. De larven der Nederlandse Chironomidae (Diptera). Autoekologie en verspreiding. – Nederl. faun. Meded. 1C. 87 pp.
Moller Pillot, H. & B. Krebs, 1981. Concept van een overzicht van de oekologie van chironomidelarven in Nederland. – Stencil, s.l., 41 pp.
Moller Pillot, H.K.M., H.J. Vallenduuk & A. bij de Vaate, 2000. Bijdrage tot de kennis over de Nederlandse Chironomidae (vedermuggen): de larven van het genus *Glyptotendipes* in West-Europa. – Lelystad, RIZA rapport 97.052. 58 pp.
Moller Pillot, H.K.M. & S.M. Wiersma, 1997. De larven van het geslacht *Einfeldia* Kieffer, 1924: nomenclatuur en tabel tot de soorten (Diptera: Chironomidae). – Nederl. faun. Meded. 7: 11-14.
Monakov, A.V., 1972. Review of studies on feeding of aquatic invertebrates conducted at the Institute of Biology of Inland waters, Academy of Science, USSR. – J. Fish. Res. Board Can. 29: 363-383.
Monakov, A.V., 2003. Feeding of freshwater invertebrates. – Kenobi Prod., Ghent. 373 pp.
Moog, O. (ed.), 1995. Fauna aquatica Austriaca. Katalog zur autökologischen Einstufung aquatischer Organismen Österreichs. – Wien, Bundesministe-rium Land- und Forstwirtschaft, loose-leaf.
Moore, J.W. , 1979. Factors influencing algal consumption and feeding rate in *Heterotrissocladius changi* Saether and *Polypedilum nubeculosum* (Meigen) (Chironomidae: Diptera). – Oecologia 40: 219-227.
Moore, J.W., 1979a. Some factors influencing the distribution, seasonal abundance and feeding of subarctic Chironomidae (Diptera). – Arch. Hydrobiol. 85: 302-325.
Morduchai-Boltovskoi, F.D., 1961. Die Entwicklung der Bodenfauna in den Stauseen der Wolga. – Verh. Internat. Verein. Limnol. 14: 647-651.
Morozova, E.E., 2005. Chironomidy roda *Cryptochironomus* (Diptera, Chironomidae) volzhskogo bassejna: ekologiya, morfologiya, kariosistematika, indikatornoe znachenie. – Avtoref. diss. Sarat. gosudarstv. Univ. 38 pp.
Morozova, E.E., 2005a. The indicator role of species of genus *Cryptochironomus* (Diptera, Chironomidae). – In: Chirov, P.A. & V.V. Anikin (eds.): Entomological and parasitological investigations in Volga region 4: 108-117. Saratov Univ. Press. (In Russian)
Mossberg, P. & P. Nyberg, 1980. Bottom fauna of small acid forest lakes. – Report Inst. Freshw. Res. Drottningholm 58: 77-87.
Müller-Liebenau, J., 1956. Die Besiedlung der Potamogeton-zone ostholsteinischer Seen. Arch. Hydrobiol. 52: 470-606.
Mundie, J.H., 1957. The ecology of Chironomidae in storage reservoirs. – Trans. R. ent. Soc. Lond. 109: 149-232.
Mundie, J.H., 1965. The activity of benthic insects in the water mass of lakes. – Proc. XIIth Intern. Congr. Entomol., London: 410.
Munsterhjelm, G., 1920. Om Chironomidernas Ägglägging och Äggrupper. – Acta Soc. Fauna Flora Fenn. 47: 1-174.
Nes, E.H. van & H. Smit, 1993. Multivariate analysis of macrozoobenthos in Lake Volkerak-Zoommeer (The Netherlands): changes in an estuary before and after closure. – Arch. Hydrobiol. 127: 185-203.
Neubert, I. & C. Frank, 1980. Response of chironomid fauna after autumn overturn in a eutrophic lake. – In: Murray, D.A. (ed.): Chironomidae: Ecology, Systematics, Cytology and Physiology. Oxford, Pergamon Press: 283-290.
Neumann, D., 1961. Der Einfluss des Eisenangebotes auf die Hämoglobinsynthese und die Entwicklung der *Chironomus*-Larve. – Z. Naturf. 16b: 820-824.
Neumann, D., M. Kramer, I. Raschke & B. Gräfe, 2001. Detrimental effects of nitrite on the development of benthic *Chironomus* larvae, in relation to their settlement in muddy sediments. – Arch. Hydrobiol. 153: 103-128.
Nijboer, R. & P. Verdonschot (red.), 2001. Zeldzaamheid van de macrofauna van de Nederlandse binnenwateren. – Werkgr. Ecol. Waterbeheer, themanr. 19: 1-77. (Basal data unpublished).
Nocentini, A., 1985. Chironomidi, 4 (Diptera: Chironomidae: Chironominae, larve). – Guide per il recogniscimento delle specie animali delle acque interne Italiane 29: 1-186.
Nolte, U., 1993. Egg masses of Chironomidae (Diptera). A review, including new observations and a preliminary key. – Ent. scand. Suppl. 43: 1-75.
Olafsson, J.S., 1992. Vertical microdistribution of benthic chironomid larvae within a selection of the littoral zone of a lake. – Neth. J.

Aquat. Ecol. 26: 397-403.
Olafsson, J.S., 1992a. A comparative study on mouthpart morphology of certain larvae of Chironomini (Diptera: Chironomidae) with reference to the larval feeding habits. – J. Zool. Lond. 228: 183-204.
Orendt, C., 1993. Vergleichende Untersuchungen zur Ökologie litoraler, benthischer Chironomidae und anderer Diptera (Ceratopogonidae, Chaoboridae) in Seen des nördlichen Alpenvorlandes. – Thesis München Univ. 315 pp.
Orendt, C., 1999. Chironomids as bioindicators in acidified streams: a contribution to the acidity tolerance of chironomid species, with a classification in sensitivity classes. – Internat. Rev. Hydrobiol. 84: 439-449.
Orendt, C., 2002. Die Chironomidenfauna des Inns bei Mühldorf (Oberbayern). – Lauterbornia 44: 109-120.
Orendt, C., 2002a. Biozönotische Klassifizierung naturnaher Flussabschnitte des nördlichen Alpenvorlandes auf der Grundlage der Zuckmücken Lebens-gemeinschaften (Diptera: Chironomidae). – Lauterbornia 44: 121-146.
Otten, J.H., 1986. Macrofauna onderzoek aan een vervuilingsgradient in de Groote Beerze met een vergelijking van kunstmatig substraat en netmonsters. – Rapp. LH Vakgr. Natuurbeheer 899: 1-80.
Otto, C.-J., 1991. Benthonuntersuchungen am Belauer See (Schleswig-Holstein): eine ökologische, phaenologische und produktionsbiologische Studie unter besonderer Berücksichtigung der merolimnischen Insekten. – Thesis Kiel. 139 pp.
Oyewo, E.A. & O.A. Saether, in prep. Revision of *Polypedilum (Pentapedilum)* Kieffer and *Ainuyusurika* Sasa et Shirasaki (Diptera: Chironomidae).
Paasivirta, L., 1974. Abundance and production of the larval Chironomidae in the profundal of a deep, oligotrophic lake in Southern Finland. – Ent. Tidskr. 95 Suppl.: 188-194.
Paasivirta, L., 1976. Suomunjärven (Lieksa) Pohjaeläimistön koostumus, biomassa ja tuotanto. – Univ. Joensuu, Karelian Inst. Publ. 18: 1-17.
Paasivirta, L. & E. Koskenniemi, 1980. The Chironomidae (Diptera) in two polyhumic reservoirs in western Finland. – In: Murray, D.A. (ed.): Chironomidae: Ecology, Systematics, Cytology and Physiology. Oxford, Pergamon Press: 233-238.
Pagast, F., 1931. Chironomiden aus der Bodenfauna des Usma-Sees in Kurland. – Folia Zool. Hydrobiol. 3: 199-248.
Pagast, F., 1934. Über die Metamorphose von *Chironomus xenolabis* Kieff., eines Schwammparasiten (Dipt.). – Zool. Anz. 105: 155-158.
Palawski, D.U., J.B. Hunn, D.N. Chester & R.H. Wiedmeyer, 1989. Interactive effects of acidity and aluminium exposure on the life cycle of the midge Chironomus riparius (Diptera). – J. Freshw. Ecol. 5: 155-162.
Palmén, E., 1962. Studies on the ecology and phenology of the Chironomids of the Northern Baltic. 1. Allochironomus crassiforceps K. – Ann. Ent. Fenn. 28: 137-168.
Palmén, E. & L. Aho, 1966. Studies on the ecology and phenology of the Chironomidae (Dipt.) of the Northern Baltic. 2. Camptochironomus Kieff. and Chironomus Meig. – Ann. Zool. Fenn. 3: 217-244.
Palomäki, R., 1989. The chironomid larvae in the different depth zones of the littoral in some Finnish lakes. – Acta Biol. Debr. Oecol. Hung. 3: 257-266.
Pankratova, V.Ya., 1964. Lichinki Tendipedid reki Oki. – Trudy Zool. Inst. AN SSSR 32: 189-207.
Pankratova, V.Ya., 1983. Lichinki i kukolki komarov podsemejstva Chironominae fauny SSSR (Diptera, Chironomidae = Tendipedidae). – Opredel. po faune SSSR 134: 1-295.
Parma, S. & C.H. Borghouts-Biersteker, 1978. Diurnal fluctuations in salinity, oxygen and chlorophyll in a shallow brackish ditch (Abstract). – Verh. Internat. Ver. Limnol. 20: 2186.
Parma, S. & B.P.M. Krebs, 1977. The distribution of chironomid larvae in relation to chloride concentration in a brackish water region of the Netherlands. – Hydrobiologia 52: 117-126.
Peeters, E.T.H.M., 1988. Hydrobiologisch onderzoek in de Nederlandse Maas. Macrofauna in relatie tot biotopen. – Rapp. LH Vakgr. Natuurbeheer. 150 pp.
Peters, A.J.G.P., R. Gijlstra & J.J.P. Gardeniers, 1988. Waterkwaliteitsbeoordeling van genormaliseerde beken met behulp van macrofauna. – STORA-rapport 88-06: 1-56 + app.
Pinder, L.C.V., 1974. The Chironomidae of a small chalk-stream in Southern England. – Ent. Tidskr. 95 Suppl.: 195-202.
Pinder, L.C.V., 1976. Morphology of the adult and juvenile stages of *Microtendipes rydalensis* (Edw.) comb. nov. (Diptera, Chironomidae). – Hydrobiologia 48: 179-184.
Pinder, L.C.V., 1978. A key to the adult males of the British Chironomidae (Diptera) the non-biting midges. – Freshw. Biol. Ass. Sci. Publ. 37: 1-169 + 189 figs.
Pinder, L.C.V., 1980. Spatial distribution of Chironomidae in an English chalk stream. – In: Murray, D.A. (ed.): Chironomidae: Ecology, Systematics, Cytology and Physiology. Oxford, Pergamon Press: 153-161.
Pinder, L.C.V., 1983. Observations on the life-cycles of some Chironomidae in southern England. – Mem. Amer. Ent. Soc. 34: 249-265.
Pinder, L.C.V., 1986. Biology of freshwater

R

Chironomidae. – Ann. Rev. Entomol. 31: 1-23.
Pinder, L.C.V. & I.S. Farr, 1987. Biological surveillance of water quality – 3. The influence of organic enrichment on the macroinvertebrate fauna of small chalk streams. – Arch. Hydrobiol. 109: 619-637.
Pinder, L.C.V. & F. Reiss, 1983. The larvae of Chironominae (Diptera: Chironomidae) of the Holarctic region. Keys and diagnoses. – In: Wiederholm, T. (ed.): Chironomidae of the Holarctic region. Keys and diagnoses. Part 1. Larvae. – Ent. Scand. Suppl. 19: 293-435.
Pinder, L.C.V. & F. Reiss, 1986. The pupae of Chironominae (Diptera: Chironomidae) of the Holarctic region. Keys and diagnoses. – In: Wiederholm, T. (ed.): Chironomidae of the Holarctic region. Keys and diagnoses. Part 2. Pupae. – Ent. Scand. Suppl. 28: 299-456.
Polukonova, N.V., 2000. The allocation of *Chironomus* Meigen species (Chironomidae, Diptera) in natural basins of Saratov. - In: Hoffrichter, E.O. (ed.): Late 20th century research on Chironomidae. Shaker Verlag, Aachen: 335-338.
Polukonova, N.V., 2005. A comparative morphological analysis of the midges *Chironomus curabilis* and *C. nuditarsis* (Chiornomidae, Diptera). 1. Preimaginal stages. – Zool. Zh. 84: 367-370. (In Russian)
Prat, N., 1978. Benthos typology of Spanish reservoirs. – Verh. Internat. Verein. Limnol. 20: 1647-1651.
Prat, N., 1978a. Quironómidos de los embalses españoles (Diptera) (2.ª parte). – Graellsia 34: 59-119.
Prat, N., 1980. Benthic populations dynamics in artificial samplers in a Spanish reservoir. – In: Murray, D.A. (ed.): Chironomidae: Ecology, Systematics, Cytology and Physiology. Oxford, Pergamon Press: 239-246.
Prat, N. & M. Rieradevall, 1992. Life cycle and production of *Cladopelma virescens* (Mg.) (Diptera: Chironomidae) in Lake Banyoles (NE Spain). – Neth. J. Aquat. Ecol. 26: 315-320.
Provost, M.W. & N. Branch, 1959. Food of chironomid larvae in Polk county lakes. – Florida Ent. 42: 49-62.
Raddum, G.G., G. Hagenlund & G.A. Halvorsen, 1984. Effects of lime treatment on the benthos of Lake Søndre Boksjø. – Report Inst. Freshw. Res. Drottningholm 61: 167-176.
Raddum, G.G. & O.A. Saether, 1981. Chironomid communities in Norwegian lakes with different degrees of acidification. – Verh. Internat. Verein. Limnol. 21: 399-405.
Rasmussen, J.B., 1985. Effects of density and microdetritus enrichment on the growth of chironomid larvae in a small pond. – Can. J. Fish. Aquat. Sci. 42: 1418-1422.
Real, M. & N. Prat, 1992. Factors influencing the distribution of chironomids and oligochaetes in profundal areas of Spanish reservoirs. – Neth. J. Aquat. Ecol. 26: 405-410.
Reiss, F., 1968. Ökologische und systematische Untersuchungen an Chironomiden (Diptera) des Bodensees. Ein Beitrag zur lakustrischen Chironomidenfauna des nördlichen Alpenvorlandes. – Arch. Hydrobiol. 64: 176-323.
Reiss, F., 1968a. Verbreitung lakustrischer Chironomiden (Diptera) des Alpengebietes. – Ann. zool. fenn. 5: 119-125.
Reiss, F., 1984. Chironomiden (Diptera, Insecta) aus dem Ampertal bei Schöngeising, Oberbayern. – Mitt. Zool. Ges. Braunau 4: 211-220.
Reiss, F., 1988. Die Gattung Kloosia Kruseman, 1933 mit der Neubeschreibung zweier Arten (Diptera, Chironomidae). – Spixiana Suppl. 14: 35-44.
Reiss, F. & E.J. Fittkau, 1971. Taxonomie und Ökologie europäisch verbreiteter *Tanytarsus*-Arten (Chironomidae, Diptera). – Arch. Hydrobiol. Suppl. 40: 75-200.
Reist, A. & J. Fischer, 1987. Experimentelle Untersuchungen zur Einwirkung von Temperatur und Besiedlungsdichte auf die Entwicklung der *Chironomus*-Arten *Ch. plumosus*, *Ch. nuditarsis* und *Ch. bernensis* (Diptera). – Zool. Jb. Syst. 114: 1-13.
Remmert, H., 1955. Ökologische Untersuchungen über die Dipteren der Nord- und Ostsee. – Arch. Hydrobiol. 51: 1-53.
Rempel, J.G., 1936. The life-history and morphology of *Chironomus hyperboreus*. – J. biol. Bd. Can. 2: 209-221.
Rieradevall, M. & N. Prat, 1989. Chironomidae from profundal samples of Banyoles Lake (NE Spain). – Acta Biol. Debr. Oecol. Hung. 3: 267-274.
Ringe, F., 1970. Einige bemerkenswerte Chironomiden (Dipt.) aus Norddeutschland. – Faun.-Ökol. Mitt. 3: 312-322.
Roback, S.S., 1974. Insects (Arthropoda: Insecta). – In: Hart, C.W. & S.L.H. Fuller (eds.): Pollution ecology of freshwater invertebrates: 313-376. Academic press, New York, London.
Rodova, R.A., 1978. Opredelitelj samok komarov-zvontsov triby Chironomini. – Izd. Nauka, Leningrad. 142 pp.
Rossaro, B., 1985. Revision of the genus *Polypedilum* Kieffer, 1912. I. Key to adults, pupae and larvae of the species known to occur in Italy (Diptera: Chironomidae). – Mem. Soc. ent. ital., Genova 62/63: 3-23.
Rossaro, B., 1987. Chironomid emergence in the Po river (Italy) near a nuclear power plant. – Ent. scand. Suppl. 29: 331-338.
Rossaro, B. & S. Mietto, 1998. Multivariate analysis using chironomid (Diptera) species. – In: Bretschko, G. & J. Helešic (eds.): Advances in river bottom ecology: 191-205. Backhuys, Leiden.
Rossaro, B., A. Solimini, V. Lencioni, L. Marziali, R. Giacchini & P. Parenti, 2007. The rela-

tionship between body size, pupal thoracic horn development and dissolved oxygen in Chironomini (Diptera: Chironomidae). – Fundam. Appl. Limnol. Arch. Hydrobiol. 169/4: 331-339.

Rusina, O.Kh., 1956. Usvoenie otmershikh vodoroslej i daphnij lichinkami *Chironomus dorsalis*. – Vopr. Ikhtiol. 6: 165-173.

Ruse, L., 2002. Chironomid pupal exuviae as indicators of lake status. – Arch. Hydrobiol. 153: 367-390.

Ruse, L., 2002a. Colonisation of gravel lakes by Chironomidae. – Arch. Hydrobiol. 153: 391-407.

Rychen Bangerter, B. & J. Fischer, 1989. Different dormancy response in the sympatric *Chironomus* species *Ch. plumosus* and *Ch. nuditarsis*. – Zool. Jb. Syst. 116: 145-150.

Ryser, H.M., H.J. Geiger, A. Scholl & W. Wülker, 1980. Untersuchungen über die Verbreitung der Zuckmücken Gattung *Chironomus* in der Schweiz, mit besonderer Berücksichtigung von drei cytologisch nicht beschriebenen Arten. - In: Murray, D.A. (ed.): Chironomidae: Ecology, Systematics, Cytology and Physiology. Oxford, Pergamon Press: 17-24.

Ryser, H.M., A. Scholl & W. Wülker, 1983. Revision der Gattung *Chironomus* Meigen (Diptera) VII: *C. muratensis* n. sp. und *C. nudiventris* n. sp., Geschwisterarten aus der *plumosus*-Gruppe. – Rev. suisse Zool. 90: 299-316.

Sadler, W.O., 1935. Biology of the midge *Chironomus tentans* Fabricius. – Memoirs Cornell University agricultural Experiment Station 173: 1-25.

Saether, O.A. , 1971. Nomenclature and phylogeny of the genus *Harnischia* (Diptera: Chironomidae). – Can. Ent. 103: 347-362.

Saether, O.A., 1977. Taxonomic studies on Chironomidae: *Nanocladius*, *Pseudochironomus* and the *Harnischia* complex. – Bull. Fish. Res. Bd. Canada 196: 1-143.

Saether, O.A., 1977a. Female genitalia in Chironomidae and other Nematocera: morphology, phylogenies, keys. – Bull. Fish. Res. Bd Can. 197: 209 pp.

Saether, O.A., 1979. Chironomid communities as water quality indicators. – Holarctic Ecol. 2: 65-74

Saether, O.A., 1979a. New name for *Beckiella* Saether, 1977 (Diptera: Chironomidae) nec *Beckiella* Grandjean, 1964 (Acari: Oribatei). – Ent. scand. 10: 315.

Saether, O.A., P. Ashe & D.A. Murray, 2000. Family Chironomidae. – In: Papp, L. & B. Darvas (eds.): Contributions to a Manual of Palaearctic Diptera. Appendix. Science Herald, Budapest: 113-334.

Saether, O.A. & M. Spies, 2004. Fauna Europaea: Chironomidae. In: Jong, H. de (ed.)(2004) Fauna Europaea: Diptera, Nematocera. Fauna Europaea version 1.1, http://www.faunaeur.org.

Saether, O.A. & A. Sundal, 1999. *Cerobregma*, a new subgenus of *Polypedilum* Kieffer, with a tentative phylogeny of subgenera and species groups within *Polypedilum* (Diptera: Chironomidae). – J. Kans. Ent. Soc. 71: 315-382.

Sankamperumal, G. & T.J. Pandian, 1991. Effect of temperature and *Chlorella* density on growth and metamorphosis of *Chironomus circumdatus* (Kieffer) (Diptera). – Aq. Insects 13: 167-177.

Scharf, B., 1972. Experimentell-ökologische Untersuchungen zur Einnischung von *Chironomus thummi thummi* und *Ch. th. piger*. - Thesis Kiel.

Scharf, B.W., 1973. Experimentell-ökologische Untersuchungen an *Chironomus thummi* und *Chironomus piger* (Diptera, Chironomidae). – Arch. Hydrobiol. 72: 225-244.

Schartau, A.K., S.J. Moe, L. Sandin, B. McFarland & G.G. Raddum, 2008. Macroinvertebrate indicators of lake acidification: analysis of monitoring data from UK, Norway and Sweden. – Aquat. Ecol. 42: 293-305.

Scheibe, M.A., 2002. Beitrag zur Artenliste der aquatischen Zuckmücken (Diptera: Chironomidae) des Taunus. – Lauterbornia 44: 99-107.

Schlee, D., 1980. Besonderheiten der Biologie und Morphologie von *Fleuria lacustris* (Diptera: Chironomidae). – Stuttg. Beitr. Naturk. A: 340: 1-23.

Schleuter, A. 1985. Untersuchung der Makroinvertebratenfauna stehender Kleingewässer des Naturparkes Kottenforst-Ville unter besonderer Berücksichtigung der Chironomidae. – Thesis Bonn Univ. 217 pp.

Schleuter, A. 1986. Die Chironomiden-Besiedlung stehender Kleingewässer in Abhängigkeit von Wasserführung und Fallaubeintrag. – Arch. Hydrobiol. 105: 471-487.

Schmale, J.C., 1999. Hydrobiologisch onderzoek Berkheide 1994 – 1995 – 1996 – 1997. N.V. Duinwaterbedrijf Zuid-Holland. 73 pp. + 80 app.

Schnabel, S. & P.-M. A. Dettinger-Klemm, 2000. Chironomiden temporärer Tümpel im Bereich der Lahnaue – faunistisch-ökologische Aspekte. – Verh. Westd. Entom. Tag 1999: 201-208.

Schreijer, M., 1983. Hydrobiologisch onderzoek op Texel. – Edam: Hoogheemr. Uitwaterende Sluizen. 204 pp.

Serra-Tosio, B. & H.Laville, 1991. Liste annotée des Diptères Chironomidés de France continentale et de Corse. – Annls Limnol. 27: 37-74.

Shcherbina, G.H., 1989. Ecology and production of monocyclic species of Chironomidae (Diptera) from Lake Vishtynetskoe of the Kaliningrad region (USSR). – Acta Biol. Debr. Oecol. Hung. 3: 295-303.

Shilova, A.I., 1958. Über die Schlüpfperiode und Generationszahl bei *Tendipes plumosus* L. im Rybinsk-Stausee. – Biull. Inst. Biol. Vodokhr. 1: 26-30. (Russian)

Shilova, A.I., 1961. Novyj rod i vid Tendipedid (Diptera, Tendipedidae). – Biull. Inst. Biol. Vodokhr. 11: 19-23.

Shilova, A.I., 1963. Metamorfoz *Lipiniella arenicola* Shilova (Diptera, Tendipedidae). – Trudy inst. biol. vodokhr. 5: 71-80.

Shilova, A.I., 1965. Metamorfoz *Parachironomus vitiosus* Goetgh. i nekotorye dannye po ego biologii (Diptera, Tendipedidae). – Trudy inst. biol. vnutr. vod 8 (11): 102-109.

Shilova, A.I., 1965a. Metamorfoz i biologiya nekotorykh vidov roda *Cryptochironomus*. – Vopr. Gidrobiol. 456-457.

Shilova, A.I., 1965b. Metamorfoz i biologiya *Stictochironomus crassiforceps* Kieff. (Diptera, Tendipedidae). – Trudy inst. biol. vnutr. vod 8 (11): 91-101.

Shilova, A.I., 1968. Materialy po biologii peristousykh komarov roda *Parachironomus* Lenz (Diptera, Chironomidae). - Trudy inst. biol. vnutr. vod 17: 104-123.

Shilova, A.I., 1973. O sezonnykh formakh *Microtendipes pedellus* de Geer (Diptera, Chironomidae). – Informatsionnyj Biulletenj Instituta Biologii vnutrennikh vod 18: 39-41.

Shilova, A.I., 1974. Stadii razvitiya *Xenochironomus xenolabis* Kieff. (Diptera, Chironomidae). – Trudy inst. biol. vnutr. vod 25: 142-153.

Shilova, A.I., 1976. Chironomidy Rybinskogo vodohranilishcha. – Izd. Nauka, Leningrad. 252 pp.

Shilova, A.I., 1980. K sitematike roda *Einfeldia* Kieff. (Diptera, Chironomidae). – Trudy inst. biol. vnutr. vod 41 (44): 162-191.

Shilova, A.I. & N.A. Shobanov, 1996. Katalog khironomid roda *Chironomus* Meigen 1803 (Diptera, Chironomidae) Rossii i byvshikh respublik SSSR. – In: Shobanov, A.N. & T.D. Zinchenko (eds.): Ekologiya, evolutsiya i sistematika khironomid. Toliatti/Borok: 28-43.

Shilova, A.I. & N.I. Zelentsov, 1972. The influence of photoperiodism on diapause in Chironomidae. – Inf. Byull. Inst. Biol. vnutr. Vod 13: 37-42. (in Russian)

Shobanov, N. A., 1989. Morfologicheskaya differentsiatsiya vidov *Chironomus* gruppy *plumosus* (Diptera, Chironomidae), lichinki. – Trudy inst. biol. vnutr. vod 56: 250-279. (In Russian)

Shobanov, N.A., 1989a. The morphological differentiation of *Chironomus* species of *plumosus* group (Diptera: Chironomidae). Larvae. – Acta Biol. Debr. Oecol. Hung. 2: 335-344.

Shobanov, N.A., 1996. Morphology and karyotype of *Chironomus anthracinus* (Diptera, Chironomidae). – Zool. Zh. 75: 1505-1516. (In Russian)

Shobanov, N.A., A.I. Shilova & S.I. Belyanina, 1996. Extent and structure of the genus *Chironomus* Meigen (Diptera, Chironomidae). Survey of the world fauna. – In: Shobanov, N.A. & T.D. Zinchenko (eds.): Ecology, evolution and systematics of of chironomids: 44-96. Tolyatti/Borok. (In Russian)

Sládeček, V., 1973. System of water quality from the biological point of view. – Arch. Hydrobiol. Beiheft 7: 1-218.

Smit, H., 1982. De Maas. Op weg naar biologische waterbeoordeling van grote rivieren. – Rapp. LH Vakgr. Natuurbeheer 667: 1-100.

Smit, H., 1995. Macrozoobenthos in the enclosed Rhine-Meuse delta. – Thesis Nijmegen Univ., 192 + XVIII pp.

Smit, H. & J.J.P. Gardeniers, 1986. Hydrobiologisch onderzoek in de Maas. Een aanzet tot biologische monitoring van grote rivieren. – H_2O 19: 314-317.

Smit, H., F. Heinis, R. Bijkerk & F. Kerkum, 1992. *Lipiniella arenicola* (Chironomidae) compared with *Chironomus muratensis* and *Ch. nudiventris*: distribution patterns related to depth and sediment characteristics, diet, and behavioural response to reduced oxygen concentrations. – Neth. J. Aquat. Ecol. 26: 431-440.

Smit, H., P. Klaren & W. Snoek, 1991. *Lipiniella arenicola* Shilova (Diptera: Chironomidae) on a sandy flat in the Rhine-Meuse estuary: Distribution, population structure, biomass and production of larvae in relation to periodical drainage. – Verh. Internat. Verein. Limnol. 24: 2918-2923.

Smit, H., J.A. van der Velden & A. Klink, 1994. Macrozoobenthic assemblages in littoral sediments in the enclosed Rhine-Meuse delta. – Neth. J. Aquat. Ecol. 28: 199-212.

Smit, H., G. van der Velde & S. Dirksen, 1996. Chironomid larval assemblages in the enclosed Rhine-Meuse Delta: spatio-temporal patterns in an exposure gradient on a tidal sandy flat. – Arch. Hydrobiol. 137: 487-510.

Smith, V.G.F. & J.O. Young, 1973. The life histories of some Chironomidae (Diptera) in two ponds on Merseyside, England. – Arch. Hydrobiol. 72: 333-355.

Sokolova, G.A., 1966. On the hibernation mode of larvae of *Limnochironomus* ex gr. *nervosus* Staeg. (Diptera, Chironomidae). – Zool. Zh. 45: 140. (Russian with English summary).

Sokolova, N. Yu., 1968. Über die Ökologie der Chironomiden im Utscha-Stausee. – Ann. zool. fenn. 5: 139-143.

Sokolova, N.Yu., 1971. Life cycles of chironomids in the Uchinskoye reservoir. – Limnologica (Berlin) 8: 151-155.

Sokolova, N.Yu., (ed.), 1983. *Chironomus plumosus* L. (Diptera, Chironomidae). Systematics, morphology, ecology, produc-

tion. – Moscow, Nauka. 309 pp. (In Russian).
Sokolova, N.Yu, A.V. Paliy & B.I. Izvekova, 1992. Biology of *Chironomus piger* Str. (Diptera: Chironomidae) and its role in the self-purification of a river. – Neth. J. Aquat. Ecol. 26: 509-512.
Soponis, A.R. & C.L. Russell, 1982. Identification of instars in some larval *Polypedilum (Polypedilum)* (Diptera: Chironomidae). – Hydrobiologia 94: 25-32.
Sorokin, Ju.I., 1968. The use of ^{14}C in the study of nutrition of aquatic animals. – Mitt. int. Ver. Limnol. 16: 1-41.
Soszka, G.J., 1975. The invertebrates on submerged macrophytes in three Masurian lakes. – Ekol. polska 23: 371-391.
Soszka, G.J., 1975a. Ecological relations between invertebrates and submerged macrophytes in the lake littoral. – Ekol. polska 23: 393-415.
Specziár, A. & L. Vöros, 2001. Long-term dynamics of Lake Balaton's chironomid fauna and its dependence on the phytoplankton production. – Arch. Hydrobiol. 152: 119-142.
Spies, M., 2000. A contribution to the knowledge of Holarctic *Parachironomus* Lenz (Diptera: Chironomidae), with two new species and a provisional key to Nearctic adult males. – Tijdschr. Entomol. 143: 125-143.
Spies, M. & O.A. Saether, 2004. Notes and recommendations on taxonomy and nomenclature of Chironomidae (Diptera). – Zootaxa 752: 1-90.
Srokosz, K., 1980. Chironomidae communities of the river Nida and its tributaries. – Acta Hydrobiol. Kraków 22: 191-215.
Steenbergen, H.A., 1993. Macrofauna-atlas van Noord-Holland: verspreidingskaarten en responsies op milieufactoren van ongewervelde waterdieren. – Prov. Noord-Holland, Dienst Ruimte en Groen. Haarlem. 650 pp.
Steinhart, M., 1999. Einflüsse der saisonalen Überflutung auf die Chironomidenbesiedlung (Diptera) aquatischer und amphibischer Biotope des Unteren Odertals. – Shaker Verlag. Aachen. 117 pp.
Šterba, O. & M. Holzer, 1977. Fauna der interstitiellen Gewässer der Sandkiessedimente unter der aktiven Strömung. – Vestnik Ceskosl. spolecn. zool. 41: 144-159.
Strenzke, K., 1959. Revision der Gattung *Chironomus* Meig. 1. Die Imagines von 15 norddeutschen Arten und Unterarten. – Arch. Hydrobiol. 56: 1-42.
Strenzke, K., 1960. Die systematische und ökologische Differenzierung der Gattung *Chironomus*. – Ann. Ent. Fenn. 26: 111-138.
Sublette, J.E., 1964. Chironomidae (Diptera) of Louisiana. I. Systematics and immature stages of some lentic chironomids of West-Central Louisiana. – Tulane Stud. Zool. 11: 109-150.
Tempelman, D., 2002. Verspreiding en ecologie van de dansmug *Glyptotendipes ospeli* in Nederland (Diptera: Chironomidae). – Nederl. faun. Meded. 17: 19-31.
Tempelman, D. (in prep.). Key to identification of 4th instar larvae of *Polypedilum* species of The Netherlands. – Draft version, 2 September 2008. Amsterdam, 10 pp.
Ten Winkel, E.H., 1987. Chironomid larvae and their foodweb relations in the littoral zone of lake Maarsseveen. – Thesis Amsterdam Univ. 145 pp.
Ten Winkel, E.H. & C. Davids, 1987. Population dynamic aspects of chironomid larvae of the littoral zone of Lake Maarsseveen I. – Hydrobiol. Bull. 21: 81-94.
Thienemann, A., 1913. Der Zusammenhang zwischen dem Sauerstoffgehalt des Tiefenwassers und der Zusammensetzung der Tiefenfauna unserer Seen. - Int. Rev. ges. Hydrobiol. 6: 243-249.
Thienemann, A., 1954. *Chironomus*. Leben, Verbreitung und wissenschaftliche Bedeutung der Chironomiden. – Binnengewässer 20: 1-834.
Thorp, J.H. & E.A. Bergey, 1981. Field experiments on interactions between vertebrate predators and larval midges (Diptera: Chironomidae) in the littoral zone of a reservoir. – Oecologia (Berlin) 50: 285-290.
Timmermans, K.R., W. Peeters & M. Tonkes, 1992. Cadmium, zinc, lead & copper in *Chironomus riparius* (Meigen) larvae (Diptera, Chironomidae): uptake and effects. – Hydrobiologia 241: 119-134.
Titmus, G. & R.M. Badcock, 1981. Distribution and feeding of larval Chironomidae in a gravel-pit lake. – Freshw. Biol. 11: 263-271.
Tokeshi, M., 1986. Population dynamics, life histories and species richness in an epiphytic chironomid community. – Freshw. Biol. 16: 431-441.
Tokeshi, M. & L.C.V. Pinder, 1985. Microhabitats of stream invertebrates on two submersed macrophytes with contrasting leaf morphology. – Holarctic Ecol. 8: 313-319.
Tokeshi, M. & L.C.V. Pinder, 1986. Dispersion of epiphytic chironomid larvae and the probability of random colonization. – Int. Revue ges. Hydrobiol. 71: 613-620.
Tokeshi, M & C.R. Townsend, 1987. Random patch formation and weak competition: coexistence in an epiphytic chironomid community. – J. Anim. Ecol. 56: 833-845.
Tolkamp, H. H., 1980. Organism-substrate relationships in lowland streams. – Thesis Wageningen. 211 pp.
Tõlp, Õ, 1956. Emajõe bentosest. – Loodusuurijate Seltsi Aastaraamat 49: 143-160. (Estonian).
Tõlp, Õ, 1958. Khironomidy reki Emajõgi i ikh znachenie v sostave donnoj fauny. – Avtoreferat dissertatsii. Tartu. 14 pp.
Tõlp, Õ, 1971. Chironomid larvae in the brack-

R

ish waters of Estonia. – Limnologica, Berlin 8: 95-97.

Tourenq, J.-N., 1975. Recherches écologiques sur les chironomides (Diptera) de Camargue. – Thesis. Toulouse. 424 pp.

Tshernovskij see Chernovskij.

Urban, E., 1975. The mining fauna in four macrophyte species in Mikolajskie lake. – Ekol. polska 23: 417-435.

Urk, G. van & A. bij de Vaate, 1990. Ecological studies in the lower Rhine in The Netherlands. – In: Kinzelbach & Friedrich (eds.): Biologie des Rheins. Limnologie aktuell 1: 131-145.

Vaate, A. bij de, 2003. Degradation and recovery of the freshwater fauna in the lower sections of the rivers Rhine and Meuse. – Thesis Wageningen. 200 pp.

Vala, J.-C., J. Moubayed & P. Langton, 2000. Chironomidae des rizières de Camargue, données faunistiques et écologiques (Diptera). – Bull. Soc. entomol. France 105: 293-300.

Vallenduuk, H., 1990. Makrofauna-onderzoek aan enkele vennen in het Natuurmonument Kampina. – Own publ., 81 pp.

Vallenduuk, 1999. Key to the larvae of *Glyptotendipes* Kieffer (Diptera, Chironomidae) in Western Europe. – Own publ., 46 pp. + app.

Vallenduuk, H.J. & H.K.M. Moller Pillot, 2007. Chironomidae larvae of the Netherlands and adjacent lowlands. General ecology and Tanypodinae. – KNNV Publ. Zeist. 144 pp.

Vallenduuk, H.J. & E. Morozova, 2005. *Cryptochironomus*. An identification key to the larvae and pupal exuviae in Europe. – Lauterbornia 55: 1-22.

Vallenduuk, H.J., S.M. Wiersma, H.K.M. Moller Pillot & J.A. van der Velden, 1995. Determineertabel voor larven van het genus *Chironomus* in Nederland. – RIZA Dordrecht, Werkdocument 95.121X. 30 pp. + app.

Värdal, H., A. Bjørlo & O.A. Saether, 2002. Afrotropical *Polypedilum* subgenus *Tripodura*, with a review of the subgenus (Diptera: Chironomidae). – Zoologica Scripta 31: 331-402.

Velde, G. van der & R. Hiddink, 1987. Chironomidae mining in *Nuphar lutea* (L.) Sm. (Nymphaeaceae). – Ent. scand. Suppl. 29: 255-264.

Velden, J.A. van der, E.M. van Dam & S. M. Wiersma, 1995. The Chironomidae (Diptera) of Lake Volkerak-Zoommeer (The Netherlands) after freshening.- Lauterbornia 21: 139-147.

Velden, J.A. van der, H.K.M. Moller Pillot, H.J. Vallenduuk & S.M. Wiersma, 1996. The occurrence of *Chironomus balatonicus* (Diptera: Chironomidae) in The Netherlands. – Entomol. Ber., Amsterdam 56: 14-15.

Verberk, W., 2008. Matching species to a changing landscape. Aquatic macroinvertebrates in a heterogeneous landscape. – Thesis Nijmegen. 150 pp.

Verdonschot, P.F.M., 1990. Ecological characterization of surface waters in the province of Overijssel (The Netherlands). Thesis. Wageningen. 255 pp.

Verdonschot, P.F.M., L.W.G. Higler, W.F. van der Hoek & J.G.M. Cuppen, 1992. A list of macroinvertebrates in Dutch water types: a first step towards an ecological classification of surface waters based on key factors. – Hydrobiol. Bull. 25: 241-259.

Verdonschot, P. & W. Lengkeek, 2006. Habitat preferences of selected indicators. – Deliverable No. 92. Euro-limpacs. 40 pp.

Verdonschot, P.F.M. & J.A. Schot, 1987. Macrofaunal community types in helocrene springs. – Ann. Rep. Res. Inst. Nature Management 1986: 85-103. Arnhem, Leersum and Texel.

Verneaux, V. & L. Aleya, 1998. Spatial and temporal distribution of chironomid larvae (Diptera: Nematocera) at the sediment-water interface in Lake Abbaye (Jura, France). – Hydrobiologia 373/374: 169-180.

Verstegen, M., 1985. De macrofauna – met name de chironomidelarven – van een twaalftal vennen in de gemeenten Boxtel, Oisterwijk en Moergestel. – Versl. Utrecht Univ. 65 pp. + bijl.

Vos, J.H., 2001. Feeding of detritivores in freshwater sediments. – Thesis. Amsterdam Univ. 140 pp.

Waajen, G.W.A.M., 1982. Hydrobiologie van veenputten in de Mariapeel en de Liesselse Peel. – LH Wageningen, sektie Hydrobiol., verslag 82-1. 67 pp.

Walshe, B.M., 1948. The oxygen requirements and thermal resistance of chironomid larvae from flowing and from still waters. – J. Exp. Biol. 25: 35-44.

Walshe, B.M., 1951. The feeding habits of certain chironomid larvae (subfamily Tendipedinae). – Proc. zool. Soc. London 121: 63-79.

Wang, X., 2000. Nuisance chironomid midges recorded from China (Diptera). – In: Hoffrichter, E.O. (ed.): Late 20th century research on Chironomidae. Shaker verlag, Aachen: 653-658.

Ward, G.M. & K.W. Cummins, 1978. Life history and growth pattern of *Paratendipes albimanus* in a Michigan headwater stream. – Ann. Ent. Soc. Am. 71: 272-284.

Ward, G.M. & K.W. Cummins, 1979. Effects of food quality on growth of a stream detritivore, *Paratendipes albimanus* (Meigen) (Diptera: Chironomidae). – Ecology 60: 57-64.

Warwick, W.F., 1975. The impact of man on the Bay of Quinte, Lake Ontario, as shown by the subfossil chironomid succession (Chironomidae, Diptera). – Verh. Internat. Verein. Limnol. 19: 3134-3141.

Webb, C.J., 1980. Modern approaches to the congruence problem in chironomid systematics. – In: Murray, D.A. (ed.): Chironomidae. Ecology, Systematics, Cytology and Physiology: 97-104. Pergamon Press, Oxford.

Webb, C.J. & A. Scholl, 1985. Identification of larvae of European species of *Chironomus* Meigen (Diptera: Chironomidae) by morphological characters. – Syst. Entom. 10: 353-372.

Webb, C.J. & A. Scholl, 1987. Comparative morphology of the larval ventromental plates of European species of *Einfeldia* Kieffer and *Chironomus* Meigen (subgenera *Lobochironomus* and *Camptochironomus*) (Diptera: Chironomidae). – Entomol. scand. Suppl. 29: 75-86.

Webb, C.J., R.S. Wilson & J.D. McGill, 1981. Ultrastructure of the striated ventromental plates and associated structures of larval Chironominae (Diptera: Chironomidae) and their role in silkspinning. – J. Zool., London 194: 67-84.

Werkgroep Hydrobiologie MEC Eindhoven, 1993. De Groote Peel als leefmilieu voor aquatische macrofauna. – M.E.C. Eindhoven, 99 pp. + bijl.

Whiteside, M.C. & C. Lindegaard, 1982. Summer distribution of zoobenthos in Grane Langsø, Denmark. – Freshw. Invertebr. Biol. 1: 2-16.

Whitman, R.L. & W.J. Clark, 1984. Ecological studies of the sand-dwelling community of an East Texas stream. – Freshw. Invertebr. Biol. 3: 59-79.

Wiederholm, T., 1979. Morphology of *Chironomus macani* Freeman, with notes on the taxonomic status of subg. *Chaetolabis* Town. (Diptera: Chironomidae). – Ent. scand. Suppl. 10: 145-150.

Wiederholm, T., 1980. Chironomids as indicators of water quality in Swedish lakes. – Acta Univ. Carolinae Biologica 1978: 275-283.

Wiederholm, T. & L. Eriksson, 1979. Subfossil chironomids as evidence of eutrophication in Ekoln Bay, Central Sweden. – Hydrobiologia 62: 195-208.

Wiederholm, T., K. Danell & K. Sjöberg, 1977. Emergence of chironomids from a small man-made lake in northern Sweden. – Norw. J. Entomol. 24: 99-105.

Wielgosz, S., 1979. The effect of wastes from the town of Olstyn on invertebrate communities in the bottom of river Lyna. - Acta Hydrobiol. 21: 149-165.

Williams, C.J., 1982. The drift of some chironomid egg masses (Diptera: Chironomidae). – Freshw. Biol. 12: 573-578.

Williams, K.A., D.W.J. Green, D. Pascoe & D.E. Gower, 1986. The acute toxicity of cadmium to different larval stages of *Chironomus riparius* (Diptera: Chironomidae) and its ecological significance for pollution regulation. – Oecologia (Berlin) 70: 362-366.

Wilson, R.S. & L.P. Ruse, 2005. A guide to the identification of genera of chironomid pupal exuviae occurring in Britain and Ireland and their use in monitoring lotic and lentic fresh waters. – Freshw. Biol. Ass., Special Publ. 13: 1-176.

Wilson, R.S. & S.E. Wilson, 1984. A survey of the distribution of Chironomidae (Diptera, Insecta) of the River Rhine by sampling pupal exuviae. – Hydrobiol. Bull. 18: 119-132.

Wotton, R.S., P.D. Armitage, K. Aston, J.H. Blackburn, M. Hamburger & C.A. Woodward, 1992. Colonization and emergence of midges (Chironomidae: Diptera) in slow sand filter beds. – Neth. J. Aquat. Ecol. 26: 331-339.

Wülker, W., 1973. Revision der Gattung *Chironomus*. III. Europäische Arten des *thummi*-Komplexes. – Arch. Hydrobiol. 72: 356-374.

Wülker, W.F., 1991. *Chironomus fraternus* sp. n. and *C. beljaninae* sp. n., sympatric sister species of the *aberratus* group in Fennoscandian reservoirs. – Entomol. Fenn. 2: 97-109.

Wülker, W.F., 1996. *Chironomus pilicornis* Fabricius, 1787 and *C. heteropilicornis* sp. n. (diptera: Chironomidae) in Fennoscandian reservoirs: karyosystematic and morphological results. – Aquat. Ins. 18: 209-221.

Wülker, W., 1999. Fennoscandian *Chironomus* species (Dipt., Chironomidae) – identified by karyotypes and compared with the Russian and Central European fauna. – Studia Dipterol. 6: 425-436.

Wülker, W., I.I. Kiknadze, I.E. Kerkis & P. Nevers, 1999. Chromosomes, morphology, ecology and distribution of *Sergentia baueri*, spec. nov., *S. prima* Proviz & Proviz, 1997 and *S. coracina* Zett., 1824 (Insecta, Diptera, Chironomidae). – Spixiana 22: 69-81.

Wülker, W. & A.M. Klötzli, 1973. Revision der Gattung *Chironomus* Meig. IV. Arten des *lacunarius*-(*commutatus*-)Komplexes. – Arch. Hydrobiol. 72: 474-489.

Wülker, W., H.M. Ryser & A. Scholl, 1983. Revision der Gattung *Chironomus* Meigen (Diptera). VIII. Arten mit Larven des *fluviatilis*-Typs (*obtusidens*-Gruppe): *C. acutiventris* n. sp. und *C. obtusidens* Goetgh. – Rev. Suisse Zool. 90: 725-745.

Wundsch, H.H., 1943. Die Seen der mittleren Havel als *Glypotendipes*-Gewässer und die Metamorphose von *Glyptotendipes paripes* Edwards. – Arch. Hydrobiol. 40: 362-380.

Wundsch, H.H., 1943a. Die Metamorphose von *Demeijerea rufipes* L. (Dipt. Tendip.). – Zool. Anz. 141: 27-32.

Young, M.R., 1973. Seasonal variation in the occurrence of the parasitic larvae of *Cryptochironomus* sp. near *pararostratus* Kieffer (Dipt.) in various freshwater molluscs. – Entomologist's mon. Mag. 110: 143-146.

Začwilichowska, K., 1970. Diptera larvae in the

R

River San and some of its tributaries. – Acta Hydrobiol., Kraków 12: 197-208.
Zavrel, J., 1926. "*Tanytarsus connectens*". – Public. Fac. Sci. Univ. Masaryk 65: 1-47.
Zelinka, M. & P. Marvan, 1961. Zur Präzisierung der biologischen Klassifikation der Reinheit fliessender Gewässer. – Arch. Hydrobiol. 10: 453-469.
Zinchenko, T.D., 1992. Long-term (30 years) dynamics of Chironomidae (Diptera) fauna in the Kuibyshev water reservoir associated with eutrophication processes. – Neth. J. Aquat. Ecol. 26: 533-542.

ACKNOWLEDGEMENTS

This book was made possible by the material or information sent by many colleagues, especially Hub Cuppen (Eerbeek), Andreas Dettinger-Klemm (Germany), Niels Evers (Limnodata), Boudewijn Goddeeris (Belgium), Ton van Haaren, Amy Storm and David Tempelman (Grontmij | Aqua Sense, Amsterdam), Hein van Kleef and Jan Kuper (Stichting Bargerveen, Nijmegen), Alexander Klink (Wageningen), Claus-Joachim Otto (Germany), Ad Kuijpers (Aqualab, Werkendam), Francien Lambregts-van de Clundert (Deltawaterlab, Breda), Peter Langton (Northern Ireland), Barend van Maanen (Waterschap Roer en Overmaas, Sittard), Susanne Michiels (Germany), Mieke Moeliker (GWL, Boxtel), Ole Sæther (Norway), Martin Spies (Germany) and Kim Vermonden (Denmark). Maria Sanabria helped with some of the field work. We received helpful comments on the text from Ronald Buskens, Mieke Moeliker and David Tempelman. The many discussions with Henk Vallenduuk contributed to a better and more complete view on several genera, such as *Cryptochironomus*, *Parachironomus* and *Chironomus*. The descriptions of the last genus would not have been possible without the cytological verification of many larvae by Iya Kiknadze (Novosibirsk). Godard Tweehuysen helped with the literature search.
Derek Middleton (Zevenaar) corrected our English and copy-edited the text.

A

INDEX

Only the first page of the description in Chapter 3 and the pages of the tables in Chapter 4 are given. Synonyms are given in italics.

OTHER AVAILABLE TITLES BY KNNV PUBLISHING

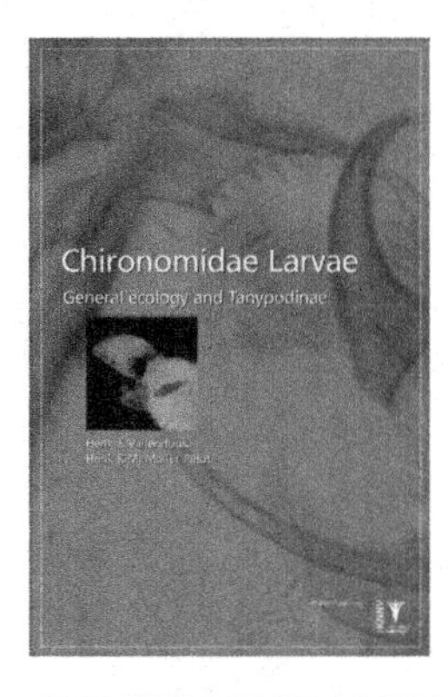

CHIRONOMIDAE LARVAE
General ecology and Tanypodinae
Henk J. Vallenduuk
Henk K.M. Moller Pillot

This book shows identification keys and additional information on the biology and ecology of Tanypodinae Larvae and contains information on the general ecology of the whole family Chironomidae Larvae.

Hardcover, 2007, 16,5 x 24 cm,
144 pp, € 79,95, English,
978-90-5011-295-8

CHIRONOMIDAE LARVAE II
Biology and Ecology of the Chironomini
Henk K.M. Moller Pillot

This 2nd part of a series on Chironomidae larvae covers the most important tribes, Chironomini and Pseudochironomini, the well-known red bloodworms.

Hardcover, 2009, 16,5 x 24 cm
274 pp, € 79,95, English
ISBN: 978-90-5011-303-8

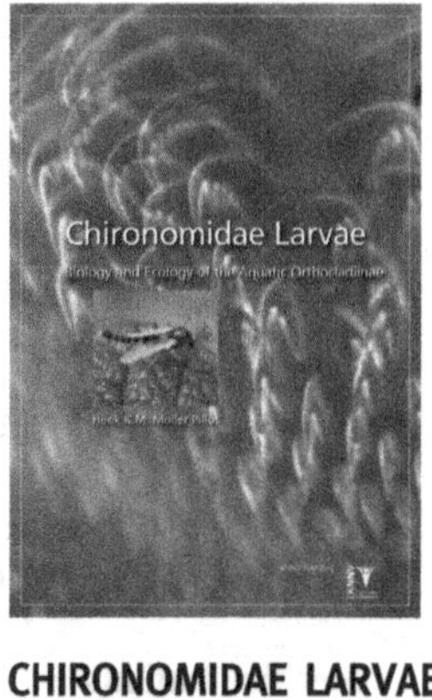

CHIRONOMIDAE LARVAE III
Biology and Ecology of the aqauatic Orthocladiinae
Henk K.M. Moller Pillot

3rd part of a series on Chironomidae larvae covers the subfamily Orthocladiinae which is especially well represented in flowing water.

Hardcover, 2013, 16,5 x 24 cm
314 pp, € 89,95, English
ISBN: 978-90-5011-459-2

DESMIDS OF THE LOWLANDS I
Mesotaeniaceae and Desmidiaceae of the European Lowlands
Peter F.M. Coesel,
Koos (J.) Meesters

This flora represents all desmid taxa known from the Netherlands and adjacent lowland areas: morphology, ecology and taxonomy. Covers more than 500 species and more than 150 additional varieties.

Hardcover + CD Rom, 2007, 16,5 x 24 cm
352 pp, € 89,95, English
ISBN: 978-90-5011-265-9

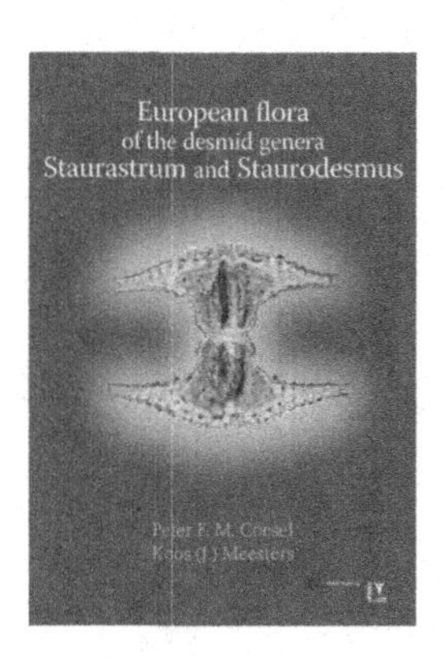

DESMIDS OF THE LOWLANDS II
European flora of the desmid genera Staurastrum and Staurodesmus
Peter F.M. Coesel,
Koos (J.) Meesters

This flora represents the European species of het desmid genera Staurastrum and Staurodesmus. Contains reliable identification keys and general information on the morphology, taxonomy, ecology and geographical distribution.

Hardcover, 2013, 16,5 x 24 cm
400 pp, € 99,95, English
ISBN: 978-90-5011-458-5

AQUATIC OLIGOCHAETA
Aquatic Oligochaeta of the Netherlands and Belgium
Ton van Haaren,
Jan Soors

This book presents a comprehensive overview, focusing on morphology, collecting and preservation, identification, and ecology and offers a new, practical key for identification to species level.

Hardcover, 2013, 19,2 x 26 cm
302 pp, € 195,-, English
ISBN: 978-90-5011-378-6

COLOPHON

Author

Henk K.M. Moller Pillot

Illustrations
Names stated under the figures
Otherwise Henk K.M. Moller Pillot

Graphic layout and design
Erik de Bruin, Varwig Design, Hengelo

Cover illustrations
Background illustration: Pheanopsectra – claws (D. Tempelman / T. van Haaren, Grontmij-AquaSense)
Front cover illustration: Chironomini larva (F.A. Bink)

This publication was financially supported by:
Stowa, Utrecht

2nd edition (POD) 2013
ISBN 978 90 5011 3038
NUR 432
www.knnvpublishing.nl

Printed in the United States
By Bookmasters